STRICT CONVEXITY AND COMPLEX STRICT CONVEXITY

PURE AND APPLIED MATHEMATICS

A Program of Monographs, Textbooks, and Lecture Notes

Contributions to *Lecture Notes in Pure and Applied Mathematics* are reproduced by direct photography of the author's typewritten manuscript. Potential authors are advised to submit preliminary manuscripts for review purposes. After acceptance, the author is responsible for preparing the final manuscript in camera-ready form, suitable for direct reproduction. Marcel Dekker, Inc. will furnish instructions to authors and special typing paper. Sample pages are reviewed and returned with our suggestions to assure quality control and the most attractive rendering of your manuscript. The publisher will also be happy to supervise and assist in all stages of the preparation of your camera-ready manuscript.

LECTURE NOTES
IN PURE AND APPLIED MATHEMATICS

1. *N. Jacobson,* Exceptional Lie Algebras
2. *L.-Å. Lindahl and F. Poulsen,* Thin Sets in Harmonic Analysis
3. *I. Satake,* Classification Theory of Semi-Simple Algebraic Groups
4. *F. Hirzebruch, W. D. Newmann, and S. S. Koh,* Differentiable Manifolds and Quadratic Forms (out of print)
5. *I. Chavel,* Riemannian Symmetric Spaces of Rank One (out of print)
6. *R. B. Burckel,* Characterization of C(X) Among Its Subalgebras
7. *B. R. McDonald, A. R. Magid, and K. C. Smith,* Ring Theory: Proceedings of the Oklahoma Conference
8. *Y.-T. Siu,* Techniques of Extension of Analytic Objects
9. *S. R. Caradus, W. E. Pfaffenberger, and B. Yood,* Calkin Algebras and Algebras of Operators on Banach Spaces
10. *E. O. Roxin, P.-T. Liu, and R. L. Sternberg,* Differential Games and Control Theory
11. *M. Orzech and C. Small,* The Brauer Group of Commutative Rings
12. *S. Thomeier,* Topology and Its Applications
13. *J. M. Lopez and K. A. Ross,* Sidon Sets
14. *W. W. Comfort and S. Negrepontis,* Continuous Pseudometrics
15. *K. McKennon and J. M. Robertson,* Locally Convex Spaces
16. *M. Carmeli and S. Malin,* Representations of the Rotation and Lorentz Groups: An Introduction
17. *G. B. Seligman,* Rational Methods in Lie Algebras
18. *D. G. de Figueiredo,* Functional Analysis: Proceedings of the Brazilian Mathematical Society Symposium
19. *L. Cesari, R. Kannan, and J. D. Schuur,* Nonlinear Functional Analysis and Differential Equations: Proceedings of the Michigan State University Conference
20. *J. J. Schäffer,* Geometry of Spheres in Normed Spaces
21. *K. Yano and M. Kon,* Anti-Invariant Submanifolds
22. *W. V. Vasconcelos,* The Rings of Dimension Two
23. *R. E. Chandler,* Hausdorff Compactifications
24. *S. P. Franklin and B. V. S. Thomas,* Topology: Proceedings of the Memphis State University Conference
25. *S. K. Jain,* Ring Theory: Proceedings of the Ohio University Conference
26. *B. R. McDonald and R. A. Morris,* Ring Theory II: Proceedings of the Second Oklahoma Conference
27. *R. B. Mura and A. Rhemtulla,* Orderable Groups
28. *J. R. Graef,* Stability of Dynamical Systems: Theory and Applications
29. *H.-C. Wang,* Homogeneous Banach Algebras
30. *E. O. Roxin, P.-T. Liu, and R. L. Sternberg,* Differential Games and Control Theory II
31. *R. D. Porter,* Introduction to Fibre Bundles
32. *M. Altman,* Contractors and Contractor Directions Theory and Applications
33. *J. S. Golan,* Decomposition and Dimension in Module Categories
34. *G. Fairweather,* Finite Element Galerkin Methods for Differential Equations
35. *J. D. Sally,* Numbers of Generators of Ideals in Local Rings
36. *S. S. Miller,* Complex Analysis: Proceedings of the S.U.N.Y. Brockport Conference
37. *R. Gordon,* Representation Theory of Algebras: Proceedings of the Philadelphia Conference
38. *M. Goto and F. D. Grosshans,* Semisimple Lie Algebras
39. *A. I. Arruda, N. C. A. da Costa, and R. Chuaqui,* Mathematical Logic: Proceedings of the First Brazilian Conference

40. *F. Van Oystaeyen,* Ring Theory: Proceedings of the 1977 Antwerp Conference
41. *F. Van Oystaeyen and A. Verschoren,* Reflectors and Localization: Application to Sheaf Theory
42. *M. Satyanarayana,* Positively Ordered Semigroups
43. *D. L. Russell,* Mathematics of Finite-Dimensional Control Systems
44. *P.-T. Liu and E. Roxin,* Differential Games and Control Theory III: Proceedings of the Third Kingston Conference, Part A
45. *A. Geramita and J. Seberry,* Orthogonal Designs: Quadratic Forms and Hadamard Matrices
46. *J. Cigler, V. Losert, and P. Michor,* Banach Modules and Functors on Categories of Banach Spaces
47. *P.-T. Liu and J. G. Sutinen,* Control Theory in Mathematical Economics: Proceedings of the Third Kingston Conference, Part B
48. *C. Byrnes,* Partial Differential Equations and Geometry
49. *G. Klambauer,* Problems and Propositions in Analysis
50. *J. Knopfmacher,* Analytic Arithmetic of Algebraic Function Fields
51. *F. Van Oystaeyen,* Ring Theory: Proceedings of the 1978 Antwerp Conference
52. *B. Kedem,* Binary Time Series
53. *J. Barros-Neto and R. A. Artino,* Hypoelliptic Boundary-Value Problems
54. *R. L. Sternberg, A. J. Kalinowski, and J. S. Papadakis,* Nonlinear Partial Differential Equations in Engineering and Applied Science
55. *B. R. McDonald,* Ring Theory and Algebra III: Proceedings of the Third Oklahoma Conference
56. *J. S. Golan,* Structure Sheaves over a Noncommutative Ring
57. *T. V. Narayana, J. G. Williams, and R. M. Mathsen,* Combinatorics, Representation Theory and Statistical Methods in Groups: YOUNG DAY Proceedings
58. *T. A. Burton,* Modeling and Differential Equations in Biology
59. *K. H. Kim and F. W. Roush,* Introduction to Mathematical Consensus Theory
60. *J. Banas and K. Goebel,* Measures of Noncompactness in Banach Spaces
61. *O. A. Nielsen,* Direct Integral Theory
62. *J. E. Smith, G. O. Kenny, and R. N. Ball,* Ordered Groups: Proceedings of the Boise State Conference
63. *J. Cronin,* Mathematics of Cell Electrophysiology
64. *J. W. Brewer,* Power Series Over Commutative Rings
65. *P. K. Kamthan and M. Gupta,* Sequence Spaces and Series
66. *T. G. McLaughlin,* Regressive Sets and the Theory of Isols
67. *T. L. Herdman, S. M. Rankin, III, and H. W. Stech,* Integral and Functional Differential Equations
68. *R. Draper,* Commutative Algebra: Analytic Methods
69. *W. G. McKay and J. Patera,* Tables of Dimensions, Indices, and Branching Rules for Representations of Simple Lie Algebras
70. *R. L. Devaney and Z. H. Nitecki,* Classical Mechanics and Dynamical Systems
71. *J. Van Geel,* Places and Valuations in Noncommutative Ring Theory
72. *C. Faith,* Injective Modules and Injective Quotient Rings
73. *A. Fiacco,* Mathematical Programming with Data Perturbations I
74. *P. Schultz, C. Praeger, and R. Sullivan,* Algebraic Structures and Applications: Proceedings of the First Western Australian Conference on Algebra
75. *L. Bican, T. Kepka, and P. Nemec,* Rings, Modules, and Preradicals
76. *D. C. Kay and M. Breen,* Convexity and Related Combinatorial Geometry: Proceedings of the Second University of Oklahoma Conference
77. *P. Fletcher and W. F. Lindgren,* Quasi-Uniform Spaces
78. *C.-C. Yang,* Factorization Theory of Meromorphic Functions
79. *O. Taussky,* Ternary Quadratic Forms and Norms
80. *S. P. Singh and J. H. Burry,* Nonlinear Analysis and Applications
81. *K. B. Hannsgen, T. L. Herdman, H. W. Stech, and R. L. Wheeler,* Volterra and Functional Differential Equations

82. *N. L. Johnson, M. J. Kallaher, and C. T. Long,* Finite Geometries: Proceedings of a Conference in Honor of T. G. Ostrom
83. *G. I. Zapata,* Functional Analysis, Holomorphy, and Approximation Theory
84. *S. Greco and G. Valla,* Commutative Algebra: Proceedings of the Trento Conference
85. *A. V. Fiacco,* Mathematical Programming with Data Perturbations II
86. *J.-B. Hiriart-Urruty, W. Oettli, and J. Stoer,* Optimization: Theory and Algorithms
87. *A. Figa Talamanca and M. A. Picardello,* Harmonic Analysis on Free Groups
88. *M. Harada,* Factor Categories with Applications to Direct Decomposition of Modules
89. *V. I. Istrăţescu,* Strict Convexity and Complex Strict Convexity: Theory and Applications

Other Volumes in Preparation

STRICT CONVEXITY AND COMPLEX STRICT CONVEXITY

Theory and Applications

VASILE I. ISTRĂŢESCU

MARCEL DEKKER, INC. New York and Basel

Library of Congress Cataloging in Publication Data

Istrăţescu, Vasile I.
Strict convexity and complex strict convexity.

(Lecture notes in pure and applied mathematics ;
v. 89)
Includes bibliographical references and indexes.
1. Banach spaces. 2. Convex domains. I. Title.
II. Series.
QA322.2.I83 1984 515.7'32 83-18849
ISBN 0-8247-1796-1

MARCEL DEKKER, INC.
270 Madison Avenue, New York, New York 10016

Current printing (last digit):
10 9 8 7 6 5 4 3 2 1

PRINTED IN THE UNITED STATES OF AMERICA

To my parents,

Paraschiva Istrăţescu and Ion(Ilie) Istrăţescu

Preface

Interest in the geometric properties of Banach spaces is due, to a great extent, to the fact that the linear topological properties, which are extremely useful in applications, are inseparably linked with a fixed geometrical object, namely, the closed unit ball of the space [i.e., the set $B_1(0) = \{x : \|x\| \leq 1\}$]. Thus we are led naturally to consider linear topological properties within the framework of a given norm on the space. The purpose of this book is to present a comprehensive survey of those properties of a Banach space related to strict convexity, together with some applications.

The book contains three chapters. The first chapter is devoted to some of the basic results of linear functional analysis; readers who have had a one-year course in functional analysis may omit this chapter. Our treatment of the subject of the book begins in Chap. 2. The class of strictly convex spaces was first defined and investigated by J. Clarkson and M. G. Krein. We present several characterizations of these spaces, using real extreme points and certain classes of spaces related to the Banach space, such as the $\ell^p(X)$ spaces. An interesting and important characterization of strictly convex spaces mentioned in the text is the one involving duality mappings (introduced by A. Beurling and A. Livingston). This was studied by many mathematicians, among whom we mention only a few: F. Browder, W. L. Bynum, and S. Gudder. Some relations between strict convexity (or some of its generalizations) and the extension of continuous linear functionals are also mentioned. Further, the problem of the strict convexity of subspaces, products, and quotient subspaces is discussed.

Since in applications it is convenient to have some concrete classes of Banach spaces which share the property of strict convexity, we present some of these. Among the most important and interesting is the class of uniformly convex spaces, introduced by J. A. Clarkson in 1936 and since then investigated and generalized in a great number of papers. Among the contributors to the study of uniformly convex spaces and related classes of Banach spaces we note V. Smulian, D. Milman, B. Pettis, R. C. James, I. Glicksberg, J. Lindenstrauss, M. I. Kadets (M. I. Kadec), P. Enflo, V. Klee, A. Lovaglia, E. Asplund, G. Nordlander, V. D. Milman, S. L. Troiansky, and V. Zizler.

Next we consider the modulus of convexity and the modulus of smoothness of a Banach space and related classes of Banach spaces. Here the basic results of J. Lindenstrauss, M. I. Kadets, and V. Milman are presented. We mention also a function, introduced by V. Milman, which facilitates the description of some classes of spaces, e.g., the uniformly convex spaces and the uniformly nonsquare spaces of James. Further, we study relations between the differentiability properties of the norm and some geometric properties of the Banach space. The problem of deciding when a Banach space has an equivalent norm with given properties is treated in the last part of this chapter, for the properties of strict convexity and uniform convexity. Here we present the theorems of V. Klee on strict convexity and the remarkable result of R. C. James and P. Enflo on uniformly convex spaces.

The chapter continues with some applications. First we give applications to the theory of approximation (the proximity operator, or the Chebyshev operator) and further to fixed point theory. Here we note that the first result in which uniform convexity plays a fundamental role was a theorem of F. Browder that a nonexpansive mapping defined on a bounded, closed, and convex subset in an uniformly convex space has a fixed point. We present results related to this theorem as well as for a class of mappings called T-maps, which were first studied by the Italian mathematician Francesco Tricomi. Also, the normal structure and geometric properties of Banach spaces and fixed point theory are discussed. Special attention is given to a new class of spaces, the so-called normed probabilistic metric spaces introduced by A. Šerstnev, which are a particular case of probabilistic metric spaces, introduced by K. Menger in 1942. A property is considered which seems to be a good candidate for strict convexity in this setting.

In their interesting study of the space of all continuous (real-valued) functions on a compact Hausdorff space, R. Arens and J. L. Kelley have determined explicitly the form of extreme points of the closed unit ball of the conjugate space. It was shown that these coincide with the set of all nontrivial multiplicative functionals. The case of linear operators between two spaces of continuous functions was first considered by C. I. Tulcea and A. I. Tulcea. The problem has attracted the attention of many people and further theorems were obtained by R. Blumenthal, F. Bonsall, J. Lindenstrauss, R. Phelps, and M. Sharir. The second chapter ends by presenting results related to those of Arens-Kelley and Tulcea.

One of the most important properties of the space of complex analytic functions is the validity of the maximum principle for elements in the space. It is quite natural to ask if this fundamental property holds for Banach-valued analytic functions. A well-known example (mentioned, for example, in the book of Hille et al.) shows that this is not the case for all Banach spaces. In the third chapter we solve the problem of characterizing those Banach spaces for which the maximum modulus holds for analytic functions with values in the space. This is achieved by defining the so-called complex extreme points. By the use of this class of points in the unit ball of a Banach space we study the class of complex strictly convex spaces (i.e., those spaces for which all elements of norm 1 are complex extreme points). Further, we present some classes of spaces which are, in some sense, analogous to some classes of Banach spaces considered in Chap. 2.

Almost all the results we include have appeared in the literature. We make no claim of encyclopedic coverage, for we have concentrated on those aspects which seem to us most interesting and significant. Also, we have attempted to provide the reader with a bibliography, which does not pretend to be complete, but which, we hope, will serve as an adequate guide to the history and the current status of the topics presented in this book.

We wish to acknowledge with thanks conversations and correspondence in functional analysis (geometric theory of Banach spaces) from which we have benefited. Out appreciation goes also to the editors of the Marcel Dekker, Inc. for their attention to this volume.

Vasile I. Istrăţescu

Contents

1
Banach Spaces

1.1 LINEAR SPACES

A fundamental notion in linear and nonlinear analysis, that of linear space over a field $\mathbb{K}$ ($\mathbb{K}$ is $\mathbb{R}$ or $\mathbb{C}$, the set of all real numbers or the set of all complex numbers, respectively) is considered in the following definition.

DEFINITION 1.1.1 The nonempty set V is a linear space over $\mathbb{K}$ if there exist two functions defined on $V \times V$ and $\mathbb{K} \times V$, denoted by + and $\cdot$, respectively, which satisfy the following properties:

1. $x + (y + z) = (x + y) + z$ (associativity)
2. $x + y = y + x$ (commutativity)
3. There exists an element 0 (zero) of V such that $x + 0 = x$
4. $0 \cdot x = 0$
5. $a \cdot (x + y) = a \cdot x + a \cdot y$
6. $a \cdot (b \cdot x) = ab \cdot x$
7. $(a + b) \cdot x = a \cdot x + b \cdot x$
8. $1 \cdot x = x$

for all x,y,z in V and a,b in $\mathbb{K}$.

Sometimes $a \cdot x$ is denoted simply as ax, and we adopt this convention in what follows. The function + is called addition and the function $\cdot$ is said to be multiplication by scalars (of $\mathbb{K}$).

We now give some examples of linear spaces.

EXAMPLE 1.1.2 We consider an arbitrary set S, with V the set of all functions defined on S with values in $\mathbb{K}$. If f,g are two elements of V, then $h = f + g$ is defined by

$$h(s) = f(s) + g(s)$$

for all s in S. For any a in $\mathbb{K}$ and f in V, the element af is the function defined as follows:

$$(af)(s) = af(s)$$

for all s in S. With the addition and multiplication by scalars defined above it is clear that V is a linear space over $\mathbb{K}$.

As a matter of terminology, we mention that a linear space is called a real linear space if $\mathbb{K} = \mathbb{R}$ and a complex space if $\mathbb{K} = \mathbb{C}$.

EXAMPLE 1.1.3 Let n be a finite integer or ∞ and consider the set V of all sequences of the form $x = (x_1, x_2, \ldots, x_n)$ if n is finite and $x = (x_1, x_2, x_3, \ldots)$ otherwise. Here x_i are elements of $\mathbb{K}$. We show that V can be structured as a linear space over $\mathbb{K}$. Indeed, if $y = (y_1, y_2, \ldots, y_n)$ is another element of V then we set as $x + y$ the element

$$(x_1 + y_1,\ x_2 + y_2,\ x_3 + y_3,\ \ldots,\ x_n + y_n)$$

and if a is an arbitrary element of $\mathbb{K}$ the element $a \cdot x$ is defined as

$$(ax_1, ax_2, ax_3, \ldots, ax_n)$$

It is obvious that V with + and $\cdot$ defined as above is a linear space over $\mathbb{K}$. We can proceed in a similar way for ∞.

REMARK 1.1.4 It is clear that Example 1.1.3 is a particular case of the space considered in 1.1.2, namely, when the set S is countable.

EXAMPLE 1.1.5 Let X be a compact Hausdorff space and C(X) be the set of all continuous complex-valued functions on X. With the natural functions + and $\cdot$, C(X) is a complex linear space over $\mathbb{C}$. If we consider only the real-valued continuous functions, then it is clear that we get a real linear space.

DEFINITION 1.1.6 If V is a linear space over $\mathbb{K}$ then a subset V_1 of V is said to be a linear subspace if V_1 is a linear space with the functions + and $\cdot$ defined exactly as in V.

EXAMPLE 1.1.7 If V is the linear space of Example 1.1.2 and s_0 is a fixed point of S then

$$V_{s_0} = \{f : f \in V,\ f(s_0) = 0\}$$

is obviously a linear subspace of V.

This is valid also for the space of Example 1.1.5 as well as for the space of Example 1.1.3. Other interesting examples of subspaces will be given later.

1.2 SETS IN LINEAR SPACES

In what follows we define various classes of subsets in linear spaces. First we note that the family of linear subspaces of a linear space is an important class of objects associated with the space. One of the most important classes of subsets of a linear space is that consisting of the so-called *convex sets*.

DEFINITION 1.2.1 Let V be a linear space over $\mathbb{K}$. A subset C of V is called convex if whenever x,y are in C, then for all t in (0,1) the element

$$z_t = tx + (1 - t)y$$

is in C.

DEFINITION 1.2.2 If E is a subset of a linear space V then conv E, the convex hull of E, is the smallest convex set containing E.

It is not difficult to see, using Zorn's lemma, that for each subset E, conv E exists.

The following proposition gives another description of conv E.

PROPOSITION 1.2.3 If E is a subset of a linear space V then

$$\text{conv } E = \left\{z : z = \sum_{i=1}^{n} a_i x_i,\ a_i \geq 0,\ \sum_{i=1}^{n} a_i = 1,\ x_i \in E\right\}$$

Proof: Let us denote by E_1 the set

$$\left\{z : z = \sum_{i=1}^{n} a_i x_i,\ a_i \geq 0,\ \sum_{i=1}^{n} a_i = 1,\ x_i \in E\right\}$$

We remark that this set is convex. Indeed, if z_1 and z_2 are two elements of E_1,

$$z_1 = a_1x_1 + \cdots + a_nx_n \qquad z_2 = b_1y_1 + \cdots + b_my_m$$

then for all t in (0,1) the element tx + (1 - t)y is in E_1, since we can write

$$tx + (1 - t)y = \sum_{i=1}^{n+m} a_i' z_i$$

with

$$a_i' = \begin{cases} ta_i & i \leq n \\ (1 - t)b_{n+m-i} & i > n \end{cases}$$

$$z_i = \begin{cases} x_i & i \le n \\ y_{i-n} & i > n \end{cases}$$

Thus the convexity of E_1 is proved. To prove that E_1 is the smallest convex set containing E, we must show that any convex set E_2 containing E contains E_1. For this it is sufficient to show that E_2 contains all the elements z of the form

$$z = a_1x_1 + \cdots + a_nx_n \qquad a_i \in [0,1],\ \Sigma\, a_i = 1,\ x_i \in E$$

For n = 2 this is obvious. Suppose that the assertion is valid for $n \le k$. We prove it for k + 1.

Let us consider $z = a_1x_1 + \cdots + a_kx_k + a_{k+1}x_{k+1}$ where a_k are positive numbers and $\Sigma_{i=1}^{n} a_i = 1$ and x_i are in E. We must show that this is in E_2. We can write

$$\begin{aligned} z &= a_1x_1 + \cdots + a_kx_k + a_{k+1}x_{k+1} \\ &= a_{k+1}x_{k+1} + (1 - a_{k+1})\left(\sum_{i=1}^{k} \frac{a_i}{1 - a_{k+1}}\right) x_i \\ &= a_{k+1}x_{k+1} + (1 - a_{k+1})z' \end{aligned}$$

where z' is in E_2 by induction. This implies clearly that z is in E_2. Thus we get that E_1 is the smallest convex set containing E, and the proposition is proved.

We give now examples of convex sets.

EXAMPLE 1.2.4 Let V be the linear space of Example 1.1.2. Then

$$C = \{f : f \in V,\ f(s) \in (0,1),\ s \in S\}$$

is obviously a convex set.

EXAMPLE 1.2.5 Let X be a compact Hausdorff space and C(X) be the space of all complex-valued continuous functions on X. The set C defined by

$$\{f : f \in C(X),\ \sup_{s \in X}|f(s)| \le 1\}$$

is clearly a convex set.

We consider now other classes of sets in linear spaces.

DEFINITION 1.2.5 Let V be a linear space over $\mathbb{K}$ and E be a subset of V. Then E is called

1. Symmetric if $E = -E = \{-x : x \in E\}$

2. Absorbing if for each x in V there exists $t_x > 0$ such that for all $|t| \leq t_x$, $tx \in E$
3. Balanced if $tE = \{tx : x \in E, |t| \leq 1\} \subseteq E\}$
4. Affine if $tE = (1 - t)E = \{tx + (1 - t)y : x,y \in E, t \in [0,1] \subseteq E\}$
5. A line through x and y if $E = \{tx + (1 - t)y : t \in \mathbb{R}\}$
6. The segment joining x and y if $E = \{tx + (1 - t)y : t \in [0,1]\}$

We now give examples of sets having some properties stated in Definition 1.2.6.

EXAMPLE 1.2.7 Let us consider V as $\mathbb{R}^2 = \{(x_1,x_2) : x_i \in \mathbb{R}\}$ ($\mathbb{R}$ the set of all real numbers) and E as the following set:

$$E = \{(x_1,x_2) : x_1^2 + x_2^2 = 1\}$$

It is easy to see that this set is symmetric and absorbing. Obviously E is not convex.

EXAMPLE 1.2.9 Let V be the space in Example 1.2.5 and E be the set defined as follows:

$$E = \{f : f \in V, f \text{ is real valued}\}$$

Then clearly E is convex and balanced.

1.3 SEMINORMS AND NORMS ON LINEAR SPACES. LOCALLY CONVEX SPACES

Let V be a linear space over $\mathbb{K}$. An important class of functions on V is considered in the following definition.

DEFINITION 1.3.1 The function

$$p : V \to \mathbb{R}$$

is called a seminorm if the following properties hold:

$$p(x + y) \leq p(x) + p(y)$$
$$p(ax) = |a|p(x)$$

for all x,y in V and all a in $\mathbb{K}$. The seminorm is said to be norm if $p(x) = 0$ if and only if $x = 0$.

We remark that the values of a seminorm are in fact in $\mathbb{R}^+$. Indeed, if we take $x = -y$ then we get

$$p(0) = 0 \leq p(x) + p(-x) = 2p(x)$$

and the assertion is proved.

Some examples of seminorms follow.

EXAMPLE 1.3.2 If V is the linear space of Example 1.1.2 and s_0 is an arbitrary but fixed point of S, the function

$$p_{s_0}(f) = |f(s_0)|$$

is obviously a seminorm on V.

EXAMPLE 1.3.3 If C(X) is the space in Example 1.1.5 then the function

$$f \to p(f) = \sup_{s \in X} |f(s)|$$

is a norm on C(X).

As is now standard, we use the following notation for a norm:

$$p(x) = \|x\|$$

and different norms will be denoted by $\|\cdot\|_1$, $\|\cdot\|_2$, $\|\cdot\|_s$, ...

If V is a linear space over $\mathbb{K}$ and p is a seminorm then the set

$$B_p(0,1) = \{x : p(x) < 1\}$$

is clearly convex, symmetric, balanced, and absorbing. The following proposition gives the connection between convex sets which are balanced and absorbing and the seminorms. This connection is expressed using the so-called *gauge* or *Minkowski functional*.

PROPOSITION 1.3.4 Let V be a linear space over $\mathbb{K}$ and E be a subset in V with the following properties:

a. E is convex.

b. E is balanced and absorbing.

Then the function p_E defined on V by the formula

$$p_E(x) = \inf\{t > 0 : x \in tE\}$$

has the following properties:

1. $p_E(x) \geq 0$.
2. $p_E(x + y) \leq p_E(x) + p_E(y)$.
3. $p_E(ax) = |a| p_E(x)$.
4. $\{x : p_E(x) < 1\} \subseteq E \subseteq \{x : p_E(x) \leq 1\}$.

Proof: Properties 1 and 3 are obvious. To prove 2, let $\varepsilon > 0$, then

$$x \in [p_E(x) + \varepsilon]E$$
$$y \in [p_E(y) + \varepsilon]E$$

and thus

$$x + y \in [p_E(x) + p_E(y) + 2\varepsilon]E$$

which follows from the convexity of E. Since ε is arbitrary we get that 2 holds. The last relations between the sets follow from the definition of p_E.

To define the very important notion of locally convex space, first we recall the notion of topological space.

DEFINITION 1.3.5 If S is an arbitrary nonempty set, then a topology on S is any collection τ of subsets of S satisfying the following properties:

1. $S \in \tau$, $\emptyset \in \tau$ (the empty set is denoted by $\emptyset$).
2. The union of an arbitrary family of elements of τ is in τ.
3. If $G_1, \ldots, G_m$ are in τ (m is finite), then $G_1 \cap G_2 \cap \cdots \cap G_m$ is in τ.

The pair (S,τ) is called a topological space. For short we say that S itself is a topological space.

DEFINITION 1.3.6 If $s \in S$, then a neighborhood of s is any subset in S containing a set V in τ such that $s \in V$.

DEFINITION 1.3.7 If S_1 and S_2 are two topological spaces and $f : S_1 \to S_2$ is a mapping defined on S_1 with values in S_2, then we say that f is continuous at the point s_1 of S_1 if for any neighborhood V^2 of $f(x_1)$ there exists a neighborhood V^1 of s_1 such that $f(s) \in V^2$ if $s \in V^1$. The function f is said to be continuous on S_1 if it is continuous at each point of S_1.

The notion of topological linear space is considered in the following definition.

DEFINITION 1.3.8 A linear space V over $\mathbb{K}$ is said to be a topological linear space if on V there exists a topology such that $V \times V$ and $\mathbb{K} \times V$ with the product topology have the property that + and $\cdot$ are continuous functions.

In this case τ is called a linear topology on V. For a detailed account of linear topological space theory we refer to the excellent texts of N. Bourbaki (1955) and A. Grothendieck (1954).

The subclass of locally convex spaces is considered in the following definition.

DEFINITION 1.3.9 A linear topology on a linear topological space V is said to be a locally convex topology if every neighborhood of 0 (the zero of V) includes a convex neighborhood of 0.

Then we have the following result.

PROPOSITION 1.3.10 If V is a locally convex space over $\mathbb{K}$ then the topology of V is determined by a family of seminorms $(p_i)_{i \in I}$.

For the proof of this assertion see the above-mentioned references.

1.4 BANACH SPACES. EXAMPLES

An important class of locally convex spaces is the class of Banach spaces, in which the family of seminorms reduces to a single norm having a special property considered below.

If

$$x \to \|x\|$$

is a norm on the linear space V then the function on $V \times V$ defined by the relation

$$(x,y) \to \|x - y\| = d(x,y)$$

defines a metric on V, i.e., d satisfies the following properties:

1. $d(x,y) \geq 0$, and $d(x,y) = 0$ if and only if $x = y$
2. $d(x,y) \leq d(x,z) + d(y,z)$
3. $d(x,y) = d(y,x)$

for all x,y,z in V.

Since the proof that d has the above properties is very simple we omit it.

DEFINITION 1.4.1 If (x_n) is a sequence of elements of V then we say that it converges to $x \in V$ if $\lim \|x_n - x\| = 0$.

The notion of Cauchy sequence is defined as follows.

DEFINITION 1.4.2 A sequence (y_n) of elements of V is called a Cauchy sequence if for any $\varepsilon > 0$ there exists an integer N_ε such that for all $n,m \geq N_\varepsilon$,

$$\|x_n - x_m\| \leq \varepsilon$$

Now the notion of Banach space is considered in the following definition.

DEFINITION 1.4.3 A linear space over $\mathbb{K}$ is called a Banach space if on V there exists a norm

$$x \to \|x\|$$

such that every Cauchy sequence of elements of V is a convergent sequence. (In other words, using the terminology of metric spaces, V with the metric d defined as above is a complete metric space.)

We give now some examples of Banach spaces. First, a linear space on which there exists a norm is called a normed space, and we have real normed spaces and complex normed spaces, respectively, according as the field $\mathbb{K}$ is $\mathbb{R}$ or $\mathbb{C}$. Also, a Banach space is a real Banach space if the field if $\mathbb{K} = \mathbb{R}$ and a complex Banach space if $\mathbb{K} = \mathbb{C}$.

EXAMPLE 1.4.5 Suppose $X = [0,1]$ and consider the following function defined on C(X), the space of all continuous $\mathbb{K}$-valued functions on X:

$$f \to \int_0^1 |f(s)|\, ds$$

which is obviously a norm on C(X). In this case C(X) is a normed linear space.

EXAMPLE 1.4.6 Let us consider $p \in (1,\infty)$ and let ℓ^p be the set of all sequences $x = (x_i)$, $x_i \in \mathbb{K}$, with the property that

$$x \to \|x\|_p = \left(\sum_{i=1}^{\infty} |x_i|^p\right)^{1/p} < \infty$$

Then, using Minkowski's inequality, we get that this is a norm on ℓ^p. (For the structure of ℓ^p as a linear space see Example 1.1.2.) Also, it is not very difficult to prove that in fact ℓ^p with the above norm is a Banach space.

EXAMPLE 1.4.7 Let us denote by ℓ^∞ the set of all bounded sequences $x = (x_i)$, $x_i \in \mathbb{K}$. With the norm

$$x \to \|x\|_\infty = \sup_i |x_i|$$

ℓ^∞ is a Banach space.

We note now some very interesting and important subspaces of ℓ^∞ (ℓ^∞ is denoted sometimes by m).

EXAMPLE 1.4.8 Let

$$c = \{x : x \in \ell^\infty,\ x = (x_i) \text{ is a convergent sequence}\}$$
$$c_0 = \{x : x \in \ell^\infty,\ x = (x_i) \text{ converges to zero}\}$$

It is clear that these are Banach spaces with the norm defined as in ℓ^∞.

EXAMPLE 1.4.9 Let (Λ,B,P) be a measure space and suppose for simplicity that $P(\Lambda) < \infty$. For p in $(1,\infty)$ consider the measurable $\mathbb{K}$-valued functions (classes of functions) defined on Λ such that $\int_\Lambda |f(s)|^p\, ds$ is finite. With the norm

$$f \to \|f\|_p = (\int_\Lambda |f(s)|^p\, dP)^{1/p}$$

the set of all such functions (classes of functions), denoted by $L^p(\Lambda,B,P)$, is a Banach space. For the proof we refer to any book on measure theory, where the reader may also find other interesting properties of these spaces (including the cases $p = 1, \infty$).

1.5 LINEAR OPERATORS ON BANACH SPACES. THE DUAL SPACE

Let us consider two Banach spaces V_1 and V_2 over $\mathbb{K}$. Among the most important functions defined on V_1 with values in V_2 are the linear operators. This class of functions is defined below.

DEFINITION 1.5.1 The function

$$T : V_1 \to V_2$$

is a linear operator if the following properties hold:

$$T(x + y) = T(x) + T(y)$$
$$T(ax) = aT(x)$$

for all x,y in V_1 and a in $\mathbb{K}$. Sometimes we write $T(x) = Tx$ for short.

If $V_2 = \mathbb{K}$ then any linear operator defined on V_1 with values in V_2 is called a linear functional.

DEFINITION 1.5.2 If V_1 and V_2 are two Banach spaces then the set of all continuous linear operators defined on V_1 with values in V_2 is denoted by $L(V_1,V_2)$; if $V_1 = V_2 = V$ then we write simply $L(V,V) = L(V)$.

DEFINITION 1.5.3 A linear operator $T : V_1 \to V_2$ is said to be bounded if

$$\|T\| = \sup_{\|x\|\le 1} \|Tx\|$$

is finite, and the number $\|T\|$ is called the norm of T. It is not difficult to see that

$$T \to \|T\|$$

is a norm on the linear space of all bounded operators on V_1.

The connection between continuity and boundedness is given in the following proposition.

PROPOSITION 1.5.4 A linear operator $T : V_1 \to V_2$ is bounded if and only if it is continuous.

Proof: It is obvious that any bounded linear operator is continuous. To prove the converse, suppose that it is not true. In this case for each n there exists an element x_n in V_1 such that

$$\|x_n\| = 1 \qquad \|Tx_n\| \geq n$$

We consider now the sequence (x_n') with $x_n' = x_n/n$. Obviously this converges to zero and $\|Tx_n'\| \geq 1$. This contradicts the continuity of T at x = 0, and thus the proposition is proved.

DEFINITION 1.5.5 The set of all continuous linear functionals on a Banach space V is denoted by V* and is called the dual space of V. The dual space of V* is called the bidual or the second conjugate of V. In general, we define V^{*n} as the dual space of $V^{*(n-1)}$.

We give now some examples of continuous linear functionals and bounded linear operators.

EXAMPLE 1.5.7 Let us consider the Banach space $C_{[0,1]}$ of all continuous real-valued functions on [0,1], with the norm

$$f \to \|f\| = \sup_t |f(t)|$$

Let t_0 be a fixed point in [0,1]. We set

$$F_{t_0}(f) = f(t_0)$$

This is obviously a continuous linear functional on $C_{[0,1]}$. It is not difficult to see that

$$\|F_{t_0}\| = 1$$

On the same space we consider the following operator, called the *Volterra operator*:

$$f \to (Tf)(x) = \int_0^x f(s)\, ds$$

It is clear that T is a bounded linear operator with $\|T\| \leq 1$.

EXAMPLE 1.5.8 If $V = \ell^p$ with p fixed in $(1,\infty)$ then for b fixed in ℓ^q $(1/p + 1/q = 1)$ we define the linear functional

$$F_b(x) = x_1 b_1 + x_2 b_2 + \cdots$$

for each x in ℓ^p. Obviously, as follows from the Hölder inequality, F_b is well defined and

$$\|F_b\| \leq \|b\|_q$$

(In fact $\|F_b\| = \|b\|_q$.)

1.6 EXTENSION THEOREMS

Let V be a Banach space over $\mathbb{K}$ and consider the space V*. It is not clear that this space contains nontrivial elements, i.e., we do not know if there exists a bounded linear functional on V which is not identically zero. The extension theorems of Hahn-Banach and Bohnenblust-Sobczyc (for real Banach spaces and complex Banach spaces, respectively) assure the nontriviality of the space V*. Moreover, these theorems have a great number of applications. In what follows we present these theorems as well as some simple applications. It is worth mentioning that these extension theorems are the only generally available methods of constructing linear operators from one Banach space to another. First we consider the case of real linear spaces.

DEFINITION 1.6.1 Let V be a linear space and V_1 be a linear subspace of V. If $f : V_1 \to \mathbb{R}$ has the property that there exists a function $F : V \to \mathbb{R}$ such that $f(x) = F(x)$ for all x in V_1 then we say that F is an extension of f (sometimes f is called the restriction of F to V_1).

THEOREM 1.6.2 (Hahn-Banach Extension Theorem) If V_1 is a real subspace of the linear real space V and $f : V_1 \to \mathbb{R}$ is a linear functional which satisfies the inequality

$$f(x) \leq p(x)$$

for all x in V_1, where p is a seminorm defined on V, then f has an extension $F : V \to \mathbb{R}$ satisfying the inequality

$$F(x) \leq p(x)$$

Proof: Let x_0 be an arbitrary but fixed element of V not in V_1 and consider the linear subspace generated by V_1 and x_0 (i.e., the smallest linear subspace of V which contains x_0 and V_1). We remark that the elements of this subspace admit a unique decomposition of the form

$$z = rx_0 + y$$

where r is a real number and y is in V_1. For any z of the above form we set

$$f_1(z) = rc + f(y)$$

with c a number which we shall determine such that

$$f_1(z) \le p(z)$$

Suppose first that such a number c exists. In this case we have

$$f_1\left(\frac{y}{r}\right) + c \le p\left(\frac{y}{r} + x_0\right) \qquad r > 0$$

$$f_1\left(\frac{y}{r}\right) + c \ge -p\left(-\frac{y}{r} - x_0\right) \qquad r < 0$$

for all y in V_1. Also

$$f(y) - p(y - x_0) \le c \le -f(x) + p(x + x_0)$$

holds for all x,y in V_1, and since

$$f(y) + f(x) = f(x + y) \le p(x + y) \le p(y - x_0) + p(x + x_0)$$

we obtain

$$m \le M$$

where

$$m = \sup \{f(y) - p(y - x_0)\}$$
$$M = \inf \{-f(x) + p(x + x_0)\}$$

It is now clear that we can choose as c any number between m and M. Thus the Hahn-Banach theorem is proved if the subspace V_1 is of special type. We remark that just from the result obtained it follows that the family of subspaces of V on which there exist extensions of f is nonempty. The proof of the Hahn-Banach theorem follows now, in the general case, by an application of Zorn's lemma. For this purpose we consider all pairs (V',f') where

V is a linear subspace of V containing V_1 and f' is an extension of f to V'. We consider a partial order on this family of pairs defined as follows: If (V',f') and (V'',f'') are two pairs we say that

$$(V',f') \le (V'',f'')$$

if V' is a linear subspace of V'' and f'' is an extension of f'. Then it is clear that this is a partial order, and it is not difficult to see that we can apply Zorn's lemma and thus we can consider a pair $(\bar{V},\bar{f})$ which is a maximal element. If $\bar{V}$ is not V then there exists an element of V, say y_0, which is not in $\bar{V}$. Applying the first part of the proof we find an extension of f to the space generated by $\bar{V}$ and y_0, say g. The pair $((\bar{V},y_0),g) \ge (\bar{V},\bar{f})$ and this contradicts the fact that $(\bar{V},\bar{f})$ is a maximal element. Thus we must have $\bar{V} = V$, and the Hahn-Banach extension theorem is proved.

Before we give the Bohnenblust-Sobczyc extension theorem (the complex case) we note three useful corollaries of the Hahn-Banach theorem.

COROLLARY 1.6.3 Let V be a real Banach space and V_1 be a subspace with $f : V_1 \to \mathbb{R}$, a continuous linear functional. Then there exists a continuous linear functional $F : V \to \mathbb{R}$ such that

$$f(x) = F(x) \qquad x \in V_1$$

$$\|f\|_{V_1} = \sup_{\substack{x \in V_1 \\ \|x\| \le 1}} |f(x)| = \|F\|$$

Proof: We consider on V the seminorm

$$x \to p(x) = \|f\|_{V_1} \|x\|$$

Then there exists an extension of f satisfying the relation $F(x) \le p(x)$. It is not difficult to see that this F satisfies our requirements.

COROLLARY 1.6.4 Let V_1 be a subspace of V, where V is a real Banach space. Then V_1 is dense in V (i.e., the closure of V_1 is V) if and only if for any f* in V* which is zero on V_1 is identically zero.

Proof: Obvious from the first part of the proof of the Hahn-Banach extension theorem.

COROLLARY 1.6.5 Let V be a real Banach space and x_0 an arbitrary element of V. Then there exists f in V* with the following properties:

$$\|f\| = 1$$

$$f(x_0) = \|x_0\|$$

Proof: We consider on the space (rx_0) generated by x_0 the linear functional

$$f(rx_0) = r\|x_0\|$$

Then f is obviously continuous, the norm of f with respect to the subspace (rx_0) is 1, and the value at x_0 is $\|x_0\|$. It is clear that the extension of f satisfies our corollary.

We consider now the complex case of the extension theorem.

THEOREM 1.6.6 (Bohnenblust-Sobczyc Extension Theorem) Let V be a complex Banach space and V_1 a subspace of V. Suppose that $f : V_1 \to \mathbb{C}$ is a continuous linear functional on V_1. Then there exists a continuous linear functional $F : V \to C$ having the following properties:

$$f(x) = F(x) \qquad x \in V_1$$

$$\|f\|_{V_1} = \|F\|$$

Proof: For any linear functional h on V_1, we can write

$$h(x) = \text{Re } h(x) + i \text{ Im } h(x) = \text{Re } h(x) - i \text{ Re } h(ix)$$

Thus we can assume that our linear functional f is of the form

$$f(x) = u(x) - iu(ix)$$

where u is a real linear functional (i.e., a linear functional on V_1 as a real linear Banach space). Then by the Hahn-Banach extension theorem there exists an extension of u, say U, with the property that $\|u\|_{V_1} = \|U\|$. Then we set

$$F(x) = U(x) - iU(ix)$$

and this satisfies our extension theorem. Indeed, it is obvious that F is an extension of f. It remains to prove that $\|f\| = \|F\|$. Obviously $\|f\| \le \|F\|$. To prove the converse inequality we note that for each x in V there exists a complex number z of modulus 1 such that $zF(x) = |F(x)|$, and thus

$$|F(x)| = F(zx) = U(zx) \le \|U\|\|zx\| = \|U\|\|x\|$$

which gives the inequality.

REMARK 1.6.7 The extension theorem for complex Banach spaces was proved independently by G. Sukhomlinov, who considered also the case of quaternionic Banach spaces.

Let V be a Banach space over $\mathbb{K}$ ($\mathbb{R}$ or $\mathbb{C}$) and consider the sequence of dual spaces (V^{*n}). Let x be an arbitrary point of V and define a linear functional on V* by

$$F_x(f) = f(x)$$

Obviously F_x is linear and

$$\|F_x\| \leq \|x\|$$

Thus it is continuous, and by Corollary 1.6.5 (this as well as the other corollaries of the extension theorem can be proved for the complex case),

$$\|F_x\| = \|x\|$$

i.e., the above mapping is isometric (preserves distances).

Thus we have a mapping defined on V with values in V**. The following class of Banach spaces was introduced by H. Hahn in connection with this mapping: A Banach space is said to be reflexive if the above mapping of V into V** is onto. The name *reflexive Banach spaces* for the class considered by H. Hahn was suggested by E. Lorch.

We note that the class of reflexive spaces includes the most familiar spaces such as ℓ^p and L^p [for p in $(1,\infty)$]. The proof of this fact will be given in the next chapter as a particular case of a more general result, the so-called Milmam-Pettis theorem on the reflexivity of the class of uniformly convex spaces of Clarkson.

Using Corollary 1.6.5 (as well as its complex version) we prove, following G. Lumer (1961), the existence of *semi-inner products* on Banach spaces.

DEFINITION 1.6.8 Let V be a complex Banach space. A mapping

$$[\ ,\ ] : V \times V \to \mathbb{C}$$

is said to be a semi-inner product on V if the following properties hold:

1. $[ax + by, z] = a[x,z] + b[y,z]$
2. $[x,by] = \bar{b}[x,y]$
3. $|[x,y]| \leq \|x\|\|y\|$
4. $[x,x] = \|x\|^2$

for all x,y,z in V and a,b in $\mathbb{C}$.

Then we have

PROPOSITION 1.6.9 On any Banach space there exist semi-inner products.

Proof: Let x be an element of V and f_x be an element of V* with the following properties:

$$\|f_x\| = \|x\|$$

$$f_x(x) = \|x\|^2$$

From Corollary 1.6.5 (as well as its complex version) we get that such a functional exists. We say that the one-dimensional space V_1 is determined by $u \in V$ if the space generated by u is V_1. For each one-dimensional subspace V_1 of V we choose an element u which determines V_1. Obviously each $v \in V_1$ is of the form v = au, and we set

$$f_v = af_u$$

The semi-inner product on V is defined as follows:

$$[x,y] = f_y(x)$$

Then it is obvious that properties 1 through 4 hold.

REMARK 1.6.10 Property 2 of semi-inner products was first considered by J. R. Giles (1967).

REMARK 1.6.11 It is not difficult to see that a Banach space is a Hilbert space (Definition 1.8.6) if there exists a semi-inner product which is linear in the second variable. (See Definition 1.8.1 for the notion of inner product on linear spaces.)

1.7 THREE BASIC THEOREMS OF LINEAR FUNCTIONAL ANALYSIS

Let V be a Banach space and $p : V \to \mathbb{R}$. The function p is said to be subadditive if for all x,y in V,

$$p(x + y) \le p(x) + p(y)$$

Concerning this class of functions we prove the following.

PROPOSITION 1.7.1 If p is subadditive and continuous at x = 0, then p is continuous on V.

Proof: Let (x_n) be a sequence converging to x. Then clearly $(x_n - x)$ converges to zero and

$$p(x) = p(x - x_n + x_n) \le p(x - x_n) + p(x_n)$$

which implies that

$$p(x) - p(x_n) \leq p(x - x_n)$$

Similarly we obtain

$$p(x_n) = p(x) \leq p(x - x_n)$$

and thus

$$|p(x) - p(x_n)| \leq p(x - x_n)$$

and this obviously implies the continuity of p at x. The proposition is proved.

For the proof of the following lemma of Zabreiko, we need Baire's theorem. [See N. Dunford and J. T. Schwartz (1958) or V. I. Istratescu (1981a).]

LEMMA 1.7.2 (P. Zabreiko, 1967) Let V be a Banach space and $p : V \to \mathbb{R}$ be a function with the following properties:

1. $\lim_{t \to 0^+} p(tx) = 0$ for each x in V.
2. For any convergent series $\Sigma_{n=1}^{\infty} x_n$, $p(\Sigma_{n=1}^{\infty} x_n) \leq \Sigma_{n=1}^{\infty} p(x_n)$.

Then p is continuous.

Proof: It is clear that p is subadditive, and thus to prove the continuity of p it suffices to prove continuity at x = 0. Let us consider $\varepsilon > 0$ and define the set

$$G_\varepsilon = \left\{x : x \in V,\ p(x) + p(-x) \leq \frac{\varepsilon}{2}\right\}$$

which is closed and symmetric. From the first property of p it follows that

$$V = \bigcup_n nG_\varepsilon$$

Thus by Baire's theorem there exists n_0 such that $n_0 G_\varepsilon$ contains a ball $B(x_0,r)$ for some $r > 0$. Since G is symmetric we can suppose without loss of generality that this is the ball $B(0,\delta(\varepsilon))$ with center at x = 0 and with a radius less than ε. Then for any x in this ball we have

$$p(x) \leq 2\varepsilon$$

Indeed, let $G'_\varepsilon = B(0,\delta(\varepsilon)) \cap G_\varepsilon$. This set is dense in $B(0,\delta(\varepsilon))$, and thus for x in $B(0,\delta(\varepsilon))$ there exists x_1 in G'_ε such that

$$\|x - x_1\| \leq \frac{\varepsilon}{2}$$

Suppose now that $x_1, \ldots, x_n$ were chosen. We choose x_{n+1} in $G'_{\varepsilon/2^{n+1}} = B(0,\delta(\varepsilon)) \cap G'_{\varepsilon/2^{n+1}}$ such that

$$\|x - x_1 - \cdots - x_n - x_{n+1}\| \leq \frac{\varepsilon}{2^{n+1}}$$

Now it is obvious that $\Sigma_1^\infty x_n$ is a convergent series, and thus by property 2 of p we get

$$p(x) \leq \sum_{i=1}^{\infty} p(x_i) = 2\varepsilon$$

This obviously implies that p is continuous at $x = 0$, and by the above proposition, p is continuous on V.

Using this lemma we prove three basic theorems of linear functional analysis.

A set M is a G_δ set if it is equal to a countable intersection of open subsets.

THEOREM 1.7.3 (Banach-Steinhaus) Let V be a Banach space and $(T_i)_{i\in I}$ be a family of elements in L(V,U), where U is another Banach space. Then either there exists $M < \infty$ such that

$$\sup_{i\in I} \|T_i\| \leq M$$

or

$$\sup_{i\in I} \|T_i x\| = \infty$$

for all x in some G_δ set of V.

Proof: We define the function p by the formula

$$p(x) = \sup_{i\in I} \|T_i x\|$$

which is clearly subadditive and lower semicontinuous. From the lower semicontinuity we get that p satisfies condition 2 in Zabreiko's lemma, and since the first property is obvious, we can apply this lemma. Now, if the first assertion of our theorem does not hold, then

$$G_n = \{x : \sup_i \|T_i x\| > n\}$$

is open, and thus the second assertion of the theorem holds for all x in $\cap_n G_n$, which is clearly a G_δ set.

REMARK 1.7.4 The result in Theorem 1.7.3 is known also as the *principle of uniform boundedness.*

For any Banach space V,

$$B_V(0,r) = \{x : x \in V, \|x\| < r\}$$

THEOREM 1.7.5 (Open Mapping Theorem) Let $T : U \to V$ be in $L(U,V)$, where U and V are Banach spaces, and suppose that T is onto [i.e., $T(U) = V$]. Then there exists $\delta > 0$ such that

$$T(B_U(0,1)) \supseteq \delta B_V(0,1)$$

Proof: We consider the following function defined on V:

$$p(x) = \inf \{\|y\| : Tx = y\}$$

It is not difficult to see that p satisfies the conditions of Zabreiko's lemma. This implies that p is continuous on X, and this obviously implies the assertion of the theorem.

THEOREM 1.7.6 (Closed Graph Theorem) Let $T : U \to V$ be a linear operator defined on the Banach space U with values in the Banach space V, and suppose that T is closed, i.e., if (x_n) converges to x and (Tx_n) converges to y then $Tx = y$. Then T is bounded.

Proof: We consider the following function on U,

$$p(x) = \|Tx\|$$

which we show satisfies the conditions of Zabreiko's lemma. The first condition obviously holds. That the second condition holds follows from the fact that T is closed. This implies that p is continuous, and this obviously is the assertion of the theorem.

1.8 HILBERT SPACES

An important class of Banach spaces is the class of Hilbert spaces which is considered below. First we define the notion of inner products on linear spaces.

DEFINITION 1.8.1 Let V be a linear space. We say that V has an inner product if there exists a function $\langle , \rangle$ defined on $V \times V$ with values in $\mathbb{C}$ satisfying the following properties:

1. $\langle ax + by, z\rangle = a\langle x,z\rangle + b\langle y,z\rangle$.
2. $\langle x,y\rangle = \overline{\langle y,x\rangle}$.

3. $\langle x,x\rangle > 0$ for all $x \in V$, $x \neq 0$.

If the space is a real linear space then property 2 is

$$\langle x,y\rangle = \langle y,x\rangle$$

PROPOSITION 1.8.2 Let V be a linear space with the inner product $\langle , \rangle$. Then the following assertions hold:

$$\langle x,ay\rangle = \bar{a}\langle x,y\rangle$$

$$\langle x,0\rangle = 0 \qquad \text{for all } x \text{ in } V$$

Proof: Obvious.

PROPOSITION 1.8.3 Let V be a linear space with an inner product $\langle , \rangle$. Then the function

$$x \to \langle x,x\rangle^{1/2} = \|x\|$$

is a norm on V.

Proof: The facts that

1. $\|x\| = 0$ if and only if $x = 0$, and $\|x\| \geq 0$
2. $\|ax\| = |a|\,\|x\|$

are obvious. We prove now that $\| \ \|$ satisfies the triangle inequality:

$$\|x + y\| \leq \|x\| + \|y\|$$

For this purpose we need the following inequality, which is also of independent interest.

LEMMA 1.8.4 (Cauchy Inequality) If V is a linear space and $\langle , \rangle$ is an inner product on V then the following inequality holds: for all x,y in V,

$$|\langle x,y\rangle| \leq \langle x,x\rangle^{1/2}\langle y,y\rangle^{1/2}$$

Proof: Let us consider a real number s with the property that

$$\langle e^{is}x,y\rangle = |\langle x,y\rangle|$$

and thus

$$0 \leq \langle e^{is}x - y,\ e^{is}x - y\rangle = \langle x,x\rangle - \langle e^{is}x,y\rangle - \langle y,e^{is}x\rangle + \langle y,y\rangle$$
$$\leq \langle x,x\rangle + \langle y,y\rangle - 2|\langle x,y\rangle|$$

which implies that

$$|\langle x,y\rangle| \leq \frac{1}{2}\left(\langle x,x\rangle + \langle y,y\rangle\right)$$

We can suppose without loss of generality that neither x nor y is the zero element of V. Then we can consider the elements

$$x_1 = \frac{x}{<x,x>^{1/2}} \qquad y_1 = \frac{y}{<y,y>^{1/2}}$$

and, according to the inequality just obtained,

$$|<x_1,y_1>| \leq 1$$

which is equivalent to the Cauchy inequality.

REMARK 1.8.5 The Cauchy inequality for the case of the L^2 spaces was first obtained by V. Buniakowsky (1859).

Now the proof that the triangle inequality is satisfied is as follows: For x,y in V, we have

$$\begin{aligned} <x + y, x + y> &= <x,x> + <y,y> + 2 \operatorname{Re} <x,y> \\ &\leq <x,x> + <y,y> + 2|<x,y>| \leq (\|x\| + \|y\|)^2 \end{aligned}$$

which implies that

$$x \to <x,x>^{1/2} = \|x\|$$

satisfies the triangle inequality.

Now we are ready to define the notion of Hilbert space.

DEFINITION 1.8.6 A Hilbert space is a Banach space on which there exists an inner product <,> such that for all $x \in V$,

$$\|x\|^2 = <x,x>$$

The norm in this case is said to be induced by an inner product.

We mention now some identities for elements of a Hilbert space, some of which characterize the inner product structure of a Banach space.

PROPOSITION 1.8.7 For any elements x,y of a Hilbert space V, the following relations hold:

1. $\|x + y\|^2 + \|x - y\|^2 = 2(\|x\|^2 + \|y\|^2)$.
2. $\|x + y\|^2 - \|x - y\|^2 = 2(<x,y> + <y,x>)$.
3. $\|x + y\|^2 - \|x - y\|^2 + i\|x + iy\|^2 - i\|x - iy\|^2 = 4<x,y>$.

Proof: Since the proof is simple it is omitted.

REMARK 1.8.8 The first identity is called the parallelogram law and is connected with the famous characterizations of Hilbert spaces in the class of Banach spaces given by Fréchet and von Neumann-Jordan [for details of this and other related problems, see V. I. Istrăţescu (1982d)].

The following proposition gives an important property of Hilbert spaces in the class of Banach spaces.

PROPOSITION 1.8.10 For any Hilbert space V and x,y in V such that

$$\|x\| \leq 1 \qquad \|y\| \leq 1$$
$$\|x - y\| \geq \varepsilon$$

there exists a positive number $\delta(\varepsilon)$ such that

$$\|x + y\| \leq 2[1 - \delta(\varepsilon)]$$

Proof: Indeed, applying the identity 1.8.7(1) we find easily that

$$\|x + y\|^2 \leq 2 - \varepsilon^2$$

and we can define

$$\delta(\varepsilon) = 1 - [1 - (\varepsilon/2)^2]^{1/2}$$

which implies that the inequality

$$\|x + y\| \leq 2[1 - \delta(\varepsilon)]$$

holds. The proposition is proved.

We give now some examples of Hilbert spaces.

EXAMPLE 1.8.11 Let ℓ^2 be the set of all sequences $x = (x_i)$ of complex numbers with the property that

$$\|x\|_2 = \left(\sum_1^\infty |x_i|^2\right)^{1/2}$$

is finite. Then ℓ^2 is a Banach space. Moreover, it is a Hilbert space with the inner product defined as follows: If $y = (y_i)$ is another element in ℓ^2 then

$$\langle x,y\rangle = x_1\bar{y}_1 + x_2\bar{y}_2 + x_3\bar{y}_3 + \cdots$$

is an inner product on ℓ^2 and the induced norm coincides with the norm defined above.

EXAMPLE 1.8.12 Let (Λ,B,P) be a measure space. For simplicity we assume that $P(\Lambda) = 1$. We consider the space L^2 of all functions (classes of functions) f such that

$$f \to \|f\|_2 = (\int_\Lambda |f(s)|^2 \, dP)^{1/2}$$

is finite. This is a Banach space. An inner product on L^2 is defined as follows: If g is another element in L^2 then we set

$$\langle f,g\rangle = \int_\Lambda f(x)\bar{g}(s) \, dP$$

which satisfies the required properties.

For more details we refer to books on measure theory.

2
Strict Convexity

The geometrical approach to the study of Banach spaces is connected with the unit ball of the space, i.e., the set $B_X(0,1) = \{x : \|x\| < 1\}$. Since Banach spaces have many topological properties that are useful in applications, the following problem is quite natural: What properties of a Banach space are expressible in isometric terms? The earliest considerations of certain aspects of this problem are in the work of J. A. Clarkson and M. G. Krein. They introduced (independently) the very important class of strictly convex spaces (also called *rotund spaces*) and gave several interesting properties of this class of Banach spaces. Also, Clarkson considered the class of uniformly convex spaces, which seems to be the most extensively studied class of Banach spaces from the above-mentioned point of view. Since the work of Clarkson and Krein the study of classes of spaces related to the above has attracted the attention of a great number of mathematicians. One reason for this is the fact that certain theorems of classical analysis as well as of modern analysis can be proved in greater generality only in such spaces. It is to be noted that these classes of Banach spaces have proved useful in many areas of mathematics, such as probability theory, function theory, harmonic analysis, fixed point theory, and operator theory. In what follows we describe these classes of Banach spaces as well as some applications.

2.1 STRICTLY CONVEX SPACES

Let V be a Banach space and

$$x \to \|x\|$$

be its norm. The class of strictly convex spaces (or rotund spaces) is defined as follows.

DEFINITION 2.1.1 A norm

$$x \to \| \ \|$$

on a Banach space is called strictly convex if whenever $\|x\| = \|y\| = 1$ and

$$\|x + y\| = 2$$

then necessarily $x = y$.

DEFINITION 2.1.2 A Banach space V is said to be strictly convex if its norm is strictly convex.

For the following characterization of strictly convex spaces we need the notion of extreme point of a convex set.

DEFINITION 2.1.3 Let C be a convex set in a Banach space V. The point z of C is said to be an extreme point for C if whenever $z = tx + (1 - t)y$ for some t in (0,1) and some x,y in C then $x = y$.

Then we have the following characterization of strictly convex spaces using extreme points.

THEOREM 2.1.4 The Banach space V is strictly convex if and only if every point z with $\|z\| = 1$ of the closed unit ball of V, $\{x : \|x\| \le 1\}$, is an extreme point.

Proof: Suppose first that V is strictly convex. We prove that every point z, $\|z\| = 1$, is an extreme point of the closed unit ball of V. If the assertion is not true, then there exists z_0, $\|z_0\| = 1$, which is not an extreme point of the closed unit ball, and thus

$$z_0 = tx_0 + (1 - t)y_0$$

where t is in (0,1) and x_0, y_0 are in the closed unit ball of V. Since $\|z_0\| = 1$ we get immediately that $\|x_0\| = \|y_0\| = 1$. Now we show that $\|x_0 + y_0\| = 2$. Indeed, in the contrary case we get that $\|x_0 + y_0\| < 2$ and

$$\begin{aligned} 1 = \|z_0\| &= \|tz_0 + (1 - t)z_0\| \\ &= \|t[tx_0 + (1 - t)y_0] + (1 - t)[tx_0 + (1 - t)y_0])\| \\ &= \|t^2x_0 + t(1 - t)(x_0 + y_0) + (1 - t)^2y_0\| \\ &< t^2 + 2t(1 - t) + (1 - t)^2 = 1 \end{aligned}$$

which is a contradiction. Thus $\|x_0 + y_0\| = 2$. This implies that $x_0 = y_0$.

Conversely, suppose that each z, $\|z\| = 1$, is an extreme point of the closed unit ball of V. We show that this implies the strict convexity of

V. Indeed, for each x,y in V with the properties

$$\|x\| = \|y\| = 1$$
$$\|x + y\| = 2$$

the point $u = (1/2)(x + y)$ has the property that $\|u\| = 1$. But u is an extreme point and thus $x = y$. Thus V is strictly convex.

The following characterization of strict convexity of a Banach space is connected with a maximum property of elements of the dual space.

THEOREM 2.1.5 The Banach space V is strictly convex if and only if for each f in V* there exists at most one point in the closed unit ball at which f attains its maximum.

Proof: Suppose first that V is a strictly convex space and let f be an arbitrary element of V*. Suppose that there exist at least two distinct points x,y of the closed unit ball such that f has the maximum value. We may assume without loss of generality that the maximum value is 1 and that the norm of f is 1. Since

$$1 = f\left(\frac{1}{2}(x + y)\right)$$

we obtain $\|x + y\| = 2$. Thus $x = y$.

For the converse, suppose that x,y are in the closed unit ball of V, $\|x\| = \|y\| = 1$, and $\|x + y\| = 2$. We must show that $x = y$. Indeed, by the extension theorem we find f in V* with the properties

$$\|f\| = 1$$
$$f(x + y) = 2$$

This obviously implies that $f(x) = f(y) = 1$ and thus, by hypothesis, that $x = y$. Thus the assertion is proved.

The following characterization may be proved in a similar way and is left to the reader.

THEOREM 2.1.6 The Banach space V is strictly convex if and only if the following equivalent assertions hold:

1. If $\|x\| = \|y\|$, $x \neq y$, then $\|x + y\| \neq 2\|x\|$.
2. If $\|x\| = \|y\|$, $x \neq y$, then $\|x + y\| < 2\|x\|$.
3. If $\|x + y\| = \|x\| + \|y\|$ then $[x,y] = \{z : z = tx + (1 - t)y,\ t \in (0,1)\}$ is a linearly dependent set in V.

For the next characterization we need the construction of new Banach spaces using the space V. For fixed p in $(1,\infty)$, $\ell^p(V)$ denotes the set of all sequences $x = (x_1, x_2, x_3, \ldots)$, with x_i in V, such that $(\Sigma_2^\infty \|x_i\|^p)^{1/p} = \|x\|$ is finite. With the operations + and $\cdot$ defined exactly as in the case of ℓ^p spaces this is clearly a linear space and even a Banach space. Another space related to these is as follows: Suppose that we have a measure space (Λ, B, P) [and we suppose for simplicity that $P(\Lambda) < \infty$]; we denote by $L^p(\Lambda, B, P, V)$ the space of all Bochner integrable functions (classes of functions) defined on Λ with values in V such that

$$\|f\|_p^p = (\int_\Lambda \|f(w)\|^p \, dP)^{1/p} < \infty$$

This is again a Banach space. For details and more information about these spaces we refer to the book by E. Hille and R. S. Philips (1957), and to papers by I. E. Leonard (1976), and S. Bochner and A. E. Taylor (1938).

The characterization of strict convexity of Banach spaces using these spaces is as follows.

THEOREM 2.1.7 The Banach space V is strictly convex if and only if for each p in $(1,\infty)$, the Banach space $\ell^p(V)$ and $L^p = L^p(\Lambda, B, P, V)$ are strictly convex spaces.

Proof: We give the proof only for the case of the ℓ^p space; the case of L^p is left to the reader. First suppose that V is strictly convex and suppose that x,y are two elements in $\ell^p(V)$ with the following properties:

$$\|x\| = \|y\| = 1$$
$$\|x + y\| = 2$$

We suppose that $x = (x_1, x_2, \ldots)$ and $y = (y_1, y_2, \ldots)$. Then, as a consequence of Minkowski's inequality, we get for each i, $\|x_i\| + \|y_i\| = \|x_i + y_i\|$, which implies that $x_i = y_i$ and thus $x = y$.

Conversely, if $\ell^p(V)$ is strictly convex and x,y are two elements of V then we can construct the elements $(x,0,0,0,\ldots)$ and $(y,0,0,0,\ldots)$, and the relations

$$\|x\| = 1 = \|y\|$$
$$\|x + y\| = 2$$

imply that

$$\|(x,0,0,0,\ldots) + (y,0,0,0,\ldots)\| = 2$$

The strict convexity of $\ell^p(V)$ gives that V is then strictly convex.

For the following characterization we need the notion of minimal points with respect to a set, a notion introduced by B. Beauzamy and B. Maurey (1977).

DEFINITION 2.1.8 Let V be a Banach space and M be a subset of V. A point x of V is called minimal with respect to M if whenever

$$\|y - m\| \leq \|x - m\| \qquad \forall\, m \in M$$

then x = y. The set of all minimal points with respect to M is denoted by min M. It is obvious that min M is a subset of M. Now the characterization of strictly convex spaces using min M is as follows.

THEOREM 2.1.9 The Banach space V is strictly convex if and only if for any points x,y of V, the points of the segment $[x,y] = \{z = tx + (1 - t)y : t \in [0,1]\}$ are minimal points with respect to the set $M = \{x,y\}$.

Proof: Suppose first that V is strictly convex and consider $z_t = tx + (1 - t)y$ in the segment [x,y]. Suppose further that for some z in V,

$$\|z - x\| \leq (1 - t)\|x - y\|$$
$$\|z - y\| \leq t\|x - y\|$$

Clearly,

$$\|z - x\| = (1 - t)\|x - y\|$$
$$\|z - y\| = t\|x - y\|$$

and this implies obviously that

$$\|x - y\| = \|x - z\| + \|z - y\|$$

From the strict convexity of V, $x - y = a(z - y)$, which implies that z is in the segment [x,y]. Thus any point of the segment [x,y] is minimal with respect to {x,y}, and the first part of the assertion is proved. Conversely, suppose that for any x,y in V, the points of the segment [x,y] are minimal points of {x,y}. We show that this implies that V is strictly convex. For this purpose we choose m = 0 and n = x + y, and the set M is {0, x + y}. If $\max(\|x\|,\|y\|) > \|x + y\|$ then clearly we have

$$\|x + y\| < \|x\| + \|y\|$$

If $\|x\|$, $\|y\|$ are less than $\|x + y\|$ then we choose the unique point on the segment [0, x + y] which has norm equal to $\|x\|$. Then by the hypothesis, this unique point, denoted by x', is minimal with respect to our set, and since $\|x'\| = \|x\|$, we have

$$\|y\| \leq \|x + y - x'\|$$

which implies that

$$\|x + y\| < \|x\| + \|y\|$$

and the theorem is proved.

For the following characterization of strict convexity we need the notion of duality mapping introduced by Beurling and Livingston (1962). For details and applications to fixed point theory see V. I. Istrăţescu (1981b).

DEFINITION 2.1.10 Let V be a Banach space and 2^{V^*} be the space of all subsets of V^*. A duality mapping $H : V \to 2^{V^*}$ is any function contained in

$$H(x) \subseteq J(x) = \{x^* : x^* \in V^*, x^*(x) = \|x\| \|x^*\|\}$$

DEFINITION 2.1.11 A function $\varphi : R^+ \to R^+$ is called a pseudogauge if it is strictly increasing and $\varphi(t) = 0$ if and only if $t = 0$. The function φ is called a gauge function if in addition it is continuous and $\varphi(t) \to \infty$ for $t \to \infty$. Using an increasing function φ with the property that $\varphi(t) = 0$ if and only if $t = 0$, we define a duality mapping by the formula

$$J_\varphi : V \to 2^{V^*}$$
$$J_\varphi(x) = \{x^* : x^* \in V^*, x^*(x) = \|x\| \|x^*\|, \|x^*\| = \varphi(\|x\|)\}$$

THEOREM 2.1.12 (S. P. Gudder and D. Strawther, 1976) For a Banach space the following assertions are equivalent:

1. V is strictly convex.
2. $J(y) \subseteq J(x)$ for $x \neq 0$ implies that $y = cx$, $c > 0$.
3. $J(y) = J(x)$ implies $y = cx$, $c > 0$.

Proof: We prove the sequence of implications

$$1 \to 2 \to 3 \to 1$$

$1 \to 2$. Suppose that $x \neq 0$ and $J(y) \subseteq J(x)$ and $y \neq cx$ for all $c > 0$. We note that we have obviously $y \neq 0$ [since in the contrary case $J(x) = J(y) = V^*$ and this implies that $x = 0$]. By the corollary to the extension theorem we get that there exists x^* in V^* such that $\|x^*\| = 1$ and $x^*(y) = \|y\|$. In this case we get $x^* \in J(y) \subseteq J(x)$. But $x/\|x\| \neq y/\|y\|$ and $x^*(y/\|y\|) = x^*(x/\|x\|) = \|x^*\|$ and thus x^* attains its norm at two different points. Thus V is not strictly convex, and this contradiction gives $1 \to 2$.

$2 \to 3$. Obvious.

$3 \to 1$. Suppose that 3 holds and V is not strictly convex. Then on the set $\{x : \|x\| = 1\}$ there exists a nontrivial segment, and we consider on this segment the points x,y,z, and w (we suppose that all points are distinct). Also we may suppose that

$$y = \frac{1}{2}(x + w) \qquad x = \frac{1}{2}(y + z)$$

We take now x^* in $J(y)$. Then we have

$$|x^*(x)| \le \|x^*\|\|x\| = \|x^*\|$$

and

$$|x^*(x) - 2\|x^*\|| = |x^*(x - 2y)| \le |x^*(-w)| \le \|x^*\|\|w\| = \|x^*\|$$
$$x^*(x) = \|x^*\|$$

and thus $x^* \in J(x)$, which gives that $J(x) \supseteq J(y)$. Similarly we show that $J(x) \subseteq J(y)$ and thus $J(x) = J(y)$. Furthermore $y = cx$ for $c > 0$, so $w = dx$ for some d. Since $\|w\| = \|x\| = 1$ we have that $w = x$ or $w = -x$. But $w \neq x$, and thus $w = -x$, and then $y = 0$. This is a contradiction, and thus $3 \to 1$.

For the characterization which follows, we need a result about the functions J.

THEOREM 2.1.13 Let V be a Banach space and φ,ψ be two functions which are continuous, positive, and increasing, and $\varphi(t) = \psi(t) = 0$ if and only if $t = 0$. Then

1. If V is strictly convex then $J_\varphi(x) \cap J_\psi(y) \neq \emptyset$ implies $y = cx$, $c > 0$.
2. If there exists $a > 0$ such that $\varphi(a) = \psi(a)$ and $J_\varphi(x) = J_\psi(y)$ imply $y = cx$ for some $c > 0$, then V is strictly convex.

Proof: (1) Suppose that V is strictly convex, $J_\varphi(x) \cap J_\psi(y) \neq \emptyset$, and $y \neq cx$ for all $c > 0$. Now any x^* in $J_\varphi(x) \cap J_\psi(y)$ is clearly not zero (in the contrary case $x = y = 0$). We have

$$x^*\left(\frac{x}{\|x\|}\right) = x^*\left(\frac{y}{\|y\|}\right)$$

and since $y \neq cx$ for all $c > 0$ we get that x^* attains its maximum at two different points of the closed unit ball. This contradicts the strict convexity of V and thus (1) holds.

(2) Suppose that the condition of (2) is satisfied and V is not strictly convex. If $x = 0$ or $y = 0$ then the other is zero. Thus we can suppose without loss of generality that both x and y are nonzero elements of V. Take x^* in $J_\psi(ay/\|y\|)$. Then x^* is in $J_\varphi(ay/\|y\|)$; thus $\|x^*\| = \varphi(a) = \psi(a) = \varphi(\|ax\|/\|x\|)$. Then x^* is in $J_\varphi(ax/\|x\|)$ which implies that the inclusion

$$J_\psi(ay/\|y\|) \subseteq J_\varphi(ax/\|x\|)$$

and similarly

$$J_\varphi(ax/\|x\|) \subseteq J_\psi(ay/\|y\|)$$

Thus the two sets are equal and this implies that $y = cx$ for some $c > 0$. By 2.1.12, V is strictly convex.

From this theorem we derive a number of results obtained by E. Torrance (1970), E. Berkson (1965), and R. Menaker (1969).

COROLLARY 2.1.14 Let φ be a pseudogauge function and V be a Banach space. Then the following assertions are equivalent:

1. V is strictly convex.
2. $J_\varphi(x) \cap J_\varphi(y) = \emptyset$ if $y \neq x$.
3. If $J_\varphi(y) \subseteq J_\varphi(x)$ then $y = cx$, $c > 0$.
4. $J_\varphi(x) \neq J_\varphi(y)$ if $x \neq y$.

Proof: First we show that $1 \to 2$. Indeed,, if $J_\varphi(x) \cap J_\varphi(y) \neq \emptyset$ then the above theorem implies that $y = cx$ for some $c > 0$. Now, if x^* is in $J_\varphi(x) \cap J_\varphi(y)$ then we have

$$\|x^*\| = \varphi(\|x\|) = \varphi(\|y\|)$$

which implies that if $x = 0$ then $y = 0$. If $x \neq 0$, from the injectivity of φ^* we get $\|x\| = \|y\|$. This implies that $c = 1$. Now $2 \to 3$ and $3 \to 4$ are obvious, and $4 \to 1$ follows from 2.1.13.

COROLLARY 2.1.15 The Banach space V is strictly convex if and only if $J(x) \cap J(y) = \emptyset$ if $x \neq y$.

COROLLARY 2.1.16 If [,] is a semi-inner product on the Banach space X then X is strictly convex if and only if

$$[x,y] = \|x\|\|y\|$$

implies $y = cx$ for some $c > 0$. Here [,] means a semi-inner product defined on X in the sense of Lumer (1961).

Proof: First suppose that X is strictly convex and for some x,y in X,

$$[x,y] = \|x\|\|y\|$$

In this case we have

$$(\|x\| + \|y\|)\|y\| \geq \|x + y\|\|y\| \geq [x + y, y] = (\|x\| + \|y\|)\|y\|$$

and we may suppose without loss of generality that $x \neq 0 \neq y$. Thus

$$\|x\| + \|y\| = \|x + y\|$$

and from the strict convexity of X, $y = cx$ for some $c > 0$. Conversely, suppose that whenever x,y in X satisfy the relation

$$[x,y] = \|x\|\|y\|$$

then necessarily $y = cx$ for some $c > 0$. We show that X is strictly convex. Indeed, in the contrary case, for some x,y in X,

$$J(y) \subseteq J(x)$$

and $y \neq cx$ for all $c > 0$. We consider $x^*(z) = [z,y]$, which is clearly in X^* and also in $J(y)$. Then x^* is in $J(x)$ and

$$x^*(x) = [x,y] = \|x\|\|y\|$$

which implies that $y = cx$ for some $c > 0$. This is a contradiction and thus X is strictly convex.

THEOREM 2.1.17 (E. Torrance, 1970) The Banach space X is strictly convex if and only if whenever $\|y + z\| \leq \|y\|$ and $[z,y] = 0$ then $z = 0$.

Here [,] means, as above, an inner product on X.

Proof: We note that the assertion of the theorem is obviously equivalent to the following assertion: X is strictly convex if and only if $[x,y] = \|y\|^2$, $x \neq y$, implies $\|x\| > \|y\|$.

Suppose first that X is strictly convex and for some x,y in X the relation

$$[x,y] = \|y\|^2$$

$(x \neq y)$ holds. From the above result we get that $y = cx$ which, by the inequality $\|x\|\|y\| \geq [x,y] = \|y\|^2$, implies that $\|x\| > \|y\|$. The converse assertion may be proved like Corollary 2.1.16 and we leave the details to the reader.

The results in 2.1.12 and 2.1.13 are given in S. P. Gudder and D. Strawther (1976), as well as the derivation of the results in 2.1.15 (R. Menaker, 1969) and 2.1.16 (E. Berkson, 1965).

The necessity condition in the next theorem was obtained by T. Husain and M. Malviya (1972) and the sufficiency part by S. P. Gudder and D. Strawther (1976). To formulate it we need the following definition.

DEFINITION 2.1.18 Let X be a Banach space and [,] be a semi-inner product on X. We say that the sequence $(x_n) \subset X$ converges weakly in the second argument to an element $x \in X$ if for any $y \in X$,

$$\lim [y,x_n] = [y,x]$$

Now the characterization of strict convexity is as follows.

THEOREM 2.1.19 The Banach space X is strictly convex if and only if the weak limits in the second argument are unique.

Proof: First we remark that the result in the theorem is equivalent to the following assertion: X is strictly convex if and only if

$$[z,x] = [z,y]$$

for all $z \in X$ implies $x = y$.

Now the proof is as follows. Suppose that X is strictly convex and let $x^*(z) = [z,x] = [z,y]$ for all $z \in X$. In this case Corollary 2.1.14 gives that $x = y$ because $x^* \in J(x) \cap J(y)$. Conversely, suppose now that X is not strictly convex. Thus we find x,y in X with the property that $x \neq y$ and $x^* \in J(x) \cap J(y)$. We define a semi-inner product on X by the formula

$$[u,v] = f_v(u)$$

where $f_v \in J(v)$, $v \neq x$, and $f_y = f_x = x^*$. In this case we get

$$[z,x] = f_x(z) = f_y(z) = [z,y]$$

and the theorem is proved.

The following characterization of strict convexity was obtained by W. V. Petryshyn (1970) and uses the concept of monotonicity and strict monotonicity of the duality mapping J_φ. The very short proof below was found by S. P. Gudder and D. Strawther (1975).

DEFINITION 2.1.20 Let φ be a positive, strictly increasing function on R^+ such that $\varphi(t) = 0$ if and only if $t = 0$. Let $J_\varphi : X \to 2^{X^*}$. We say that J_φ is strictly monotone if for any $x \neq y$ and every $f \in J_\varphi(x)$, $g \in J_\varphi(y)$,

$$\text{Re } (f - g)(x - y) > 0$$

THEOREM 2.1.21 (W. V. Petryshyn, 1970) The Banach space X is strictly convex if and only if J_φ is strictly monotone.

Proof: First we remark that the following identity holds: for any $f \in J_\varphi(x)$, $g \in J_\varphi(y)$,

$$\begin{aligned}\text{Re } (f - g)(x - y) &= \|f\|\|x\| - \text{Re } f(y) - \text{Re } g(x) + \|g\|\|y\| \\ &= [(\|f\| - \|g\|)(\|x\| - \|y\|)] + [\|f\|\|y\| - \text{Re } f(y)] \\ &\quad + [\|g\|\|x\| - \text{Re } g(x)]\end{aligned}$$

Now since each term in square brackets is nonnegative, $\text{Re } (f - g)(x - y) \geq 0$, and $\text{Re } (f - g)(x - y) = 0$ if and only if $\text{Re } f(y) = \|f\|\|y\|$, $\text{Re } g(x) = \|g\|\|x\|$, and $(\|f\| - \|g\|)(\|x\| - \|y\|) = 0$.

Suppose that Re $f(y) = \|f\|\|y\|$ and thus

$$[\text{Im } f(y)]^2 + \|f\|^2\|y\|^2 = [\text{Im } f(y)]^2 + [\text{Re } f(y)]^2 = |f(y)|^2 \leq \|f\|^2\|y\|^2$$

This implies that Re $f(y) = \|f\|\|y\|$ if and only if $f(y) = \|f\|\|y\|$, and the same assertion holds for x, i.e., Re $g(x) = \|g\|\|x\|$ if and only if $g(x) = \|g\|\|x\|$. Now if $\|f\| = \|g\|$ then $\varphi(\|x\|) = \|f\| = \|g\| = \varphi(\|y\|)$ and since φ is supposed to be strictly increasing, we get that $\|x\| = \|y\|$. Thus $(\|f\| - \|g\|)(\|x\| - \|y\|) = 0$ if and only if $\|f\| = \|g\|$ and $\|x\| = \|y\|$. In this case from 2.1.14 we see that X is not strictly convex if and only if there exists $x \neq y$ and $f \in X^*$ such that

$$f \in J_\varphi(x) \cap J_\varphi(y)$$

i.e., if and only if $f(x) = \|f\|\|x\|$, $f(y) = \|f\|\|y\|$, $\|f\| = \varphi(\|x\|)$, $\|f\| = \varphi(\|y\|)$, or there exists $x \neq y$, $f \in J_\varphi(x)$, $g \in J_\varphi(y)$ with Re $(f - g)(x - y) = 0$, if and only if J_φ is not strictly monotone. The theorem is proved.

We present now a characterization of the strict convexity property of a Banach space in connection with the extension of functionals.

As we know, every continuous functional defined on a subspace of a Banach space X has an extension to the entire space (with the same norm). But for a given functional there may exist many functionals on X (with the same norm) which extend the given functional. Using the notion of strict convexity we can characterize the class of Banach spaces for which the extension is unique. For this purpose we need the following notion.

DEFINITION 2.1.22 A real Banach space X is said to have the property (L_r) if whenever $x \neq y$ are in X and $\|x\| = \|y\| = 1$ then there exists a real number s such that

$$\|sx + (1 - s)y\| < 1$$

The following result gives a characterization of the class of real Banach spaces for which any functional defined on a subspace has a unique extension (preserving the norm) to the entire space.

THEOREM 2.1.23 A real Banach space X has the property that every continuous linear functional defined on a subspace of X has a unique norm-preserving linear extension to X if and only if X* is strictly convex.

For the proof we need the following characterization of strict convexity using the property (L_r).

THEOREM 2.1.24 A real Banach space is strictly convex if and only if it has the property (L_r).

Proof: If X is strictly convex then it has the property (L_r) for $s = 1/2$. Suppose now that X has the property (L_r) and is not strictly convex. This means that for some x,y in X, $\|x\| = \|y\| = 1$, and for all t in (0,1), $\|tx + (1 - t)y\| = 1$. Let $x^* \in X^*$, $\|x^*\| = 1$ and $x^*((1/2)(x + y)) = (1/2)(\|x + y\|)$. Then it is clear that $x^*(x) = x^*(y) = 1$ and thus for all $s \in \mathbb{R}$,

$$\|sx + (1 - s)y\| \geq |x^*(sx + (1 - s)y)| = 1$$

and the property (L_r) is not satisfied. Thus we proved the theorem.

Now the proof of Theorem 2.1.23 is as follows.

Suppose first that X has the property that X* is strictly convex. Then we show that any continuous linear functional defined on a subspace of X has a unique norm-preserving extension to a linear functional on X. If this is not true, then we find a subspace X_1 of X and a linear functional f on X_1 [and we may suppose without loss of generality that the norm of the functional is 1 (with respect to X_1)] that has two distinct norm-preserving extensions to the entire space X. Let us denote these functionals by F_1 and F_2. In this case for all $s \in \mathbb{R}$,

$$F_s = sF_1 + (1 - s)F_2$$

is again a continuous linear functional on X which is an extension of f. Since X* is assumed to be strictly convex, for some $s \in \mathbb{R}$,

$$\|F_s\| < 1$$

But this obviously is a contradiction since

$$\|F_s\| = 1$$

Conversely, suppose that X has the property that all norm-preserving extensions of continuous functionals are unique. We show now that this implies that X* is strictly convex. Consider two elements f,g in X* with $\|f\| = \|g\| = 1$,

$$X_1 = \text{Ker } (f - g) = \{x : f(x) = g(x)\}$$

and consider $f_1 = f|X_1$ (the restriction of f to X_1). This is obviously a continuous linear functional and there exists, by hypothesis, a unique norm-preserving linear extension to X of f_1. We denote this extension by f_1^*. Since f and f_1^* are both extensions of f_1 we must have $\|f_1^*\| < 1$. Let us consider now x_0 in $X - X_1$ with the property that $f(x_0) \neq f_1^*(x_0)$. Since the equation in s,

$$sf(x_0) + (1 - s)g(x_0) = f_1^*(x_0)$$

has a unique solution, say s_0, we have

$$s_0 f + (1 - s_0)g = f_1^*$$

since we may suppose without loss of generality that $X = \text{span}(X_1, x_0)$. This implies that

$$\|s_0 f + (1 - x_0)g\| < 1$$

and by Theorem 2.1.24, X* is strictly convex.

We present now an extension of this result to complex Banach spaces. First we give

DEFINITION 2.1.25 (P. R. Beesack, R. Hughes, and M. Ortel, 1979) A complex Banach space X is said to have the property (L_c) [the property (L) in the notation of Beesack, Hughes, and Ortel] if whenever x,y are in X, $\|x\| = \|y\| = 1$, there exists a complex number z_0 such that

$$\|z_0 x + (1 - z_0 y)\| < 1$$

THEOREM 2.1.26 (P. R. Beesack, R. Hughes, and M. Ortel, 1979) A complex Banach space is strictly convex if and only if it has the property (L_c).

Proof: It is clear that if X is strictly convex then it has the property (L_r) for $s = 1/2$. Conversely, suppose that X has the property (L_c). We show that it is a strictly convex space. To prove this it is sufficient to show that if X is not strictly convex then the (L_c) property is not satisfied.

Suppose now that there exist x,y in X with the property that $\|x\| = \|y\| = 1$ and for all t in (0,1), $\|tx + (1 - t)y\| = 1$. Let x* be in X*, having the following properties:

$$\|x^*\| = 1$$

$$x^*\left(\frac{1}{2}(x + y)\right) = \frac{1}{2}\|x + y\| = 1$$

This implies that $x^*(x) = x^*(y) = 1$, and thus for all complex numbers z,

$$\|zx + (1 - z)y\| \geq |x^*(zx + (1 - z)y)| = 1$$

which shows that the property (L_c) does not hold. The theorem is proved.

Using the characterization of strict convexity for complex Banach spaces we can prove the following result on the uniqueness of norm-preserving extension for continuous linear functionals.

THEOREM 2.1.27 (A. E. Taylor, 1939) The complex Banach space X has the property that any continuous linear functional defined on a subspace of X has a unique norm-preserving extension to an element of X^* if and only if X^* is strictly convex.

Proof: Similar to proof of Theorem 2.1.27.

REMARK 2.1.28 The above characterization of strict convexity using the properties (L_r) and (L_c) can be extended easily to characterize k-strict convexity, a notion which we will consider in Sec. 2.8.

2.2 PRODUCT AND QUOTIENT SPACES AND STRICT CONVEXITY

In what follows we consider the problem of strict convexity of the product of two (or a finite number) of strictly convex spaces, as well as the problem of strict convexity of the quotient spaces of strictly convex spaces. Results related to this problem were obtained essentially by V. Klee (1959).

Let us suppose that $X_1, \ldots, X_n$ are strictly convex spaces and consider the space $X_1 \times \cdots \times X_n$. For this we consider the complex n-dimensional space $\mathbb{C}^n$ with a norm $|\ |$ which is monotone in each argument, i.e., for $z = (z_1,\ldots,z_n)$ and $z^- = (z_1^-,\ldots,z_n^-)$, then if $|z_i| \leq |z_i^-|$ then $|z| \leq |z^-|$. In this case we have

THEOREM 2.2.1 The space $X_1 \times \cdots \times X_n$ is a strictly convex Banach space with the norm

$$\|x\| = |\,\|x_1\|,\|x_2\|,\ldots,\|x_n\|\,|$$

if and only if the norm $|\ |$ is strictly convex on $\mathbb{C}^n$.

Proof: Since the proof that the product is a Banach space under this norm is easy we omit it. We prove the second assertion. Indeed, suppose that for some $x = (x_1,\ldots,x_n)$ and $y = (y_1,\ldots,y_n)$ we have the relations:

1. $\|x\| = \|y\| = 1$
2. $\|x + y\| = \|x\| + \|y\| = 2$

Since the norm on $\mathbb{C}^n$ is supposed to be strictly convex and monotone in each argument we obtain the following relations:

$$\|x_i + y_i\| = \|x_i\| + \|y_i\| \qquad i = 1, 2, \ldots, n$$

In this case we get that for each i there exists c_i such that $y_i = c_i x_i$ and condition 1 implies that all $c_i = 1$. Thus $x = y$. Since the converse assertion is obvious the theorem is proved.

REMARK 2.2.2 The fact the number of spaces is finite is essential for the validity of the theorem. Indeed, if we consider the Banach space $X_i = \mathbb{C}$ for each i and on the product space the norm

$$\|x\| = \|x_1\| + \|x_2\| + \cdots + \|x_n\| + \cdots = \sum_{n=1}^{\infty} \|x_n\|$$

then the subspace of the product space consisting of all $x = (x_n)$ with finite norm is the Banach space ℓ^1, and we see easily that this is not strictly convex.

REMARK 2.2.3 We note an example of a norm on $\mathbb{C}^n$ that has the properties required in Theorem 2.2.1. Let $p \in (1,\infty)$ and for each $x_1, \ldots, x_n$ set

$$|(x_1,\ldots,x_n)| = (a_1|x_1|^p + \cdots + a_n|x_n|^p)^{1/p}$$

where a_i are strictly positive real numbers.

We consider now the problem of when the strict convexity of X is transmitted to the quotient subspaces. First we recall that if X is a Banach space and X_1 is a closed subspace of X then the quotient space X/X_1 is the space whose elements are the translates of X_1, with the norm

$$\|x + X_1\| = \|\hat{x}\| = \inf\{\|x + y\| : y \in X_1\}$$

The mapping

$$x \to \hat{x}$$

defined on X with values in X/X_1 is the so-called canonical projection of X onto X/X_1 and is denoted m (sometimes π). We remark that the norm of x is exactly $d(x,X_1)$ where $d(x,X_1)$ is the distance from x to the subset X_1. Also, we have the isomorphism between X/X_1 and $X_1 = \{x^* : x^* \in X^*, x^*(x_1) = 0, x_1 \in X_1\}$. We set $B_X(0,1) = \{x : \|x\| \leq 1\}$ for any Banach space X.

LEMMA 2.2.4 For any Banach spaces X,Y the following assertions are equivalent:

1. There exists a closed subspace X_1 of X such that Y is equivalent (i.e., isometrically isomorphic) to X/X_1.

2. There exists a linear mapping $T : X \to Y$ such that $T(B_X(0,1)) = B_Y(0,1)$.

Proof: Consider a subspace X_1 of X and let $S_{X/X_1}(0,1)$ and m be the canonical map of X onto X/X_1, i.e., $mx = x + X_1$. It is clear that $mB_X(0,1) \subseteq B_{X/X_1}(0,1)$. If assertion 1 holds then for $x + X_1$ in $B_{X/X_1}(0,1)$ there exists some $x_1 \in X_1$ with $x + x_1 \in B_X(0,1)$. But $x + x_1 + X_1 = x + X_1$, which proves that $1 \to 2$.

Suppose now that 2 holds, i.e., there exists T with the above property, and set $X_1 = \text{Ker } T = \{x : Tx = 0\}$ and we remark that T and the canonical map have the same kernel, and then mT^{-1} is a linear map of Y onto X/X_1. We show that this is in fact an isometry. For this it is sufficient to show that for all $x \in X$, mx is in $B_{X/X_1}(0,1)$ for some x_1 with $x + x_1$ in $B_X(0,1)$ if and only if Tx is in $B_Y(0,1)$. Suppose that $mx \in B_{X/X_1}(0,1)$ and for some $x_1 \in X_1$, $x + x_1 \in B_X(0,1)$. Then $Tx = T(x + x_1) \in B_Y(0,1)$. Now, if Tx is in $B_Y(0,1)$ then $Tx = Ty$ for some $y \in B_X(0,1)$, and thus $mx = my \in B_{X/X_1}(0,1)$, and the lemma is proved.

THEOREM 2.2.5 (V. Klee, 1959) If the Banach space X is strictly convex and X_1 is a reflexive subspace then X/X_1 is strictly convex.

Proof: We consider the quotient subspace X/X_1 and denote by m the canonical map of X onto X/X_1. From the reflexivity of X_1 we have that $mB_X(0,1) = \bar{B}_{X/X_1}(0,1)$ (closure). Indeed, the inclusion $mB_X(0,1) \subseteq \bar{B}_{X/X_1}(0,1)$ is obvious. Let X' be an element in $\bar{B}_{X/X_1}(0,1)$ and thus $\inf\{\|x'\| : x' \in X'\} = 1$, which gives that

$$d(\bar{B}_X(0,1),X') = 0 \qquad d(B_X(0,1),2\bar{B}_X(0,1) \cap X') = 0$$

Since the first set in the last relation is weakly closed, and the second, being a subset of X/X_1, is weakly compact, then they must have a common point. We denote this point by x and thus $mx = X'$. Now the assertion of the theorem follows since strict convexity is preserved by continuous linear mappings.

The following example shows that the property of a Banach space's being strictly convex is not transmitted, in general, to the quotient subspaces of that space. We recall the notion of summability. Let S be an arbitrary nonempty set and $P_b(S)$ be the family of all finite subsets of S.

DEFINITION 2.2.6 The family $(c_s)_{s\in S}$ of complex numbers is said to be summable and has the sum c if and only if for each $\varepsilon > 0$ there exists a set F_ε in $P_b(S)$ such that

$$\left|\sum_{s\in F} c_s - c\right| < \varepsilon$$

for all $F \supset F_\varepsilon$, $F \in P_b(S)$.

Now the example is as follows.

EXAMPLE 2.2.7 (V. Klee, 1959) We consider the space L^1_S consisting of all families $(c_s)_{s\in S}$ such that $(|c_s|)_{s\in S}$ is summable. If we set as the norm of $(c_s)_{s\in S}$ the sum of $(|c_s|)_{s\in S}$ then it is not difficult to prove that this is a Banach space. We choose the set S to be infinite and then L^1_S admits an equivalent norm which is strictly convex such that every Banach space with density character less than or equal to card S is equivalent to a quotient of L^1_S. To prove this assertion, we consider for each $n = 1, 2, 3, \ldots,$ a continuous strictly convex function g_n on $[0,1]$ with the property that

$$\frac{n}{n+1}\, t \le g_n(t)t \le t$$

We consider S to be covered by a sequence of pairwise disjoint subsets (S_n), each of the same cardinality as S, and set $g_s = g_n$ for all $s \in S_n$. We consider now the set U defined as follows:

$$U = \{(c_s)_{s\in S} : (g_s c_s) \in S_{L^1_S}(0,1)\}$$

This is clearly a convex subset, and the gauge (the Minkowski functional of U) is a norm on L^1_S which is equivalent with the original one, since

$$B_{L^1_S}(0,1) \subset U \subset 2B_{L^1_S}(0,1)$$

But each point of the boundary of Cl U is extreme and thus (L^1_S, p_U), p_U being the gauge of U, is a strictly convex space. Suppose now that X is a Banach space whose density character is less than or equal to card S.

We consider a map ξ defined on S with values in a dense subset of $B_X(0,1)$, with the property that

$$\frac{n}{n+1}\xi S_n \subset B_X(0,1)$$

Now define a mapping on L^1_S (the norm on this space is p_U) as follows: we set

$$x \to Tx = c_s \xi(s) \in X$$

Obviously this is a linear mapping and TU contains $\bar{B}_X(0,1)$. We show now that $TU = B_X(0,1)$. Indeed, let us consider Tx with $x \in U$. Since

$$\|Tx\| \le \sum_n \left(\sum_{s \in S_n} |c_s| \|\xi(s)\| \right) \le \sum_{n, s \in S_n} \frac{n}{n+1} |c_s| \le \sum_{s \in S} g_s |c_s| = 1$$

the assertion is proved, as well as the assertion of the example.

Now we consider the relation between strict convexity and certain approximation properties of Banach spaces. We recall some notions. [The reader interested in this area may consult the book of I. Singer (1967, 1970).]

Let X be a Banach space and X_1 be a closed linear subspace of X. The subspace X_1 is called an existence subspace if for each $x \in X$ there exists an element x_1 of X_1 such that

$$\|x - x_1\| = \inf \{\|x - y\| : y \in X_1\}$$

The element x_1 with this property is said to be the best approximation of x in X_1 and we note that using the cannonical map associated with X_1 this may be written as

$$\|x - x_1\| = \|mx\|$$

For the characterization of strict convexity using existence subspaces we need the following lemma.

LEMMA 2.2.8 Let X be a strictly convex space and X_1 be a closed subspace of X. Then no convex subset of the closed unit ball of X/X_1 contains more than one element mx such that x has a best approximation in X_1.

Proof: Suppose that the assertion of the lemma is not true. Thus we find x,y with the property that mx and my are elements of a convex subset of X/X_1 whose norm is less or equal to 1, and x and y have best approximations in X_1, say the elements x_1 and y_1, respectively. Now we find f in

$(X/X_1)^*$ with the property that

$$f(mx) = f(my) = 1$$

Then m^*f is in the closed unit ball of $X_1^- = \{x^* : x^* \in X^*, x^*(x_1^-) = 0,$ $x_1^- \in X_1\}$ and $m^*f(x - x_1) = m^*f(y - y_1) = 1$. This implies that $x - x_1$ and $y - y_1$ are in a convex subset of the closed unit sphere of X, and since this is strictly convex, the closed convex subsets of the unit sphere are single points. Thus $x - x_1 = y - y_1$, which implies that $mx = my$, and the lemma is proved.

From this lemma we get the following characterization of strict convexity in quotient spaces; the details are left to the reader.

THEOREM 2.2.9 Let X be a strictly convex space and X_1 be a closed subspace of X. The space X/X_1 is strictly convex if and only if for every mx of the unit sphere of X/X_1 with the property that x has no best approximation in X_1, mx is an extreme point of the closed unit ball of X/X_1.

Another characterization of strict convexity of X is as follows, using existence subspaces and the strict convexity of some quotient subspaces.

THEOREM 2.2.10 Let X be a Banach space. Then the following assertions are equivalent:

1. X is strictly convex.
2. For every existence subspace X_1 of X, X/X_1 is strictly convex.
3. There exists a natural number $n \leq \dim X - 2$ such that for every n-dimensional subspace X_1 of X, X/X_1 is strictly convex.

Proof: The assertion $1 \to 2$ follows from Theorem 2.2.9, and the assertion $2 \to 3$ is obvious. We prove now that $3 \to 1$. Suppose on the contrary that X is not strictly convex. Thus there exist x_1 and x_2, $\|x_1\| = \|x_2\| = 1$, and x^* in X^*, such that $\|x^*\| = 1$, $x^*(x_1) = x^*(x_2) = 1$. Since $x^{*-1}(0)$ contains an n-dimensional subspace of X_1 whose intersection with the subspace generated by $x_1 - x_2$ reduces to $\{0\}$, the quotient space X/X_1 is not strictly convex since mx_1 and mx_2 are distinct points of the unit sphere of X/X_1. Indeed, the fact that they are distinct follows from the definition of X_1, and we remark that they are in the convex set

$$m^{*-1}(x_1) = m^{*-1}(x_2)$$

so the theorem is proved.

2.3 INTERPOLATION SPACES AND STRICT CONVEXITY

In what follows we consider the problem of strict convexity of some interpolation spaces introduced by J. L. Lions and J. Peetre (1964). Since we shall need some information about these important classes of spaces, we give the basic definitions; for further details we refer to the Lions-Peetre paper.

We suppose that A_0 and A_1 are Banach spaces which are subspaces of a Hausdorff topological linear space A, and

$$p_i : A_i \to A \qquad i = 0, 1$$

$$p_i(x) = x$$

are continuous mappings. An interpolation space is a Banach space B with the property that

$$A_0 \cap A_1 \subset B \subset A_0 + A_1$$

The norm on A_0 is denoted by $\| \|_{A_0}$ and the norm on A_1 by $\| \|_{A_1}$. A norm on $A_0 \cap A_1$ is defined by the formula: If $a = a_0 + a_1$, $a_i \in A_i$, then

$$\|a\| = \max \{\|a_0\|_{A_0}, \|a_1\|_{A_1}\}$$

A norm on $A_0 + A_1$ is defined as follows: If $a = a_0 + a_1$, $a_i \in A_i$, then

$$\|a\| = \inf \{\|a_0\|_{A_0} + \|a_1\|_{A_1} : a = a_0 + a_1\}$$

It is obvious that $A_0 \cap A_1$ and $A_0 + A_1$ with the norms defined above are Banach spaces.

Let $\xi_0, \xi_1 \in \mathbb{R}$ with the property that $\xi_0\xi_1 < 0$ and let $p_i \in [1,\infty)$, $i = 0, 1$. We consider the space $W(p_0,\xi_0,A_0, p_1,\xi_1,A_1)$ of all functions f (classes of functions) defined on $\mathbb{R}$ with values in A such that:

1. The function $s \to e^{\xi_0 s} f(s) \in L^{p_0}(A_0) = \{g : \mathbb{R} \to A : \|g\|_{L^{p_0}(A_0)} = (\int_{-\infty}^{\infty} \|g(s)\|^{p_0} ds)^{1/p_0} < \infty\}$

2. The function $s \to e^{\xi_1 s} f(s) \in L^{p_1}(A_1) = \{h : \mathbb{R} \to A : \|h\|_{L^{p_1}(A_1)} = (\int_{-\infty}^{\infty} \|h(s)\|^{p_1} ds)^{1/p_1} < \infty\}$

A norm on $W(p_0,\xi_0,A_0,p_1,\xi_1,A_1)$ is defined as follows:

$$f \to \|f\|_{W(p_0,\xi_0,A_0,p_1,\xi_1,A_1)} = \max\{\|e^{\xi_0 s} f(s)\|_{L^{p_0}(A_0)}, \|e^{\xi_1 s} f(s)\|_{L^{p_1}(A_1)}\}$$

Since $\xi_0\xi_1 < 0$ the integral $\int_{-\infty}^{\infty} f(s)\,ds$ converges (in $A_0 + A_1$) we can consider the space $S(p_0,\xi_0,A_0,p_1,\xi_1,A_1)$, defined as the set

$$\{\int_{-\infty}^{\infty} f(s)\,ds : f \in W(p_0,\xi_0,A_0,p_1,\xi_1,A_1)\}$$

A norm on this space is defined as follows:

$$a = \int_{-\infty}^{\infty} f(s)\,ds \to \|a\|_{S(p_0,\xi_0,A_0,p_1,\xi_1,A_1)}$$

$$= \inf_{a=\int_{-\infty}^{\infty} f(s)ds} \max\{\|e^{\xi_0 s} f(s)\|_{L^{p_0}(A_0)}, \|e^{\xi_1 s} f(s)\|_{L^{p_1}(A_1)}\}$$

and we call $S(p_0,\xi_0,A_0,p_1,\xi_1,A_1)$ with this norm a Lions-Peetre interpolation space.

There exist several equivalent definitions of these spaces and we give some of them below, since for the study of some problems it seems appropriate to use one or another definition. We remark that in the above definition we have used the spaces L^p with respect to the Lebesgue measure ds on R. We can use the space $\ell^p(B)$ of all sequences of elements $x = (x_i)_{-\infty}^{\infty}$ of a Banach space B with the property that

$$\|x\|_{\ell^p(B)} = \left(\sum_{-\infty}^{\infty} \|x_i\|^p\right)^{1/p} < \infty$$

As in the definition of $S(p_0,\xi_0,A_0,p_1,\xi_1,A_1)$, we consider the space $w(p_0, \xi_0,A_0,p_1,\xi_1,A_1)$ of all sequences $u = (u_n)_{-\infty}^{\infty}$ of elements u_n in A with the property that

$$(e^{\xi_0 n} u_n) \in \ell^{p_0}(A_0)$$

$$(e^{\xi_1 n} u_n) \in \ell^{p_1}(A_1)$$

with the norm on this space defined by the formula

$$\|u\|_{w(p_0,\xi_0,A_0,p_1,\xi_1,A_1)} = \max \{\|(e^{\xi_0 n} u_n)\|_{\ell^{p_0}(A_0)}, \|(e^{\xi_1 n} u_n)\|_{\ell^{p_1}(A_1)}\}$$

We remark that the condition $\xi_0\xi_1 < 0$ implies that $\Sigma_{-\infty}^{\infty} u_n$ converges in $A_0 + A_1$, and thus we can consider the space $s(p_0,\xi_0,A_0,p_1,\xi_1,A_1)$ of all elements $a = \Sigma u_n$ with (u_n) in $w(p_0,\xi_0,A_0,p_1,\xi_1,A_1)$. The norm on this space is defined as follows: If $a = \Sigma_{-\infty}^{\infty} u_n$,

$$\|a\|_{s(p_0,\xi_0,A_0,p_1,\xi_1,A_1)} = \inf_{a=\Sigma_{-\infty}^{\infty} u_n} \max \{\|(e^{\xi_0 n} u_n)\|_{\ell^{p_0}(A_0)}, \|(e^{\xi_1 n} u_n)\|_{\ell^{p_1}(A_1)}\}$$

We consider now the space denoted $\underline{S}(p_0,\xi_0,A_0,p_1,\xi_1,A_1)$ of all a in $A_0 + A_1$, $a = a_0(s) + a_1(s)$ ($s \in \mathbb{R}$ and almost every ds, where ds is the Lebesgue measure), such that

$$e^{\xi_0 s} a_0(s) \in L^{p_0}(A_0) \qquad e^{\xi_1 s} a_1(s) \in L^{p_1}(A_1)$$

with the norm on this space defined as follows:

$$\|a\|_{\underline{S}(p_0,\xi_0,A_0,p_1,\xi_1,A_1)} = \inf_{a=a_0(s)+a_1(s)} \max \{\|e^{\xi_0 s} a_0(s)\|_{L^{p_0}(A_0)}, \|e^{\xi_1 s} a_1(s)\|_{L^{p_1}(A_1)}\}$$

The discrete analogue of this space is denoted by $\underline{s}(p_0,\xi_0,A_0,p_1,\xi_1,A_1)$. The basic result of Lions and Peetre on these spaces is that they are identical and all the above norms are equivalent. Thus in topological problems it is immaterial what norm is used; for some geometrical properties certain norms are appropriate.

Now we give a result concerning the strict convexity of interpolation spaces.

THEOREM 2.3.1 (B. Beauzamy, 1976) Suppose that the Banach spaces A_0 and A_1 are as above and the following properties hold:

1. $A_0 \subset A_1$, the embedding has the norm 1, A_0 is dense in A_1.
2. $B_{A_0}(1) = \{x : \|x\|_{A_0} \le 1\}$ is weakly compact in A_1.
3. A_0 or A_1 is strictly convex.

Then the space $(A_0,A_1)_{\theta,p} = s(p_0,\xi_0,A_0,p_1,\xi_1,A_1)$ is strictly convex, where $p_0 = p_1 = p$ is in $(1,\infty)$ and $\theta = \xi_0/(\xi_0 - \xi_1)$ is in $(0,1)$.

For the proof we need the following lemma.

LEMMA 2.3.2 Let A_0,A_1 be as in Theorem 2.3.1. Then

1. The following norm is equivalent with the norm introduced above:

$$\|u\| = \left(\sum_{-\infty}^{\infty} \inf_{u=u_0+u_1} \max\{\|e^{\xi_0 n} u_0\|^p_{A_0}, \|e^{\xi_1 n} u_1\|^p_{A_1}\}\right)^{1/p} = \left(\sum_{-\infty}^{\infty} \|u\|^p_n\right)^{1/p}$$

where $\| \ \|_n$ denotes the Minkowski functional of the convex set

$$e^{-\xi_0 n} B_{A_0}(1) + e^{-\xi_1 n} B_{A_1}(1)$$

2. For each n and for each u there exist $u_0 \in A_0$, $u_1 \in A_1$, such that:

$$u = u_0 + u_1$$

$$\max\{\|e^{\xi_0 n} u_0\|_{A_0}, \|e^{\xi_1 n} u_1\|_{A_1}\} = \inf_{\substack{u=\bar{u}_0+\bar{u}_1 \\ \bar{u}_i \in A_i}} \max\{\|e^{\xi_0 n} \bar{u}_0\|_{A_0}, \|e^{\xi_1 n} \bar{u}_1\|_{A_1}\}$$

3. If A_0 is dense in A_1 then u_0 and u_1 have the property that

$$\|e^{\xi_0 n} u_0\|_{A_0} = \|e^{\xi_1 n} u_1\|_{A_1}$$

Proof: The proof given below shows that assertions (1) and (2) hold without the condition of density of A_0 in A_1 [hence the formulation of (3)]. We consider the following norm:

$$\|u\|^- = \inf_{u=u_0^n+u_1^n} \max\{\|e^{\xi_0 n} u_0^n\|_{\ell^p(A_0)}, \|e^{\xi_1 n} u_1^n\|_{\ell^p(A_1)}\}$$

and we remark that it is equivalent to the norm

$$\|u\|_- = \left(\inf_{u=u_0^n+u_1^n} \{\|e^{\xi_0 n} u_0^n\|^p_{\ell^p(A_0)} + \|e^{\xi_1 n} u_1^n\|^p_{\ell^p(A_1)}\}\right)^{1/p}$$

$$= \left[\inf_{u=u_0^n+u_1^n} \sum_{-\infty}^{\infty} (\|e^{\xi_0 n} u_0^n\|^p_{A_0} + \|e^{\xi_1 n} u_1^n\|^p_{A_1})\right]^{1/p}$$

which gives

$$\|u\|_- = \left(\sum_{-\infty}^{\infty} \inf_{u=u_0+u_1} \{\|e^{\xi_0 n} u_0\|^p_{A_0} + \|e^{\xi_1 n} u_1\|^p_{A_1}\}\right)^{1/p} = \left(\sum_{-\infty}^{\infty} \|u\|^p_n\right)^{1/p}$$

and assertion (1) is proved.

To prove assertion (2), first we note that for each n, $\| \ \|_n$ is equivalent to the norm of A_1 because of the following relation:

$$e^{-\xi_0 n} B_{A_1}(1) \subset e^{-\xi_0 n} B_{A_0}(1) + e^{-\xi_1 n} B_{A_1}(1) \subset (e^{-\xi_0 n} + e^{-\xi_1 n}) B_{A_1}(1)$$

Further, we have that for each n,

$$e^{-\xi_0 n} B_{A_0}(1) + e^{-\xi_1 n} B_{A_1}(1)$$

is closed in A_1 (as a sum of two sets, one weakly compact and the other closed). We consider the number

$$k = \inf\{t \geq 0 : u \in t[e^{-\xi_0 n} B_{A_0}(1) + e^{-\xi_1 n} B_{A_1}(1)]\}$$

and we prove that

$$u \in k[e^{-\xi_0 n} B_{A_0}(1) + e^{-\xi_1 n} B_{A_1}(1)]$$

From the definition of k there exists a sequence (t_n) with the following properties:

$$\lim t_n = k$$

$$u \in t_n[e^{-\xi_0 n} B_{A_0}(1) + e^{-\xi_1 n} B_{A_1}(1)]$$

Thus we have (we can suppose without loss of generality that $k \neq 0$) that $\lim (u/t_n) = u/k$, and so

$$u \in k[e^{-\xi_0 n} B_{A_0}(1) + e^{-\xi_1 n} B_{A_1}(1)]$$

Then there exists a_i in

$$e^{-\xi_i n} B_{A_i}(1) \qquad i = 0, 1$$

such that

$$\frac{u}{k} = a_0 + a_1$$

If we set

$$u_0 = ka_0 \qquad u_1 = ka_1$$

we have

$$\|e^{-\xi_0 n} u_i\| \le k \qquad i = 0, 1$$

and assertion (2) is proved.

Assertion (3) is that if u_0 and u_1 are elements with the properties

$$u = u_0 + u$$

$$\max\{\|e^{-\xi_0 n} u_0\|_{A_0}, \|e^{\xi_1 n} u_1\|_{A_1}\} = \inf_{u=u_1^*+u_2^*} \max\{\|e^{\xi_0 n} u_0^*\|_{A_0}, \|e^{\xi_1 n} u_1^*\|_{A_1}\}$$

then

$$\|e^{\xi_0 n} u_0\|_{A_0} = \|e^{\xi_1 n} u_1\|_{A_1}$$

Suppose that this is not true. We distinguish two cases:

Case 1 For some $\varepsilon_0 > 0$,

$$\|e^{\xi_0 n} u_0\|_{A_0} > \|e^{\xi_1 n} u_1\|_{A_1} + \varepsilon_0$$

Case 2 For some $\eta_0 > 0$,

$$\|e^{\xi_1 n} u_1\|_{A_1} > \|e^{\xi_0 n} u_0\|_{A_0} + \eta_0$$

In the first case we consider the pair

$$(v_0^t, v_1^t) \qquad t \in (0,1)$$

where

$$v_0^t = (1 - t)u_0 \qquad v_1^t = u_1 + tu_0$$

Clearly,

$$u = v_0^t + v_1^t$$

and

$$\|e^{\xi_1 n} v_1^t\|_{A_1} \le \|e^{\xi_1 n} u_1\|_{A_1} + t\|e^{\xi_1 n} u_0\|_{A_1} \le \|e^{\xi_1 n} u_1\|_{A_1} + \frac{\varepsilon_0}{2}$$

if

$$t < \frac{\varepsilon_0}{2\|e^{\xi_1 n} u_0\|_{A_1}}$$

In this case we have

$$\|e^{\xi_0 n} v_0^t\|_{A_0} \le (1 - t)\|e^{\xi_0 n} u_0\|_{A_0} < \|e^{\xi_0 n} u_0\|_{A_0}$$

and this contradicts the property of the pair u_0, u_1.

In the second case we use essentially the density of A_0 in A_1 (because u_1 is not in A_0). For any a_0 in A_0 we have

$$\|e^{\xi_0 n}(a_0 + u_0)\|_{A_0} \le \|e^{\xi_0 n} u_0\|_{A_0} + \|e^{\xi_0 n} a_0\|_{A_0} \le \|e^{\xi_0 n} u_0\|_{A_0} + \frac{\eta_0}{2}$$

if

$$\|a_0\|_{A_0} \le \frac{e^{-\xi_0 n}\eta_0}{2}$$

Let

$$C = \left\{a^* : a^* \in A_1, \|a^*\|_{A_0} \le \frac{\varepsilon_0}{2} e^{-\xi_0 n}\right\}$$

To finish the proof for case 2, it suffices to show that for some b in C,

$$\|e^{\xi_1 n}(u_1 - b)\|_{A_1} < \|e^{\xi_1 n} u_1\|_{A_1}$$

Suppose that this is not true; thus for all b in C we have

$$\|u_1 - b\|_{A_1} \ge \|u_1\|_{A_1}$$

We consider the set

$$C_1 = \{u_1 - a^* : a^* \in C\} \subset A_1$$

and we remark that

$$C_1 \cap \{a_1 : a_1 \in A_1, \|a_1\|_{A_1} \le \|u_1\|_{A_1}\} = \emptyset$$

We can separate the two sets by a hyperplane. The half-space containing C_1 is invariant under homotheties with center at u_1 (and positive ratio), and we see that it contains the set $\{u_1 - a_0 : a_0 \in A_0\}$. But this obviously contradicts the density of A_0 (in A_1). The lemma is proved.

Now the proof of Theorem 2.3.1 is as follows:

Suppose that the assertion of the theorem is false. Then we find u,v in the space with the property

$$\|u\| = \|v\| = \left\|\frac{1}{2}(u + v)\right\| = 1$$

We consider the elements u_0^n, u_1^n, v_0^n, v_1^n corresponding to u and v as given by Lemma 2.3.2(2). In this case we have

$$1 = \frac{1}{2}(u + v) = \left(\sum_{-\infty}^{\infty} \max\left\{\left\|e^{\xi_0 n}\frac{u_0^n + v_0^n}{2}\right\|_{A_0}^p, \left\|e^{\xi_1 n}\frac{u_1^n + v_1^n}{2}\right\|_{A_1}^p\right\}\right)^{1/p}$$

$$\le \left(\sum_{-\infty}^{\infty} \max\left\{\frac{1}{2}(\|e^{\xi_0 n}u_0^n\|_{A_0}^p + \|e^{\xi_0 n}v_0^n\|_{A_0}^p), \frac{1}{2}(\|e^{\xi_1 n}u_1^n\|_{A_1}^p + \|e^{\xi_1 n}v_1^n\|_{A_1}^p)\right\}\right)^{1/p}$$

and from Lemma 2.3.2,

$$\|e^{\xi_0 n}u_0^n\|_{A_0} = \|e^{\xi_1 n}u_1^n\|_{A_1}$$

$$\|e^{\xi_0 n}v_0^n\|_{A_0} = \|e^{\xi_1 n}v_1^n\|_{A_1}$$

for all n. Since

$$\|u\| = \|v\| = 1$$

we must have

$$\sum_{-\infty}^{\infty} \frac{1}{2}(\|e^{\xi_0 n}u_0^n\|_{A_0}^p + \|e^{\xi_1 n}v_0^n\|_{A_0}^p) = 1$$

which gives

$$\sum_{-\infty}^{\infty} \max\left\{\left\|e^{\xi_0 n}\frac{u_0^n + v_0^n}{2}\right\|_{A_0}^p, \left\|e^{\xi_1 n}\frac{u_1^n + v_1^n}{2}\right\|_{A_1}^p\right\} = 1$$

and thus

$$\sum_{-\infty}^{\infty} \left\| e^{\xi_0 n} \frac{u_0^n + v_0^n}{2} \right\|_{A_0}^p = \sum_{-\infty}^{\infty} \left\| e^{\xi_1 n} \frac{u_1^n + v_1^n}{2} \right\|_{A_1}^p = 1$$

Thus we consider the elements

$$(e^{\xi_0 n} u_0^n) \qquad (e^{\xi_0 n} v_0^n) \qquad \left(e^{\xi_0 n} \frac{1}{2} (u_0^n + v_0^n)\right)$$

and

$$(e^{\xi_1 n} u_1^n) \qquad (e^{\xi_1 n} v_1^n) \qquad \left(e^{\xi_1 n} \frac{1}{2} (u_1^n + v_1^n)\right)$$

which are in $\ell^p(A_0)$ and in $\ell^p(A_1)$, respectively. All have norm equal to 1, which is impossible since at least one of the spaces $\ell^p(A_0)$ or $\ell^p(A_1)$ is strictly convex by Theorem 2.1.7. The theorem is proved.

2.4 UNIFORMLY CONVEX SPACES

The class of uniformly convex spaces was introduced by J. A. Clarkson (1936), who proved that the spaces ℓ^p and L^p are in this class [for p in $(1,\infty)$]. Another interesting property discovered by Clarkson for the class of uniformly convex spaces was the so-called Radon-Nikodým property. Numerous papers have since appeared in which this class or related classes of Banach spaces were studied. Among the contributors in this area we mention D. P. Milman, B. J. Pettis, V. Klee, A. Lovaglia, V. Smulian, J. Lindenstrauss, and M. Kadets. In what follows we present this class of Banach spaces together with some basic properties.

DEFINITION 2.4.1 The Banach space X is said to be uniformly convex if and only if for any $\varepsilon > 0$ there exists $\delta(\varepsilon) > 0$ such that whenever $\|x\| = \|y\| = 1$ and $\|x - y\| \geq \varepsilon$ then $\|x + y\| \leq 2[1 - \delta(\varepsilon)]$.

We now give some equivalent formulations of this notion which may prove useful.

THEOREM 2.4.2 The Banach space X is uniformly convex if and only if for any sequences (x_n), (y_n) of elements of X with the properties

$$\|x_n\| \to 1 \quad \|y_n\| \to 1$$

$$\lim \|x_n + y_n\| = 2$$

then $\lim (x_n - y_n) = 0$.

It is obvious that this is equivalent to the following formulation.

THEOREM 2.4.3 The Banach space X is uniformly convex if and only if for any sequences $(x_n),(y_n)$ of elements of X satisfying the following properties:
(1) $\|x_n\| = \|y_n\| = 1$
(2) $\lim \|x_n + y_n\| = 2$
then $\lim (x_n - y_n) = 0$.

Proof: We prove Theorem 2.4.3. First we show that the condition is necessary. Suppose that $(x_n),(y_n)$ have properties (1) and (2) and that the sequence $(x_n - y_n)$ does not converge to zero. Then there exists $\varepsilon_0 > 0$ and a sequence of integers (n_k) such that

$$\|x_{n_k} - y_{n_k}\| \geq \varepsilon_0$$

Since the space X is supposed to be uniformly convex, for this ε_0 there exists $\delta(\varepsilon_0) > 0$ such that

$$\|x_{n_k} + y_{n_k}\| \leq 2[1 - \delta(\varepsilon_0)]$$

and this obviously contradicts property (2).

Now we prove that the condition is sufficient. Suppose that X satisfies the conditions of Theorem 2.4.3 and that X is not uniformly convex. Then for some $\varepsilon > 0$ and for each $\delta = 1/n$ there exists elements x_n and y_n with the following properties:

(i) $\|x_n\| = \|y_n\| = 1$
(ii) $\|x_n + y_n\| \geq 2(1 - 1/n)$
(iii) $\|x_n - y_n\| \geq E$

Obviously (iii) contradicts the hypothesis since (ii) gives $\lim \|x_n + y_n\| = 2$. Thus X must be uniformly convex.

From just the definition of a uniformly convex space we have the following result:

THEOREM 2.4.4 Every uniformly convex space is strictly convex.

We give now an important property of closed convex sets in uniformly convex spaces.

THEOREM 2.4.5 Let X be a uniformly convex space and C be a closed, bounded, and convex set in X. Then C has a unique element u_0 with the property that

$$\|u_0\| = \inf \{\|u\| : u \in C\}$$

(This u_0 is called the element with minimal norm.)

Proof: Let $d = \inf \{\|u\| : u \in C\}$. If $d = 0$, from the fact that C is closed, the assertion about the existence of u_0 is obvious. Thus we can suppose that $d > c$. In this case from the definition of d it follows that

$$d = \lim \|x_n\| \qquad x_n \in C$$

Since C is supposed to be convex, for all n,m,

$$\frac{1}{2}(x_n + x_m) \in C$$

and thus

$$\|\frac{1}{2}(x_n + x_m)\| \geq d$$

If we set

$$y_n = \frac{x_n}{d}$$

then we have that the sequence (y_n) has the properties

$$\lim \|y_n\| = 1$$

$$\lim_{n,m\to\infty} \|y_n + y_m\| = 2$$

Set

$$z_n = \frac{y_n}{\|y_n\|}$$

Then

1. $\|z_n\| = 1$
2. $\lim \|z_n + z_m\| = 2$

We remark that property 2 means that for any $\eta > 0$ there exists an integer N_η such that if $n,m \geq N_\eta$ then

$$\big|\|z_n + z_m\| - 2\big| \leq \eta$$

Now, since X is supposed to be uniformly convex we get that (z_n) is a Cauchy sequence and this implies that (x_n) is a Cauchy sequence. If $u_0 = \lim x_n$ then it is clear that u_0 is an element with minimal norm. Now we show that this element is unique. Indeed, in the contrary case there exists another element, say v_0, with the same property. Then clearly

$$tu_0 + (1 - t)v_0$$

are elements with norm equal to $\|u_0\|$. But this contradicts the strict convexity of X (Theorem 2.4.4).

Now we consider the problem of reflexivity of uniformly convex Banach spaces. The fact that every uniformly convex space is reflexive was proved independently by D. P. Milman (1938) and B. J. Pettis (1939). We mention that there exist now many elegant proofs of the reflexivity of uniformly convex Banach spaces. Some of them are very short but make use of some advanced results of Banach space theory. One of the simplest proofs seems to be that of J. R. Ringrose (1959). In what follows we offer two proofs.

THEOREM 2.4.6 (Milman-Pettis) Every uniformly convex Banach space is reflexive.

First proof: Let us consider $x^{**} \in X^{**}$ and suppose that $\|x^{**}\| = 1$. From the definition of the norm we conclude that there exists a sequence (x_n^*) with the properties

$$\|x_n^*\| = 1$$
$$|x^{**}(x_n^*)| \geq 1 - \frac{1}{n} \qquad n = 1, 2, 3, \ldots$$

Now for each fixed n we can apply Helly's theorem and find the element x_n such that

$$\|x_n\| \leq 1 + \frac{1}{n}$$
$$x^{**}(x_k^*) = x_k^*(x_n) \qquad k = 1, 2, 3, \ldots, n$$

From this we have that the sequence (x_n) converges to 1 and also that

$$\lim \|(x_n + x_m)\| = 2$$

This implies that the sequence (x_n) is a Cauchy sequence (see the proof of Theorem 2.4.5). Let $x = \lim x_n$ and let y^* be an arbitrary element in X^*. We can apply Helly's theorem for $y^*, x_1^*, \ldots, x_n^*$, and thus we have the relations

$$x^{**}(x_k^*) = x_k^*(x) \qquad k = 1, 2, 3, \ldots, n$$
$$x^{**}(y^*) = y^*(x)$$

This gives that x^{**} is the image of x in X^{**} and the reflexivity is proved.

Second proof: We use the deep result of R. C. James (1957, 1972) characterizing reflexivity: The Banach space X is not reflexive if and only if there exists x^* in X^* which does not achieve its norm.

Let us consider arbitrary x^* in X^*. From the definition of the norm of x^* we get that there exists a sequence (x_n), $\|x_n\| = 1$, such that $x^*(x_n) \to 1$ (we suppose without loss of generality that $\|x^*\| = 1$). Now we show that this sequence (x_n) is a Cauchy sequence. Indeed, in the contrary case, for some $\varepsilon > 0$ and a sequence (n_k) of integers we have the inequality

$$\|x_{n_p} - x_{n_q}\| \geq \varepsilon$$

Since X is supposed to be uniformly convex, for this ε we find $\delta(\varepsilon) > 0$ such that

$$\|x_{n_p} + x_{n_q}\| \leq 2[1 - \delta(\varepsilon)]$$

In this case we have

$$1 = \lim x^*(\frac{1}{2}(x_{n_p} + x_{n_q})) \leq 2[1 - \delta(\varepsilon)]$$

which is clearly a contradiction. Thus the sequence (x_n) is a Cauchy sequence, and if $x = \lim x_n$ then clearly $x^*(x) = 1$. By James' theorem X must be reflexive.

We present now an interesting characterization of uniformly convex spaces using some intrinsic functions defined on the unit sphere $S_1 = \{x : \|x\| = 1\}$ of X. We consider φ to be a strictly convex and strictly increasing function on $[0,2]$ [such a function is continuous on $(0,2)$].

THEOREM 2.4.7 (W. L. Bynum, 1971) The Banach space X is uniformly convex if and only if for any function φ as above with $\varphi(1) = 1$, for each t in $(0,1]$,

$$a(t) = \inf \{\varphi(\|x + ty\|) + \varphi(\|x - ty\|) - 2 : \|x\| = \|y\| = 1\}$$

is strictly positive.

Proof: First we show that the above condition is necessary. Thus, if X is uniformly convex and φ is a function as above we must show that the function $a(t) > 0$ for all t in $(0,1]$. Suppose that this is not true and thus for some t_0 in $(0,1]$ we have $a(t_0) = 0$. From the definition of $a(t)$ there exist sequences (x_n) and (y_n) with the properties

$$\|x_n\| = \|y_n\| = 1$$
$$\lim_{n\to\infty} (\varphi(\|x_n + t_0 y_n\|) + \varphi(\|x_n - t_0 y_n\|) = 2$$

But φ is supposed to be convex and nondecreasing with $\varphi(1) = 1$, and thus we have

$$2 \leq 2\varphi\left(\frac{\|x_n + t_0 y_n\| + \|x_n - t_0 y_n\|}{2}\right) = \varphi(\|x_n + t_0 y_n\|) + \varphi(\|x_n - t_0 y_n\|) \to 2$$

Thus the strict convexity of φ implies that

$$\lim \big| \|x_n + t_0 y_n\| - \|x_n - t_0 y_n\| \big| \to 0$$

The function φ^{-1} is continuous at $t = 1$ and thus

$$\|x_n + t_0 y_n\| + \|x_n - t_0 y_n\| \to 2$$

X being uniformly convex, by Theorem 2.4.2 we have $2t_0 = \|x_n + t_0 y_n - (x_n - t_0 y_n)\| \to 0$, which is a contradiction.

Conversely, suppose that X has the property of the theorem. We show that this implies the uniform convexity of X. For each fixed x,y in X, $\|x\| = \|y\| = 1$, the function

$$f_{x,y}(t) = \varphi(\|x + ty\|) + \varphi(\|x - ty\|) - 2$$

is obviously convex and nondecreasing, with $f_{x,y}(t) \geq 0$, $f_{x,y}(0) = 0$. Since

$$a(t) = \inf_{x,y} f_{x,y}(t)$$

we obtain that $a(t)$ is positive and nondecreasing on $[0,1]$, and for all x,y in X with $\|x\| \geq \|y\|$, $x \neq 0$, the following inequality holds:

$$\varphi\left(\frac{\|x + y\|}{\|x\|}\right) + \varphi\left(\frac{\|x - y\|}{\|x\|}\right) - 2 \geq a\left(\frac{\|y\|}{\|x\|}\right)$$

From this we have for all u,v with $\|u\| = \|v\| = 1$ and $\|u - v\| \leq \|u + v\|$ that

$$2\varphi\left(\frac{2}{\|u + v\|}\right) - 2 \geq a\left(\frac{\|u - v\|}{\|u + v\|}\right) \geq a\left(\frac{\|u - v\|}{2}\right) \qquad (*)$$

Suppose now that (x_n) and (y_n) are two sequences in X with the properties

$$\|x_n\| = \|y_n\| = 1 \qquad \lim \|x_n + y_n\| = 2$$

We may assume (without loss of generality) that for n sufficiently large the following inequality holds:

$$\|x_n - y_n\| \le \|x_n + y_n\|$$

From the continuity of φ at $t = 1$ we obtain

$$\lim a(x_n - y_n) = 0$$

which gives further that

$$\lim \|x_n - y_n\| = 0$$

and thus by Theorem 2.4.3 the space X is uniformly convex.

We present now two characterizations of uniform convexity using the duality mappings. (See Definition 2.1.11.)

THEOREM 2.4.8 (W. L. Bynum, 1971) The Banach space X is uniformly convex if and only if for each t in (0,2] the function

$$b(t) = \inf \{1 - x^*(y) : \|x\| = \|y\| = 1, \|x - y\| \ge t, x^* \in J(x)\}$$

is positive. Here J is a duality mapping defined on X.

Proof: Suppose first that X is uniformly convex and let x, y, and x^* be as in the theorem. In this case, we have

$$1 - x^*(y) = 2 - x^*(x + y) \ge 2 - \|x + y\| = 2 - \delta(\|x - y\|)$$

and thus the condition is necessary.

Conversely, suppose that the condition of the theorem is satisfied. We must show that X is uniformly convex. Indeed, in the contrary case there exist sequences (x_n) and (y_n) with the properties:

1. $\|x_n\| = \|y_n\| = 1$
2. $\lim \|x_n + y_n\| = 2$
3. $\|x_n - y_n\| \ge t$

We set

$$k_n = \|x_n + y_n\|^{-1} \qquad u_n = k_n(x_n + y_n)$$

and for u_n^* in $J(u_n)$, x_n^* in $J(x_n)$, y_n^* in $J(y_n)$,

$$2 - \|x_n + y_n\| = 1 - u_n^*(x_n) + 1 - u_n^*(y_n) \ge b(\|x_n - u_n\|)$$

Now we remark that

$$\min \{\|x_n - u_n\|, \|y_n - u_n\|\} \ge tk_n - |1 - 2k_n|$$

and thus if n is sufficiently large we have

$$\|x_n - u_n\| \geq \frac{t}{4} \qquad \|y_n - u_n\| \geq \frac{t}{4}$$

which gives that

$$2 - \|x_n + y_n\| \geq 2b(\frac{t}{4})$$

This contradicts property 2 of the sequences (x_n) and (y_n). Thus the condition is sufficient.

For the formulation of the next characterization we need the following.

DEFINITION 2.4.9 Let X be a Banach space and J be a duality mapping from X into X*. We say that J is uniformly monotone if for each t in (0,2],

$$c(t) = \inf \{(x^* - y^*)(x - y) : \|x\| = \|y\| = 1,\ \|x - y\| \geq t,$$
$$x^* \in J(x),\ y^* \in J(y)\}$$

is strictly positive.

THEOREM 2.4.10 (W. L. Bynum, 1971) The Banach space X is uniformly convex if and only if the duality mapping J from X into X* is uniformly monotone.

Proof: Suppose that X is a uniformly convex Banach space and x,y are in X, $\|x\| = \|y\| = 1$, x* in J(x), y* in J(y). Then

$$(x^* - y^*)(x - y) = 2 - y^*(x + y) + 2 - x^*(x + y) \geq 2(2 - \|x + y\|)$$

which clearly is strictly positive by uniform convexity of X.

Now we prove that the condition is sufficient. Suppose that the duality mapping J is uniformly monotone and that X is not uniformly convex. According to Theorem 2.4.8 the function b(t) is zero at some t in (0,2]. Thus there exist sequences (x_n),(y_n) with the following properties:

$$\|x_n\| = \|y_n\| = 1 \qquad \|x_n - y_n\| \geq t \qquad \text{for some } t \in (0,2]$$
$$x_n^* \in J(x_n) \qquad 1 - x_n^*(y_n) \to 0$$

Now

$$1 - x_n^*(y_n) = 2 - \|x_n + y_n\| \geq 0$$

which implies that

$$\lim \|x_n + y_n\| = 2$$

We may assume without loss of generality that for all n, $x_n + y_n \neq 0$, and thus we can set

$$k_n = (\|x_n + y_n\|)^{-1} \qquad u_n = k_n(x_n + y_n)$$

and for u_n^* in $J(u_n)$ we have

$$u_n^*(x_n + y_n) = \|x_n + y_n\| - 2$$

But

$$\|u_n^*\| = 1 = \|x_n\| = \|y_n\|$$

which implies that

$$\lim u_n^*(x_n) = 1$$

and thus

$$(u_n^* - x_n^*)(u_n - x_n) = 1 - k_n - k_n x_n^*(y_n) + 1 - u_n^*(x_n) \to 0$$

Exactly as in the proof of Theorem 2.4.8, if n is sufficiently large,

$$\|x_n - u_n\| \geq \frac{t}{4} \qquad (u_n^* - x_n^*)(u_n - x_n) \geq c(\frac{t}{4})$$

This is a contradiction and the theorem is proved.

We present now a characterization of uniformly convex spaces obtained by A. R. Blass and C. V. Stanojevič (1976) using the so-called *partial Mielnik spaces*. The notion of partial Mielnik space is considered in the following.

DEFINITION 2.4.11 A partial Mielnik space is a pair (S,p) where S is a nonempty set and p is a function on S × S with values in [0,1] satisfying the following properties:

1. p(a,b) = 1 if and only if a = b.
2. p(a,b) = p(b,a) for all a,b in S.
3. There exists a maximal family of elements $(b_\alpha)_{\alpha \in I}$, I an index set, such that $p(b_\alpha, b_\beta) = 0$ for $\alpha \neq \beta$, and for each a in S the family of numbers $(p(a,b_\alpha)_{\alpha \in I})$ is summable and the sum is in [0,1].

The characterization of uniformly convex spaces using partial Mielnik spaces is as follows.

THEOREM 2.4.12 (A. R. Blass and C. V. Stanojevič, 1976) The Banach space X is uniformly convex if and only if there exists a continuous, strictly increasing function f on [0,2] with values in [0,1] such that $f(0) = 0$, $f(2) = 1$, and (X,p) is a partial Mielnik space, where p is defined by the formula:

$$p(x,y) = f(\|x + y\|)$$

For the proof we need the following lemma.

LEMMA 2.4.13 If $g : [0,2] \to [0,1]$ is a monotone nondecreasing function with $g(0) = 0$, $g(2) = 1$, and g continuous, then there exists $g_1 : [0,2] \to [0,1]$ which is continuous, strictly increasing, $g_1(0) = 0$, $g_1(2) = 1$, and

$$g_1(t) \le g(t)$$

on [0,2].

Proof: We define the function g_1 by the formula

$$g_1(t) = \int_{s=0}^{t} \frac{t - s}{2 - s}\, dg(s)$$

and it is not difficult to see that this function has all properties stated in the lemma.

Now the proof of Theorem 2.4.12 is as follows.

Suppose first that X is uniformly convex and suppose further that f is a continuous, strictly increasing function on [0,2] with values in [0,1] satisfying the conditions $f(0) = 0$, $f(2) = 1$. We note that for any f with these properties, Axioms 1 and 2 for a Mielnik space are satisfied, since X is strictly convex. We show now that we can choose f also satisfying Axiom 3. For this it is sufficient to remark that Axiom 3 is satisfied if f has the property

$$f(\|x + y\|) + f(\|x - y\|) \ge 1$$

for all $\|x\| = \|y\| = 1$. We show that such an f may be constructed. First we set

$$h(t) = \sup \{\|x + y\| : \|x\| = \|y\| = 1,\ \|x - y\| \ge t\} \qquad 0 \le t \le 2$$

We note that h is monotone nonincreasing with $h(0) = 2$. From the uniform convexity of X we get that $h(t) \le 2$, and $h(2) = 0$ for $\|x - y\| = 2$ if and only if $x = -y$. Also, $t = 2$ is a continuity point for h. We set

$$g(t) = 1 - \frac{h(t)}{2}$$

and this satisfies the hypothesis of Lemma 2.4.13. Denote by g_1 the functions constructed in that lemma for our g. Clearly the function $\min \{f,g_1\} = p$ gives (X,p) as a Mielnik space.

Conversely, suppose that X is a Banach space such that for some p, (X,p) is a Mielnik space. We show that this implies that X is uniformly convex. Indeed, let $(x_n),(y_n)$ be two sequences with the properties

$$\|x_n\| = \|y_n\| = 1$$
$$\lim \|x_n + y_n\| = 2$$

We must show that $\lim (x_n - y_n) = 0$. Now p is derived from a function f such that f is defined on [0,2],, has values in [0,1], is continuous and strictly increasing, f(0) = 0, and f(2) = 1. Thus we obtain that any family $(a_\alpha)_{\alpha \in I}$ of elements of X satisfies the property $p(a_\alpha, a_\beta) = 0$, $\alpha \neq \beta$, and the family that is maximal (for the partial order relation by inclusion) is of the form $(a,-a)_{a \in X}$. We note that the existence of such a family follows easily from Zorn's lemma. Since f(2) = 1 we obtain

$$\lim f(\|x_n - y_n\|) = 0$$

and f^{-1} being continuous, this implies that

$$\lim (x_n - y_n) = 0$$

Then Theorem 2.4.3 asserts that X is uniformly convex.

For the formulation of the following theorem we need the notion of modulus of convexity of a Banach space, introduced by J. Clarkson.

DEFINITION 2.4.14 Let X be a Banach space. The modulus of convexity of X is the function $\delta_X : [0,2] \to [0,1]$ defined by the formula

$$\delta_X(t) = \inf \{1 - \tfrac{1}{2}(\|x + y\|) : \|x\| = \|y\| = 1, \|x - y\| \geq t\}$$

From the definition we have the following result.

PROPOSITION 2.4.15 Let X be a Banach space. Then the following assertions hold:

1. The modulus of convexity is a nondecreasing function.
2. X is uniformly convex if and only if for each $t > 0$, $\delta_X(t) > 0$.

If (X_n) is a sequence of Banach spaces then for each p in $\mathbb{R}$ we can consider the space of all sequences $x = (x_n)$, x_n in X_n, such that

$$\|x\|_p = (\Sigma \|x_n\|^p)^{1/p} < \infty$$

If p is in $(1,\infty)$ then we get that $\ell^p(X_n)$ is a Banach space. It is quite natural to ask about uniform convexity of this space. The answer to this question is given in the following result.

THEOREM 2.4.16 (M. M. Day, 1944) The Banach space $\ell^p(X_n)$, p in $(1,\infty)$, is uniformly convex if and only if all X_n have the same modulus of convexity [i.e., the same function $t \to \delta_{X_n}(t) = h(t)$] and are uniformly convex.

Proof: First we show that the condition is necessary. Indeed, for this purpose it is sufficient to remark that each X_n is isomorphic and isometric with the subspace of $\ell^p(X_n)$ consisting of all elements x of the form

$$(0,0,\ldots,x_n,0,\ldots) \qquad X_n \text{ in the nth place}$$

and the assertion is obvious.

Conversely, suppose that all X_n have the same modulus of convexity and are uniformly convex. Let x,y be in $\ell^p(X_n)$ with the following properties:

$$\|x\| = \|y\| = 1$$
$$\|x - y\| \geq t$$

We suppose further that $x = (x_n)$, $y = (y_n)$, with x_n, y_n in X_n. We distinguish two cases:

Case 1. For all n, $\|x_n\| = \|y_n\|$. In this case, if $\|x_n\| = b_n$ and $\|x_n - y_n\| = r_n$, using the uniform convexity of X_n (as well as the fact that the modulus of convexity is the same for all X_n), we find that

$$\|x_n + y_n\| \leq 2\left(1 - \frac{r_n}{b_n}\right)b_n$$

which implies

$$\|x + y\| = (\Sigma \|x_n + y_n\|^p)^{1/p} \leq 2\left[\sum_{n=1}^{\infty} \left(1 - \frac{r_n}{b_n}\right)^p\right]^{1/p}$$

Since

$$\|x_n - y_n\| \le \|x_n + y_n\|$$

we have

$$r_n \le 2b_n$$

We decompose the set of integers into two subsets as follows:

$$E_1 = \left\{n : \frac{r_n}{b_n} > \frac{\varepsilon}{4}\right\} \qquad E_2 = C_{E_1}$$

(E_2 = complement of E_1). From the definition of E_2 we obtain that for each n in E_2,

$$\varepsilon b_n \ge 4r_n$$

and thus

$$1 = \left(\sum_1^\infty b_n^p\right)^{1/p} \ge \left(\sum_{n\in E_2} b_n^p\right)^{1/p} \ge \frac{4}{\varepsilon}\left(\sum_{n\in E_2} r_n^p\right)^{1/p}$$

or

$$\sum_{n\in E_2} r_n^p \le \left(\frac{\varepsilon}{4}\right)^p$$

Also,

$$\left(\sum_{n\in E_1} r_n^p\right)^{1/p} \ge \left(\sum_n r_n^p - \sum_{n\in E_2} r_n^p\right)^{1/p} \ge \left[\varepsilon^p - \left(\frac{\varepsilon}{4}\right)^p\right]^{1/p} \ge 3\,\frac{\varepsilon}{4}$$

which implies that

$$a = \left(\sum_{n\in E_1} b_n^p\right)^{1/p} \ge 3\,\frac{\varepsilon}{4}$$

Using the inequality just obtained for x + y we have

$$\begin{aligned}\|x + y\| &\le 2\left[1 - \delta\left(\frac{\varepsilon}{4}\right)\right]^p\left(\sum_{n\in E_1} b_n^p + \sum_{n\in E_2} b_n^p\right)^{1/p}\\ &\le 2\left\{\left[1 - \delta\left(\frac{\varepsilon}{4}\right)\right]^p a^p + (1 - a^p)\right\}^{1/p}\\ &\le 2\left\{1 - \left[1 - \delta\left(\frac{\varepsilon}{4}\right)\right]^p a^p\right\}^{1/p} \le 2\left\{1 - \left[1 - \delta\left(\frac{\varepsilon}{4}\right)\right]^p\left(\frac{3\varepsilon}{8}\right)^p\right\}^{1/p}\end{aligned}$$

We remark that the coefficient of a^p is positive and is a function of ε only, and the right-hand side is of the form $2[1 - \delta_1(\varepsilon)]$, with $\delta_1(\varepsilon) > 0$

if $\varepsilon > 0$. Thus in this case the assertion is proved.

Case 2. Not all $\|x_n\| = \|y_n\|$. We consider the space ℓ^p formed by the numerical sequences. As we shall show later (Theorem 2.7.11) this is a uniformly convex space; let δ^* be its modulus of convexity. Suppose that

$$\|x + y\| > 2[1 - \delta^*(\varepsilon)] \qquad \varepsilon \in (0,2]$$

In this case we have

$$2[1 - \delta^*(\varepsilon)] \leq (\Sigma \|x_n + y_n\|^p)^{1/p} \leq [\Sigma(\|x_n\| + \|y_n\|)^p]^{1/p} \leq 2$$

The sequences $(\|x_n\|)$, $(\|y_n\|)$ are in ℓ^p and thus, according to the definition of the modulus of convexity, we have

$$(\Sigma |\|x_n\| - \|y_n\||^p)^{1/p} \leq \varepsilon_1$$

for some $\varepsilon_1 \in [0,2]$. Consider now the sequence (z_n) where

$$z_n = \frac{\|x_n\|}{\|y_n\|} y_n$$

Obviously

$$\|x_n\| = \|z_n\|$$

for all n. We remark that the following inequality holds:

$$\|z - y\| = [\Sigma \; |\|y_n\| - \|x_n\||^p]^{1/p} < \varepsilon_1 \qquad z = (z_n)$$

Now for the elements x and z we can apply the result obtained for case 1, and thus

$$\|x - z\| \leq \frac{\varepsilon_1}{2}$$

if

$$\|x + z\| \geq 2\left[1 - \delta\left(\frac{\varepsilon}{2}\right)\right]$$

and also,

$$\|x + z\| \geq \|x + y\| - \|y - z\| > 2\left\{[1 - \delta^*(\varepsilon_1)] - \frac{\varepsilon_1}{2}\right\}$$

But we can take ε_1 in the interval $(0,\varepsilon/2)$, with the property that

$$\delta^*(\varepsilon_1) + \frac{\varepsilon_1}{2} < \delta\left(\frac{\varepsilon}{2}\right)$$

and thus

$$\|x + y\| \geq 2[1 - \delta^*(\varepsilon_1)]$$

which gives that

$$\|x + z\| > 2\left[1 - \delta\left(\frac{\varepsilon}{2}\right)\right]$$

and thus

$$\|x - z\| \leq \frac{\varepsilon}{2}$$

In this case we have that

$$\|x - y\| \leq \|x - z\| + \|z - y\| \leq \frac{\varepsilon}{2} + \frac{\varepsilon}{2} = \varepsilon$$

We denote $\delta^*(\varepsilon_1)$ (which is a function of ε) by $\delta_2(\varepsilon)$; this is clearly strictly positive for $\varepsilon > 0$. Then we have that

$$\|x - y\| \leq \varepsilon \qquad \text{for } \|x\| = \|y\| = 1$$

gives

$$\|x + y\| \leq 2[1 - \delta_2(\varepsilon)]$$

and this implies the uniform convexity of $\ell^p(X_n)$.

As we know, every uniformly convex space is strictly convex and reflexive. The problem naturally arises as to how large is the class of Banach spaces which are isomorphic to uniformly convex spaces. The following example shows that this class is strictly narrower than the class of reflexive strictly convex spaces.

EXAMPLE 2.4.17 (M. M. Day, 1944) Let (X_n) be a sequence of Banach spaces with the property that $\dim X_n = n$, $n = 1, 2, 3, \ldots$. We identify each X_n with the space of all sequences $x = (x_1, \ldots, x_n)$ equipped with the norm

$$\|x\| = \max_i |x_i|$$

Since each X_n is finite dimensional, $\ell^p(X_n)$ for p in $(1,\infty)$ is reflexive and clearly is a separable Banach space. We consider the following norm on $\ell^p(X_n)$:

$$\|x\|_1 = \left[\|x\|^2 + \sum_{i=1}^{\infty} \frac{1}{2^i}|x_i^* 2(x)|\right]^{1/2}$$

where $\{x_i^*\}_1^\infty$ is any sequence in the dual space of $\ell^p(X_n)$ which is dense in the closed unit ball. Then using Minkowski's inequality and the property of the sequence (x_n^*) it is not difficult to see that $(\ell^p(X_n), \|\ \|_1)$ is a strictly convex space, and clearly it is reflexive. We remark that the above construction of the norm $\|\ \|_1$ goes back to Clarkson (1936). We show now that there exists a new equivalent norm $|\ |$ on $\ell^p(X_n)$ such that it is uniformly convex with respect to this norm. Let x,y be in $\ell^p(X_n)$ and ε in (0,2], and assume

$$|x| = |y| = 1$$
$$|x - y| > \varepsilon$$

We set

$$\delta(\varepsilon) = \inf_{|x|=|y|=1} \left\{1 - \frac{1}{2}|x + y|\right\}$$

The space $\ell^p(X_n)$ is uniformly convex with respect to the norm $|\ |$ if and only if $\delta(\varepsilon) > 0$ for $\varepsilon > 0$. Since the norm $|\ |$ is equivalent to the norm $\|\ \|_1$ we may suppose without loss of generality that for some $M \geq 1$,

$$\|x\|_1 \leq |x| \leq M\|x\|_1$$

Let n be fixed and consider the unit ball $B_n(1)$ of X_n, which contains all points of X_n of norm less than or equal to 1/M. Set a = 1/M; then

$$(\varepsilon_1 a, \varepsilon_2 a, \ldots, \varepsilon_n a) \in B_n$$

for all $\varepsilon_j = \pm 1$. We consider now the elements

$$b = (a,a,\ldots,a,a) \qquad b' = (a,\ldots,a,-a)$$

Then

$$b - b' = (0,0,\ldots,0,2a)$$
$$b + b' = 2a,\ldots,2a,0)$$

We have the following inequalities, taking into account the relation between the norms:

$$|b - b'| \geq \|b - b'\|_1 \geq 2a$$
$$|b + b'| \leq 2[1 - \delta(2a)]$$

which implies that

$$|(a,\ldots,a,0)| < 1 - \delta(2a)$$

or

$$\left\| \left(\frac{a}{1 - \delta(2a)}, \ldots, \frac{a}{1 - \delta(2a)}, 0 \right) \right\| < 1$$

We can repeat this n - 2 times, and since $1 - \delta(2a) < 1$ [$\delta(\varepsilon)$ decreases], we have that

$$\delta\left(\frac{a}{1 - \delta(2a)}\right)^j \geq \delta(2a)$$

for $j = 1, 2, \ldots, n - 1$. If $b_n = (a/[1 - \delta(2a)]^{n-1}, 0, \ldots, 0)$ we get that $|b_n| \geq 1$ and $|b_n| = \|b_n\|_1 \geq 1$ for large n. But this is a contradiction, and thus the space $\ell^p(X_n)$ is not isomorphic to any uniformly convex space although it is reflexive.

The following result gives a method to obtain from a given uniformly convex space new uniformly convex spaces.

THEOREM 2.4.18 Let X be a uniformly convex Banach space and X_1 be a closed linear subspace of X. Then the quotient space X/X_1 is uniformly convex.

Proof: We remark in passing that every closed subspace of a uniformly convex space is obviously uniformly convex. Now the proof of the theorem is as follows. Let $\hat{x}$ and $\hat{y}$ be two elements of X/X_1 with the properties that $\|\hat{x}\| = 1$, $\|\hat{y}\| = 1$, and $\|\hat{x} - \hat{y}\| \geq \varepsilon$. From the definition of the norm of the quotient space there exist x and y in X that are in $\hat{x}$ and $\hat{y}$, respectively, and $\|x\| \leq 1 + \lambda$, $\|y\| \leq 1 + \lambda$, with λ arbitrarily close to 0. Since

$$\|x - y\| \geq \|\hat{x} - \hat{y}\| \geq \varepsilon$$

from the uniform convexity of X we get

$$\left\|\frac{\hat{x} + \hat{y}}{2}\right\| \leq \left\|\frac{x + y}{2}\right\| \leq 1 - \delta\left(\frac{\varepsilon}{1 + \lambda}\right)(1 + \lambda)$$

which implies that

$$\delta_1(\varepsilon) = \lim \delta\left(\frac{\varepsilon}{1 + \lambda}\right) \geq \delta\left(\frac{\varepsilon}{2}\right)$$

Now $\delta_1(\varepsilon)$ may be a function that is smaller than the modulus of convexity of X/X_1. Since the above function is strictly positive for $\varepsilon > 0$, X/X_1 is thus uniformly convex.

The following theorems show that uniform convexity is determined by certain classes of subspaces. For the proofs we refer to the original paper.

THEOREM 2.4.19 (M. M. Day, 1944) The Banach space X is uniformly convex if and only if all the two-dimensional subspaces have a common modulus of convexity.

THEOREM 2.4.20 (M. M. Day, 1944) The Banach space X is uniformly convex if and only if all the two-dimensional quotient spaces of X have a common modulus of convexity.

2.5 UNIFORM CONVEXITY AND INTERPOLATION SPACES

In what follows we consider the relation between the uniform convexity of interpolation spaces and the uniform convexity of factors (i.e., of the spaces A_0 and A_1). (See Sec. 2.3 for notation and terminology.)

THEOREM 2.5.1 (B. Beauzamy, 1976) Let A_0 and A_1 be two Banach spaces continuously embedded in a Hausdorff topological linear space. Then if either A_0 or A_1 is uniformly convex, so is the space $S(p,\xi_0,A_0,p,\xi_1,A_1)$, where p is in $(1,\infty)$ and $\theta = \xi_0(\xi_0 - \xi_1)$ is in $(0,1)$.

Proof: We consider two elements u,v in the space with the following properties:

$$\|u\| = \|v\| = 1$$
$$\|u - v\| \geq \varepsilon$$

Let $\gamma > 0$. According to the definition of the norm in our space, we find u(t) and v(t) having the properties

$$\int_{-\infty}^{\infty} u(t)\,dt = u \qquad \int_{-\infty}^{\infty} v(t)\,dt = v$$

$$\left.\begin{aligned} \|e^{\xi_0 t} u(t)\|_{L^p(A_0)} &\leq 1 + \gamma \qquad & \|e^{\xi_1 t} u(t)\|_{L^p(A_1)} &\leq 1 + \gamma \\ \|e^{\xi_0 t} v(t)\|_{L^p(A_0)} &\leq 1 + \gamma \qquad & \|e^{\xi_1 t} v(t)\|_{L^p(A_1)} &\leq 1 + \gamma \end{aligned}\right\} \qquad (*)$$

Since

$$\varepsilon \leq \left\|\frac{u - v}{2}\right\| \leq \left\|e^{\xi_0 t}\,\frac{u(t) - v(t)}{2}\right\|_{L^p(A_0)}^{1-\theta} \left\|e^{\xi_1 t}\,\frac{u(t) - v(t)}{2}\right\|_{L^p(A_1)}^{\theta}$$

if we set

$$u^*(t) = \frac{u(t)}{1 + \gamma} \qquad v^*(t) = \frac{v(t)}{1 + \gamma}$$

then the above inequalities imply

$$\begin{aligned} \|e^{\xi_0 t}[u^*(t) - v^*(t)]\|_{L^p(A_0)} &\geq 2\left(\frac{\varepsilon}{1+\gamma}\right)^{1/(1-\theta)} \\ \|e^{\xi_1 t}[u^*(t) - v^*(t)]\|_{L^p(A_1)} &\geq 2\left(\frac{\varepsilon}{1+\gamma}\right)^{1/\theta} \end{aligned} \qquad (**)$$

If $\delta_p^0(\varepsilon), \delta_p^1(\varepsilon)$ are the moduli of convexity of the spaces $L^p(A_0)$ and $L^p(A_1)$, respectively, then the inequalities in (*) and (**) imply that

$$\left\|e^{\xi_0 t}\frac{u^*(t) + v^*(t)}{2}\right\|_{L^p(A_0)} \leq 1 - \delta_p^0\left[2\left(\frac{\varepsilon}{1+\gamma}\right)^{1/(1-\theta)}\right]$$

$$\left\|e^{\xi_1 t}\frac{u^*(t) + v^*(t)}{2}\right\|_{L^p(A_1)} \leq 1 - \delta_p^1\left[2\left(\frac{\varepsilon}{1+\gamma}\right)^{1/\theta}\right]$$

Now if γ is sufficiently small such that

$$\frac{2}{(1+\gamma)^{1/(1-\theta)}} \geq 1 \qquad \frac{2}{(1+\gamma)^{1/\theta}} \geq 1$$

we get

$$\delta_p^0\left(\frac{2}{(1+\gamma)^{1/(1-\theta)}}\right) \geq \delta_p^0(\varepsilon^{1/(1-\theta)})$$

$$\delta_p^1\left(\frac{2}{(1+\gamma)^{1/\theta}}\right) \geq \delta_p^1(\varepsilon^{1/\theta})$$

Further, we have

$$\frac{1}{2}\|u + v\| \leq (1+\gamma)\left\|e^{\xi_0 t}\frac{u^*(t) + v^*(t)}{2}\right\|_{L^p(A_0)}^{1-\theta}\left\|e^{\xi_1 t}\frac{u^*(t) + v^*(t)}{2}\right\|_{L^p(A_1)}^{\theta}$$

$$\leq (1+\gamma)[1 - \delta_p^0(\varepsilon^{1/(1-\theta)})^{1-\theta}][1 - \delta_p^1(\varepsilon^{1/\theta})^{\theta}]$$

and γ being arbitrary, this implies that

$$\|u + v\| \leq 2[1 - \delta_p^0(\varepsilon^{1/(1-\theta})^{1-\theta}][1 - \delta_p^1(\varepsilon^{1/\theta})^{\theta}]$$

For $\varepsilon > 0$ this is equivalent to the function

$$\varepsilon \to 1 - [(1-\theta)\delta_p^0(\varepsilon^{1/(1-\theta)}) + \theta\delta_p^1(\varepsilon^{1/\theta})]$$

which is clearly a modulus of convexity of our interpolation space. This form of modulus obviously implies the assertion of the theorem.

2.6 SOME GENERALIZATIONS OF UNIFORM CONVEXITY

The generalizations of uniformly convex spaces may be viewed as the study of those properties of Banach spaces lying between strict convexity and uniform convexity. These properties can be classified, in some sense and to some extent, as either localizations or directionalizations of uniform convexity. In what follows we give some of these generalizations as well as some properties of spaces in these classes.

DEFINITION 2.6.1 (V. L. Šmulian, 1939b, 1940) The Banach space X is said to be weakly uniformly convex (WUC) if the following property holds: for any sequences $(x_n),(y_n)$ such that

$$\|x_n\| = \|y_n\| = 1 \qquad \lim \frac{1}{2}(\|x_n + y_n\|) = 1$$

it follows that

$$(x_n - y_n) \rightharpoonup 0$$

where $\rightharpoonup$ means weak convergence.

DEFINITION 2.6.2 (V. L. Šmulian, 1939b, 1940) The Banach space X is said to be weak* uniformly convex (W*UC) if the following property holds: if $(x_n^*),(y_n^*)$ are two sequences in X* such that

$$\|x_n^*\| = \|y_n^*\| = 1 \qquad \lim \frac{1}{2}(\|x_n^* + y_n^*\|) = 1$$

then $(x_n^* - y_n^*)$ is weak* convergent to 0. We write this as $x_n^* - y_n^* \xrightarrow{w^*} 0$.

DEFINITION 2.6.3 (A. R. Lovaglia, 1955) The Banach space X is said to be locally uniformly convex (LUC) if for $\varepsilon > 0$ given and x in X, $\|x\| = 1$, there exists $\delta(x,\varepsilon) > 0$ such that for all $y \in X$ with $\|y\| = 1$, $\|y - x\| \geq \varepsilon$, the following inequality holds:

$$\|x - y\| \leq 2[1 - \delta(x,\varepsilon)]$$

DEFINITION 2.6.4 The Banach space X is said to have the Radon-Riesz property if the following holds: if (x_n) is a sequence of elements of X such that

$$\lim \|x_n\| = \|x\|$$
$$(x_n - x) \rightharpoonup 0$$

Then

$$\lim \|x_n - x\| = 0$$

More generally we can define the following class of Banach spaces.

DEFINITION 2.6.5 The Banach space is said to have the Radon-Riesz property of order k if the following holds: If (x_n) is a sequence of elements of X such that

1. $\lim \|x_n\| = \|x\|$
2. For each x* in X*

$$\lim_{n_i \to \infty} x^*((x_{n_1} + \cdots + x_{n_k}) - x) = 0$$

then

$$\lim \|x_n - x\| = 0$$

Obviously for k = 1 we have the Radon-Riesz property.

DEFINITION 2.6.6 (K. W. Anderson, 1960) The Banach space X is called midpoint locally uniformly convex (MLUC) if whenever x is in X and (x_n), (y_n) are sequences in X such that

$$\|x\| = 1 \qquad \lim \|x_n\| = \lim \|y_n\| = 1 \qquad \lim \|2x - (x_n + y_n)\| = 0$$

then

$$\lim \|x_n - y_n\| = 0$$

DEFINITION 2.6.7 (V. I. Istrăţescu, 1981a) The Banach space X is said to be weakly midpoint locally uniformly convex (WMLUC) if whenever x is in X and (x_n), (y_n) are sequences in X such that

$$\|x\| = 1 \qquad \lim \|x_n\| = \lim \|y_n\| = 1 \qquad \lim \|2x - (x_n + y_n)\| = 0$$

then

$$(x_n - y_n) \xrightarrow{w} 0$$

DEFINITION 2.6.8 (V. I. Istrăţescu, 1981a) The Banach space X is said to be weak* midpoint locally uniformly convex (W*MLUC) if the following property holds: if x* is in X* and (x_n^*), (y_n^*) are two sequences in X* such that

$$\|x^*\| = 1 \qquad \lim \|x_n^*\| = \lim \|y_n^*\| = 1 \qquad \lim \|2x^* - (x_n^* + y_n^*)\| = 0$$

then

$$(x_n^* - y_n^*) \xrightarrow{w^*} 0$$

DEFINITION 2.6.9 (A. L. Garkavi, 1962) The Banach space X is called uniformly convex in every direction (UCED) if whenever $z \neq 0$ is in X and $(x_n), (y_n)$ are two sequences in X such that

$$\lim \|x_n\| = \lim \|y_n\| = 1 \qquad \lim \|x_n + y_n\| = 2 \qquad x_n - y_n = a_n z$$

then

$$\lim a_n = 0$$

DEFINITION 2.6.10 (M. A. Smith, 1977a) The Banach space X is called uniformly convex in weakly compact sets of directions (UCWC) if whenever z is in X and $(x_n), (y_n)$ are two sequences in X such that

$$\lim \|x_n\| = \lim \|y_n\| = 1 \qquad \lim \|x_n + y_n\| = 2 \qquad \lim (x_n - y_n) = z$$

(the weak limit), then $z = 0$.

DEFINITION 2.6.11 (R. C. James, 1964) The Banach space X is called uniformly nonsquare if there exists a positive number r such that there do not exist x,y in X with the properties

$$\|x\| \leq 1 \qquad \|y\| \leq 1 \qquad \|x + y\| \geq 2(1 - r) \qquad \|x - y\| < 2(1 - r)$$

DEFINITION 2.6.12 (I. Glicksberg et al., 1958) The Banach space X is called k-uniformly convex (k-UC) if whenever (x_n) is a sequence in X with $\|x_n\| = 1$ and

$$\lim_{n_i \to \infty} \frac{1}{k} (x_{n_1} + \cdots + x_{n_k}) = 1$$

then

$$\lim (x_n - x_m) = 0$$

(The number k is any integer ≥ 2.)

DEFINITION 2.6.13 (V. I. Istrăţescu, 1981b) The Banach space X is called k-locally uniformly convex (k-LUC) if whenever $x = \|x\| = 1$ is in X and $\varepsilon > 0$, there exists $\delta(x,\varepsilon) > 0$ such that for all $y_1, \ldots, y_k$ in X with $\|y_i\| = 1$, $\|y_i - x\| \geq \varepsilon$, the following inequality holds:

$$\|x + y_1 + \cdots + y_k\| \leq (k + 1)[1 - \delta(x,\varepsilon)]$$

For $k = 2$ we obtain the locally uniformly convex spaces of Lovaglia.

DEFINITION 2.6.14 (V. I. Istrăţescu, 1981a) The Banach space is said to be uniformly convex of order k (UC-k) if X is k-locally uniformly convex in the sense of Definition 2.6.13 and $\delta(x,\varepsilon)$ is independent of x, $\|x\| = 1$.

DEFINITION 2.6.15 (V. I. Istrăţescu, 1981a) The Banach space X is said to be weakly k-locally uniformly convex (k-WLUC) if whenever x is in X, $\|x\| = 1$, and (x_n) is a sequence in X such that

$$\lim \|x_n\| = 1$$

$$\lim_{n_i \to \infty} \|x_{n_1} + \cdots + x_{n_k} + x\| = k + 1$$

then

$$x_n \rightharpoonup x$$

DEFINITION 2.6.16 (M. A. Smith, 1977a) Let X be a Banach space and $\mathcal{F}$ be a nonempty family of subsets of X 0. For every A in $\mathcal{F}$ and ε in (0,2] set

$$\delta(\varepsilon,A) = \inf\{\delta(\varepsilon,z) : z \in A\}$$

where

$$\delta(\varepsilon,z) = \inf\{1 - \frac{1}{2}\|x + y\| : \|x\| = \|y\| = 1,\ x - y = az,\ \|x - y\| \geq \varepsilon\}$$

The space X is said to be $UC_{\mathcal{F}}$ if $\delta(\varepsilon,A) > 0$ for every A in $\mathcal{F}$ and any ε in (0,2].

DEFINITION 2.6.17 (L. P. Vlasov, 1967) The Banach space is said to be an (M)-space if the following property holds: if x is in X, $\|x\| = 1$, and (x_n) is a sequence such that

$$\|x_n\| = 1 \qquad \lim \|x_n + x\| = 2$$

then (x_n) is relatively compact in X.

DEFINITION 2.6.18 The Banach space X is said to be a (WM)-space if the following property holds: if x is in X, $\|x\| = 1$ and (x_n) is a sequence satisfying

$$\|x_n\| = 1 \qquad \lim \|x_n + x\| = 2$$

then (x_n) has a weakly convergent subsequence.

DEFINITION 2.6.19 (F. Sullivan, 1979) The Banach space X is said to be S_k-UC if for each $\varepsilon > 0$ there exists $\delta(\varepsilon) > 0$ such that for all $x_1, \ldots,$

x_k, $\|x_i\| = 1$, with the property that $\|x_1 + \cdots + x_k\| > k - \delta(\varepsilon)$, the following inequality holds:

$$\sup_{\substack{(x^*,\ldots,x^*_{k-1}) \\ \|x^*_i\| = 1}} \begin{vmatrix} 1 & 1 & \cdots & \cdots & \cdots & 1 \\ x^*_1(x_1) & x^*_2(x_2) & \cdots & \cdots & \cdots & x^*_1(x_k) \\ \cdots & \cdots & \cdots & \cdots & \cdots & \cdots \\ x^*_{k-1}(x_1) & x^*_{k-1}(x_2) & \cdots & \cdots & \cdots & x^*_{k-1}(x_k) \end{vmatrix} < \varepsilon$$

It is not difficult to see that S_2-UC spaces are exactly the uniformly convex spaces of Clarkson.

We note that the above list of classes of spaces does not exhaust the classes of Banach spaces considered in connection with uniformly convex spaces. The references at the end of the book may be useful in giving more information about related classes of Banach spaces.

In what follows we give some examples of Banach spaces in the classes defined above, as well as results which show some relations among these classes. First we mention the following obvious property.

PROPOSITION 2.6.20 Every uniformly convex space is LUC.

THEOREM 2.6.21 (A. R. Lovaglia, 1955) For each n let X_n be a Banach space which is LUC, and consider the Banach space $\ell^p(X_n)$, $p \in (1,\infty)$. Then $\ell^p(X)$ is LUC.

Proof: We consider $x = (x_i)$ with $\|x\| = 1$ and let $(y_n) \subset \ell^p(X_n)$ with the properties that $\|y_n\| = 1$ and $\lim \|x + y_n\| = 2$. We consider further the following associated elements:

$$x^*_n = (0,\ldots,0,x_{n+1},x_{n+2},\ldots)$$
$$y_{m,n} = (0,\ldots,0,y^m_{n+1},y^m_{n+2},\ldots)$$

where

$$y_p = (y^p_1,y^p_2,\ldots,y^p_{n+1},y^p_{n+2},\ldots)$$

To prove that $\ell^p(X_n)$ is LUC we must show that $\lim \|x - y_n\| = 0$. Using Minkowski's inequality we have

$$\|x + y_m\| = \left(\sum_{n=1}^{\infty} \|x_n + y^m_n\|^p\right)^{1/p} = \left(\sum_{n=1}^{k} \|x_n + y^m_n\|^p + \sum_{n=k+1}^{\infty} \|x_n + y^m_n\|^p\right)^{1/p}$$
$$= \left(\sum_{n=1}^{k} \|x_n + y^m_n\|^p + \|x^*_k + y_{k,n}\|^p\right)^{1/p}$$

$$\leq \left[\sum_{n=1}^{k} (\|x_n + y_n^m\|)^p + (\|x_k^* + y_{k,m}\|)^p\right]^{1/p}$$

$$\leq \left[\sum_{n=1}^{k} (\|x_n\|^p + \|x_k^*\|^p)\right]^{1/p} + (\|y_n^m\|^p + \|y_{m,k}\|^p)^{1/p}$$

$$\leq \|x\| + \|y_m\| = 2$$

Since $\lim \|x + y_m\| = 2$ this inequality implies that

$$\lim_{n\to\infty} \left[\sum_{n=1}^{k} (\|x_n\| + \|y_n^m\|)^p + (\|x_k^*\| + \|y_{m,k}\|)^p\right]^{1/p} \leq 2$$

This gives further that $(\|y_n^m\|)$ is a bounded sequence for each n, so that by the diagonal method, we may choose a subsequence of elements such that for each n, $\lim \|y_n^m\|$ exists; we denote it by $\{a_n\}$. Since

$$1 = \|y_{p_n}\| = \left(\sum_{n=1}^{\infty} \|y_n^{p_n}\|^p\right)^{1/p}$$

we get

$$\lim_{p_n\to\infty} \|y_{p_n,k}\|^p = 1 - \sum_{n=1}^{k} a_n^p = b_k^p$$

and

$$\lim_{p_j\to\infty} [(\|x_n^*\| + \|y_n^{p_j}\|)^p + (\|x_k^*\| + \|y_{p_j,k}\|)^p]^{1/p}$$

$$= [(\|x_n^*\| + a_n)^p + (\|x_k^*\| + b_k)^p]^{1/p} \leq 2$$

But

$$\left(\sum_{n=1}^{k} \|x_n\|^p + \|x_k^*\|^p\right)^{1/p} + \left(\sum_{n=1}^{k} a_n^p + b_k^p\right)^{1/p} \leq 2$$

which gives

$$a_n = \|x_n\| \qquad b_n = \|x_n^*\|$$

for $n = 1, 2, 3, \ldots$. These relations may be written as follows:

$$\lim_{j\to\infty} \|y_n^{p_j}\| = \|x_n\| \qquad \lim_{j\to\infty} \|y_{p_j,k}\| = \|x_k^*\|$$

Now, if $\ell^p(X_n)$ is not LUC then there exists a (sub)sequence (p_j) with the property that $\|x - y_{p_j}\| \geq \varepsilon > 0$ for some ε. But we have

$$
\begin{aligned}
0 < \varepsilon \leq \|x - y_m\| &= \left(\sum_{n=1}^{k} \|x_n - y_n^m\|^p + \sum_{n=k+1}^{\infty} \|x_n - y_n^m\|^p \right)^{1/p} \\
&\leq \left(\sum_{n=1}^{k} \|x_n - y_n^m\|^p + \|x_k^* - y_{m,k}\|^p \right)^{1/p} \\
&\leq \left(\sum_{n=1}^{k} \|x_n - y_n^m\|^p \right)^{1/p} + \|x_k^* - y_{m,k}\| \\
&\leq \left(\sum_{n=1}^{k} \|x_n - y_n^m\|^p \right)^{1/p} + \|x_k^*\| + \|y_{m,k}\|
\end{aligned}
$$

which implies that

$$\left(\sum_{n=1}^{k} \|x_n - y_n^m\|^p \right)^{1/p} \geq \varepsilon - (\|x_k^*\| + \|y_{m,k}\|)$$

But

$$\lim \|y_{m,k}\| = \|x_k^*\|$$

and

$$\lim \|x_k^*\| = 0$$

which gives that for some k and m_0, for all $m \geq m_0$,

$$\|x_k^*\| + \|y_{m,k}\| < \varepsilon$$

Then for k and $m \geq m_0$,

$$\sum_{n=1}^{k} \|x_n - y_n^m\|^p \geq s^p > 0$$

Thus we find n_0 in [1,k] and a (sub)sequence of m's with the property that

$$\|x_{n_0} - y_{n_0}^m\| \geq t > 0$$

Since

$$\|x_{n_0} + y_{n_0}^m\| \leq \|x_{n_0}\| + \|y_{n_0}^m\|$$

and

$$\lim_{m\to\infty} \|x_{n_0} + y^m_{n_0}\| = 2\|x_{n_0}\|$$

we have that $x_{n_0} \neq 0$. This implies that for some (sub)sequence of m's, $y^m_{n_0} \neq 0$ and

$$\liminf_{m\to\infty} \left\| \frac{x_{n_0}}{\|x_{n_0}\|} - \frac{y^m_{n_0}}{\|y^m_{n_0}\|} \right\| = \liminf_{m\to\infty} \left\| \frac{x_{n_0}}{\|x_{n_0}\|} - \frac{y^m_{n_0}}{\|y^m_{n_0}\|} \right\| \geq \frac{t}{\|x_{n_0}\|} \geq t > 0$$

But the space X_{n_0} is supposed to be LUC and thus

$$\lim_{m\to\infty} \left\| \frac{x_{n_0}}{\|x_{n_0}\|} + \frac{y^m_{n_0}}{\|y^m_{n_0}\|} \right\| = 2 - \delta_{n_0} \qquad \delta_{n_0} = \delta_{n_0}(x_{n_0}, t)$$

which gives further that

$$\limsup_{m\to\infty} \|x_{n_0} + y^m_{n_0}\| \leq (2 - \delta_{n_0})\|x_{n_0}\|$$

But

$$\|x + y_m\| \leq \left(\sum_{n=1}^{k} \|x_n + y^m_n\|^p \right)^{1/p} + \|x^*_k\| + \|y_{m,k}\|$$

so we have

$$\limsup_{m\to\infty} \|x + y_m\| \leq \left(\sum_{n=1}^{k} \|x_n + y^m_n\|^p \right)^{1/p} + 2\|x^*_k\|$$

$$\leq [(2 - \delta_{n_0})^p \|x_{n_0}\|^p + 2^p \|x_{n_0}\|^p]^{1/p} + 2\|x^*_k\|$$

and for $k \to \infty$,

$$\limsup \|x + y_m\| < [(2 - \delta_{n_0})\|x_{n_0}\|^p + \sum_{\substack{n=1 \\ n \neq n_0}}^{k} 2^p \|x_n\|^p]^{1/p}$$

$$< \left(\sum_{1}^{\infty} 2^p \|x_n\|^p \right)^{1/p} = 2$$

which contradicts the fact that

$$\lim \|x + y_m\| = 2$$

Thus $\ell^p(X_n)$ is LUC.

REMARK 2.6.22 For an appropriate choice of X_n, $\ell^p(X_n)$ is not uniformly convex (see Theorem 2.4.16).

The following example shows that the class of locally uniformly convex spaces is narrower than the class of strictly convex spaces. First we note the following property of LUC spaces.

PROPOSITION 2.6.23 Every LUC space is strictly convex.

Proof: Since the proof is simple it is omitted.

EXAMPLE 2.6.24 (A. R. Lovaglia, 1955) Let $C_{[0,1]}$ be the Banach space of all continuous functions on [0,1] with the norm

$$\|f\|_1 = \left[\|f\|^2 + \sum_{n=1}^{\infty} \frac{1}{2^{2n}} |f(t_n)|^2\right]^{1/2}$$

where (t_i) is a dense set in [0,1], and

$$\|f\| = \sup \{|f(t)| : t \in [0,1]\}$$

Using the density of (t_i) and Minkowski's inequality we obtain easily that $(C_{[0,1]}, \| \|_1)$ is a strictly convex space. We note that the norm $\| \|_1$ was defined by Clarkson in 1935. We show that $(C_{[0,1]}, \| \|_1)$ is not LUC. For this purpose we consider the function $x(t)$ defined by

$$x(t) = \frac{3^{1/2}}{2}$$

and the sequence of functions $(x_n(t))$,

$$x_n(t) = \begin{cases} \dfrac{3^{1/2}}{2} & \dfrac{1}{n} < t < 1 \\ \dfrac{3^{1/2}}{2t} & 0 < t < 1/n \end{cases}$$

These are continuous functions and we remark that

$$\|x_n\|_1 = \|x\|_1 = 1 \qquad \lim \|x + x_n\|_1 = 2$$
$$\lim \|x - x_n\|_1 = 3^{1/2}$$

From these relations it is clear that $(C_{[0,1]}, \| \|_1)$ is not a LUC space.

The following result is similar to that in Theorem 2.6.21 and may be proved in a similar way. To formulate it we need some notation, which is now given. We consider a Banach space X which consists of sequences of real numbers $x = (x_n)$ with the following property: if $y = (y_n)$ is an arbitrary sequence of real numbers and for some $x \in X$, $x = (x_i)$, the following relations hold:

$$|x_i| \geq |y_i| \qquad i = 1, 2, 3, \ldots$$

then y is in X. We suppose further that on X there exists a norm, denoted by

$$x \to N(x)$$

such that the following properties hold:

1. (X,N) is LUC.

2. N is strictly increasing in each component, i.e., if $|x_i| \leq |y_i|$ then $N(x) \leq N(y)$, where $x = (x_i)$, $y = (y_i)$.

3. For each $x \in X$, $\lim N(0,0,0,\ldots,x_{n+1},x_{n+2},\ldots) = 0$.

Let (X_n) be a sequence of Banach spaces, and consider the space $m((X_n))$ of all sequences $x = (x_n)$, $x_n \in X_n$, with the property that $N((\|x_n\|)) < \infty$. We denote $m((X_n))$ with this norm by $(m((X_n)),N)$. Then we have

THEOREM 2.6.25 If each X_n is LUC then $(m((X_n)),N)$ is LUC.

Now we give some results concerning the class of Banach spaces with the Radon-Riesz property. We note that the property was proved by Fr. Riesz for the case of L^p spaces for p in $(1,\infty)$.

THEOREM 2.6.26 (R. Vyborny, 1956) If X is LUC then X has the Radon-Riesz property.

Proof: Suppose on the contrary that this assertion is not true. Then there exists x and (x_n) in X such that

$$\lim \|x_n\| = \|x\|$$

$$x_n \rightharpoonup x$$

and (x_n) does not converge to x. In this case, for some $\varepsilon > 0$ and for some sequence (n_k),

$$\|x_{n_k} - x\| \geq \varepsilon$$

We may suppose without loss of generality that $\|x\| = \|x_n\| = 1$. Now, by the Hahn-Banach theorem there exists x^* in X^* such that

$$\|x^*\| = 1 = x^*(x)$$

From the weak convergence we get

$$2 = \lim x^*(x + x_n) \le \|x + x_n\|$$

and since X is LUC we find $\delta(x,\varepsilon) > 0$ such that

$$\|x + x_n\| \le 2[1 - \delta(x,\varepsilon)]$$

Then we obtain

$$2 \le \limsup \|x + x_n\| \le 2[1 - \delta(x,\varepsilon)]$$

This is a contradiction and the theorem is proved.

PROPOSITION 2.6.27 If the Banach space X is LUC then it is k-locally uniformly convex.

The following theorem may be proved similarly to Theorem 2.6.25.

THEOREM 2.6.28 If X is a k-locally uniformly convex Banach space then X has the Radon-Riesz property.

We prove now some equivalent formulations of the property that a Banach space is a WUC space or a W*UC space.

THEOREM 2.6.29 (V. Zizler, 1968) The Banach space X is a WUC space if and only if the following property holds: if $(x_n),(y_n)$ are two sequences in X, $\max \{\|x_n\|,\|y_n\|\} \le 1$, n = 1, 2, 3, ..., and $\lim (1/2)(\|x_n + y_n\|) = 1$, then $(x_n - y_n) \rightharpoonup 0$.

Proof: We remark that the condition is sufficient. We show now that it is necessary. This will follow from the assertion: if y^*, $\|y^*\| = 1$, and $\varepsilon > 0$ are given, then there exists $\delta(y^*,\varepsilon) > 0$ such that

$$|y^*(x - y)| \ge \varepsilon \max \{\|x\|,\|y\|\}$$

implies

$$\|x + y\| \le 2[1 - \delta(y^*,\varepsilon)] \max \{\|x\|,\|y\|\}$$

We note that we may suppose that $\|x\| \ge \|y\|$ and $\|x\| = 1$. If $|y^*(x - y)| \ge \varepsilon$ with $\varepsilon \in (0,1)$, let $\|y\| \ge 1 - r$ and let $z = y/\|y\|$. Then we have

$$y^*(x - z) = y^*(x - y) + y^*(y - z)$$

which implies that

$$|y^*(x - z)| \ge \varepsilon - |y^*(y - z)|$$

and thus

$$|y^*(y - x)| = \varepsilon - r$$

In this case

$$\|x + z\| \le 2[1 - \delta(y^*,\varepsilon)]$$

and thus

$$\|x + y\| \le \|x + z\| + \|y - z\| \le 2[1 - \delta(y^*,\varepsilon)] + r$$

If

$$\|y\| \le 1 - r$$

then

$$\|x + y\| \le 2\left(1 - \frac{r}{2}\right)$$

and thus

$$\|x + y\| \le 2 \max\left\{[1 - \delta(y^*,\varepsilon)] + \frac{r}{2},\ 1 - \frac{r}{2}\right\}$$

Thus by proper choice of r the right-hand side of this last inequality can be made less than 1, and the assertion of the theorem is proved.

In a similar way we prove the following result.

THEOREM 2.6.30 (V. Zizler, 1968) The Banach space X is W*UC if and only if the following property holds: if $(x_n^*),(y_n^*)$ are two sequences in X* with $\|x_n^*\| \le 1$, $\|y_n^*\| \le 1$, and $\lim \|x_n^* + y_n^*\| = 2$, then $x_n^* - y_n^* \xrightarrow{w^*} 0$.

We mention the following characterizations of WUC and W*UC Banach spaces in terms of sequences.

THEOREM 2.6.31 (V. Zizler, 1971) The Banach space X is WUC if and only if the following property holds: if (x_n) and (y_n) are sequences in X, (x_n) bounded, and

$$2(\|x_n\|^2 + \|y_n\|^2) - (\|x_n + y_n\|)^2 \to 0$$

then

$$(x_n - y_n) \rightharpoonup 0.$$

THEOREM 2.6.32 (V. Zizler, 1971) The Banach space X is W*UC if and only if the following property holds: if $(x_n^*),(y_n^*)$ are sequences in X*, (x_n^*) bounded, and

$$2\|x_n^*\|^2 + 2\|y_n^*\|^2 - (\|x_n^* + y_n^*\|)^2 \to 0$$

then

$$(x_n^* - y_n^*) \xrightarrow{w^*} 0$$

Now we present some results about the class of Banach spaces introduced by A. L. Garkavi, the so-called spaces which are uniformly convex in every direction (UCED). We mention that these spaces are important in approximation theory since they are exactly those Banach spaces in which every bounded set has at most one Cebyshev center (if K is a subset of a Banach space X then the Cebyshev centers of K are the elements c in K with the property that

$$\sup_{k \in K} \|c - k\| = \inf_{s \in X} \sup_{t \in K} \|s - t\|$$

We mention that an excellent detailed study of Garkavi's UCED spaces is contained in Zizler's paper (1971). The papers of R. C. James et al. (1971) and M. A. Smith (1977a,b) also contain interesting results about this and related classes of Banach spaces.

THEOREM 2.6.33 (M. M. Day et al., 1971) For a Banach space X the following assertions are equivalent:

(1) X is UCED.

(2) For every nonzero z in X and $\varepsilon > 0$ there exists $\delta(z,\varepsilon) > 0$ such that $|a| < \varepsilon$ if $\|x\| = \|y\| = 1$, $x - y = az$, and $\|x + y\| = 2[1 - \delta(z,\varepsilon)]$.

(3) If $(x_n),(y_n)$ are sequences in X such that

a. $\|x_n\| \le 1$ $\quad\|y_n\| \le 1$ $\quad n = 1, 2, 3, \ldots$

b. $\lim (x_n - y_n) = z$

c. $\lim \|x_n + y_n\| = 2$

then $z = 0$.

(4) For any fixed p in $[2,\infty)$, then for no nonzero z in X is there a bounded sequence (x_n) in X with the property that

$$2^{p-1}(\|x_n + z\|^p + \|x_n\|^p) - 2\|x_n + z\|^p \to 0$$

(5) For each nonzero element z of X there exists δ such that if $\|x\| \le 1$ and $\|x + z\| \le 1$, then $\|x + (1/2)z\| < 1 - \delta$.

Proof: Since, in a sense property (2) is a reformulation of the UCED property, we have that (1) is equivalent to (2). Suppose now that (2) holds and (x_n), (y_n), and z are as in (3), and that $z \neq 0$. Let us consider $k_n = \min \{1, \|x_n - z\|^{-1}\}$ and let

$$u_n = k_n x_n \qquad v_n = k_n(x_n - z)$$

Since k_n is in [0,1], from the properties a and b in (3) we get that $\lim k_n = 1$. This implies that the following relations hold:

$$\|u_n\| = 1 \qquad \|v_n\| = 1$$
$$u_n - v_n = k_n z$$
$$\lim \|u_n + v_n\| = 2$$

We set

$$s_n = u_n + a_n z \qquad t_n = v_n - b_n z$$

with a_n, b_n positive numbers such that for all n,

$$\|s_n\| = \|t_n\| = 1$$

Then we have the inequalities

$$0 \le a_n \le \frac{2}{\|z\|} \qquad 0 \le b_n \le \frac{2}{\|z\|}$$

Clearly we have

$$t_n - s_n = (k_n + a_n + b_n)z$$

with

$$\liminf (k_n + a_n + b_n) \ge 1$$

But we have also

$$\begin{aligned} s_n + t_n &= u_n + v_n + (a_n - b_n)z = k_n x_n + k_n(k_n(x_n - z) + (a_n - b_n)z \\ &= x_n + y_n + (a_n - b_n)(x_n - y_n) \\ &\quad + [(k_n - 1)x_n + (k_n x_n - k_n z - y_n) + (a_n - b_n)(z - x_n + y_n)] \end{aligned}$$

and if $a_n \ge b_n$ and $\|x_n + y_n\| > 2 - \delta$ then

$$\begin{aligned} \|x_n + y_n + (a_n - b_n)(x_n - y_n)\| &= \|(1 + a_n - b_n)(x_n + y_n) \\ &\quad - 2(a_n - b_n)y_n\| \\ &> (1 + a_n - b_n)(2 - \delta) - 2(a_n - b_n) \\ &= 2 - \delta(1 + a_n - b_n) \end{aligned}$$

We remark that the expression in brackets in the formula for $t_n + s_n$ tends to 0 for $n \to \infty$. Reasoning similarly in the case $b_n \le a_n$, we obtain that

$$\lim \|t_n + s_n\| = 2$$

and the triple $((t_n),(s_n),z)$ shows that X does not satisfy the UCED property.

Suppose now that (3) holds; we prove that (4) holds. Suppose on the contrary that it is not true. Then we find $z \neq 0$ in X and a bounded sequence (x_n) such that

$$2^{p-1}(\|x_n + z\|^p + \|x_n\|^p) - \|2x_n + z\|^p \to 0$$

Since z is not zero the sequence $(\|x_n\|)$ is bounded away from zero, and we may assume that $\lim \|x_n\| = 1$. Now we have the following inequality. For $a \geq b \geq 0$, $p \in [2,\infty)$,

$$(a - b)^p + (a + b)^p \leq 2^{p-1}(a^p + b^p)$$

Applying this inequality we get

$$\begin{aligned} 2^{p-1}(\|x_n + z\|^p + \|x_n\|^p) - \|2x_n + z\|^p &\geq 2^{p-1}(\|x_n + z\|^p + \|x_n\|^p) \\ &\quad - (\|x_n + z\| + \|x_n\|)^p \\ &\geq (\|x_n + z\| - \|x_n\|)^p \end{aligned}$$

which implies obviously that

$$\lim (\|x_n + z\| - \|x_n\|) = 0$$
$$\lim \|x_n + z\| = 1$$

Using the condition satisfied by the sequence (x_n) and the element z we see that

$$\lim \|2x_n + z\| = 2$$

Now we set

$$u_n = \frac{x_n}{\|x_n\|} \qquad v_n = \frac{x_n + z}{\|x_n + z\|}$$

and for the triple $((u_n),(v_n),z)$ this gives $z = 0$. This contradiction proves the assertion.

Now suppose that (4) holds. If x and z are as in (5) we remark that for some δ in $(0,1/2)$ the following inequality holds:

$$2^{p-1}(\|x + z\|^p + \|x\|^p) - \|2x + z\|^p > 2^p p\delta$$

for $\|x\| \le 1$, and thus

$$\|2x + z\|^p < 2^p - 2^p p\delta$$

which gives

$$\|x + \frac{1}{2} z\|(1 - p\delta)^{1/p} < 1 - \delta$$

Thus (5) is satisfied.

Suppose now that (5) holds; we show that this implies condition (2). Let z in X be a nonzero element and $\varepsilon > 0$. Then there exists r such that

$$\|u + \frac{1}{2} \varepsilon z\| < 1 - r \qquad r > 0$$

for all $\|u\| \le 1$, $\|u + \varepsilon z\| \le 1$. Suppose now that x,y are as in condition (2), i.e., $\|x\| = \|y\| = 1$, $x - y = az$. Now if $|a| \ge \varepsilon$ let $u = -(\text{sign } a)x$. Then clearly,

$$\|u\| = 1 \qquad \|u + |a| z\| = \|y\| = 1$$

and

$$\|u + \varepsilon z\| = \|(1 - |\frac{\varepsilon}{a}|)u + |\frac{\varepsilon}{a}|(u + |a| z)\| \le 1$$

Since (5) holds we must have

$$\|u + \frac{1}{2} \varepsilon z\| < 1 - r$$

Then

$$\begin{aligned}\|\frac{1}{2}(x + y)\| = \|x - \frac{1}{2} az\| = \|u + \frac{1}{2}|a| z\| &\le \frac{|a|}{2|a| - \varepsilon}\|u + \frac{1}{2}\varepsilon z\| \\ &+ \frac{|a| - \varepsilon}{2|a| - \varepsilon}\|u + |a| z\| < |a|\frac{1 - r}{2|a| - \varepsilon} \\ &+ \frac{|a| - \varepsilon}{2|a| - \varepsilon} \\ &= \frac{(2 - r)|a| - \varepsilon}{2|a| - \varepsilon}\end{aligned}$$

If we take $\delta = r/2$ then

$$\|x + y\| < 2 - \delta$$

Thus X is UCED.

The following result gives a method to obtain Banach spaces which are UCED.

THEOREM 2.6.34 (V. Zizler, 1971) Let X be a Banach space with the following property: there exists a UCED Banach space Y and an injective continuous linear operator $T : X \to Y$. Then there exists on X an equivalent norm, say $\| \|^*$, such that $(X, \| \|^*)$ is UCED.

Proof: We define the new norm on X by the formula

$$\|x\|^* = (\|x\|^2 + \|Tx\|^2)^{1/2}$$

where the norms of x and Tx are computed in X and Y, respectively. Let z be an element of X and suppose that the following relation holds:

$$2(\|x_n + z\|^{*2} + \|x_n\|^{*2}) - \|2x_n + z\|^{*2} \to 0$$

where (x_n^*) is a bounded sequence. From the form of the norm * we obtain

$$Tx_n = 2(\|x_n + z\|^2 + \|x_n\|^2) - \|2x_n + z\|^2 + 2(\|Tx_n + Tz\|^2 + \|Tx_n\|^2)$$

$$- \|2Tx_n + Tz\|^2 \to 0$$

Now, T being continuous and (x_n) bounded, (Tx_n) is bounded in Y. From the last relation we conclude, since it is a sum of two sequences of positive real numbers, that both converge to zero. Since $\{\|Tx_n\|\}$ is bounded, we have

$$2(\|Tz\|^{-1}\|Tx_n + Tz\|^2 + \|Tz\|^{-1}\|Tx_n\|)^2 - \|Tz\|^{-1}\|2Tx_n + Tz\|^2 \to 0$$

Since Y is supposed UCED, for the direction $Tz/\|Tz\|$ we obtain a contradiction, and thus $(X, \| \|^*)$ is UCED.

COROLLARY 2.6.35 (V. Zizler, 1971) Let X be a Banach space and suppose that X* has a countable dense subset. Then on X there exists an equivalent norm $\| \|_1$ such that $(X, \| \|_1)$ is UCED.

Proof: Let (x_n^*) be the countable dense set in X* (and we may suppose that each x_n^* has the property that $\|x_n^*\| = 1$. Then for each x in X we set

$$Tx = x_n^* \frac{(x)}{2^n}$$

which maps X into ℓ^2. Then the corollary follows from the above theorem.

Another way to find UCED spaces is given in the following result.

PROPOSITION 2.6.36 (V. Zizler, 1971) Let (X_n) be a family of Banach spaces. Then $\ell^2(X_n)$ is UCED if and only if each X_n is UCED.

Proof: This follows easily from Theorem 2.6.33(4) for $p = 2$.

We mention now the following interesting result obtained by M. Smith concerning the products of UCED spaces.

DEFINITION 2.6.37 Let S be an index set. A full function space V on S is a Banach space consisting of real-valued functions defined on S with the following property: if f is a function in V and g is any real-valued function such that

$$|f(s)| \geq |g(s)|$$

for each s in S, then g is in V.

REMARK 2.6.38 The space co X considered in Theorem 2.6.26 is a full function space on $N = \{1,2,3,\ldots\}$.

Suppose now that for each s in S we have a Banach space B_s. Then we consider the product of B_s over V, defined as follows.

DEFINITION 2.6.39 The product of $(B_s)_{s \in S}$ over V is the space of all functions f defined on S such that for each $s \in S$, f(s) is an element of B_s and the following property holds: for each f in the product of $(B_s)_{s \in S}$ over V the function

$$f^* : s \to \|f(s)\| \qquad (*)$$

where the norm is that of B_s, is an element of V. We denote the product of $(B_s)_{s \in S}$ over V as $(P_V B_S)$, and a norm on this space may be defined by the formula

$$\|f\| = \|f^*\|_V$$

where f^* is the function defined in (*).

Then we have the following result.

THEOREM 2.6.40 (M. A. Smith, 1977b) The space $(P_V B_S)$ is UCED in the direction z if each B_s and X is UCED and the order interval $[0,\{\|z(s)\|\}]$ is compact.

Proof: Refer to Smith's paper.

For the connection between UCED spaces and the class of Banach spaces with normal structure, see Sec. 2.11.

We give below some examples of Banach spaces that illustrate the distinctions between various generalizations of uniformly convex spaces.

EXAMPLE 2.6.41 (M. A. Smith, 1978a) We give an example of a Banach space which is LUC and not UCED. For this purpose consider the space ℓ^2 of all sequences $x = (x_i)$ with

$$\|x\|_2^2 = \sum_1^\infty |x_i|^2 < \infty$$

Since ℓ^2 is a linear subspace of c_0 we can consider on ℓ^2 the following norm: if $x = (x_i)$ is in ℓ^2 we enumerate the support of x, (n_k), so that

$$|x_{n_k}| \geq |x_{n_{k+1}}|$$

for $k = 1, 2, 3, \ldots$. We consider now the mapping

$$x \to x^* = (x_n^*)$$

where

$$x_n^* = \begin{cases} \dfrac{x_{n_k}}{2^k} & \text{if } n = n_k \text{ for some } k \\ 0 & \text{otherwise} \end{cases}$$

Let

$$x \to \|x\|_1 = \|x^*\|_2$$

Then a result of Rainwater (1969) (see Sec. 2.9, Theorem 2.9.27) shows that c_0 with the norm $\| \ \|_1$ is LUC. We define on ℓ^2 the following norm: let x be in ℓ^2 and

$$x \to \|x\|^- = \|u_x^*\|_1$$

with

$$u_x = \left(\frac{1}{2}\|x\|_2, x_1, x_2, x_2, \ldots, \underbrace{x_n, \ldots, x_n}_{n \text{ times}}, x_{n+1}, \ldots\right)$$

It is clear that this norm is equivalent to the norm $\| \ \|_2$ of ℓ^2, and we show that ℓ^2 with this norm is LUC. Indeed, if $x, (x_n)$ are in ℓ^2 and

$$\|x\|^- = 1 \qquad \|x_n\|^- = 1 \qquad \lim \|x_n + x\|^- = 2$$

let u_x, u_{x_n} be the corresponding elements defined as above. Then we have

$$\|u_x\|_1 = 1 \qquad \lim \|u_{x_n}\|_1 = 1 \qquad \lim \|u_x + u_{x_n}\|_1 = 2$$

Since the norm $\| \ \|_1$ is LUC we obtain that (x_n) converges weakly to x, and since

$$\lim \|x_n\|_2 = \|x\|_2$$

we get

$$\lim \|x_n - x\|_2 = 0$$

since ℓ^2 has the Radon-Riesz property (this is almost obvious).

Now we show that $(\ell^2, \| \ \|^-)$ is not UCED. Indeed, consider the following sequences $(x_n),(y_n)$:

$$x_n = e_1 + e^{1/2} e_n$$

$$y_n = 3^{1/2} e_n$$

with

$$e_i = (0,0,\ldots,0,1,0,\ldots)$$

(1 in the ith place). We remark that

$$\lim \|x_n\|^- = 1 \qquad \lim \|y_n\|^- = 1 \qquad \lim \|x_n + y_n\|^- = 1$$

and since

$$x_n - y_n = e_1$$

the space is not UCED.

EXAMPLE 2.6.42 (M. A. Smith, 1978a) We give now an example of a Banach space which is WUC and not MLUC. Again consider the space ℓ^2 and define a new equivalent norm on ℓ^2 by the formula:

$$x \to \|x\|_1 = \max \{|x_1|, \|(0,x_2,x_3,\ldots)\|_2\}$$

where

$$x = (x_1,x_2,x_3,\ldots)$$

Further we set

$$x \to \|x\|_1^* = (\|x\|_1^2 + \|Tx\|_2^2)^{1/2}$$

where

$$Tx = (x_1, ax_2, a_3x_3, \ldots, a_nx_n)$$

and (a_n) is a sequence of positive numbers converging to zero. It is clear that $\|x\|_1^*$ is equivalent to $\|x\|_2$, and since ℓ^2 is WUC and T is obviously bounded and injective, the norm $\|x\|_1^*$ is WUC.

Now we show that ℓ^2 with the norm $\|\ \|_1^*$ is not MLUC. For this we consider $a = 1/2^{1/2}$ and the sequences $(x_n), (y_n)$, defined as follows:

$$x_n = a(e_1 + e_n) \qquad y_n = a(e_1 - e_n)$$

If $x = ae_1$ then the triple $(x,(x_n),(y_n))$ does not satisfy the condition in the definition of MLUC since

$$\|x\|_1^* = 1 \qquad \lim \|x_n\|_1^* = \lim \|y_n\|_1^* = 1 \qquad \lim \|2x - (x_n + y_n)\|_1^* = 0$$

and

$$\lim \|x_n - y_n\|_1^* = 2^{1/2}$$

EXAMPLE 2.6.43 (M. A. Smith, 1978a) We give an example of a Banach space which is UCED and not WLUC. We consider on ℓ^2 the following equivalent norms to $\|\ \|_2$:

$$x \to \|x\|_1 = |x_1| + \|(0,x_2,x_3,\ldots)\|_2$$
$$x \to \|x\|^- = (\|x\|_1^2 + \|Tx\|_2^2)^{1/2}$$

where

$$Tx = (a_2x_2, a_3x_3, \ldots)$$

and (a_n) is a sequence of positive numbers converging to zero. It is clear that $(\ell^2, \|\ \|^-)$ is UCED, by Theorem 2.6.34. If we set $x = e_1$, $x_n = e_n$, then clearly we have the following relations:

$$\|x\|^- = 1 \qquad \|x_n\|^- = 1 \qquad \lim \|x + x_n\|^- = 2$$

and obviously (x_n) converges weakly to 0. Thus the space is not WLUC.

We remark that $(\ell^2, \|\ \|^-)$ has the Radon-Riesz property, and being UCED it is strictly convex. Thus we have an example of a space which is in the class (HR) (i.e., the common part of the class of spaces which are strictly convex and the class of Banach spaces which have the Radon-Riesz property).

EXAMPLE 2.6.44 (M. A. Smith, 1978a) The following is an example of a Banach space which is LUC and not WUC. We consider the space ℓ^1 of all sequences $x = (x_i)$ with $\|x\| = \Sigma_1^\infty |x_i| < \infty$. On ℓ^1 we define the following equivalent norm:

$$x \to (\|x\|^2 + \|x\|_2^2)^{1/2} = \|x\|^-$$

where $\| \|_2$ is the ℓ^2 norm. From the form of the norm $\| \|^-$ it is not difficult to see that $(\ell^1, \| \|^-)$ is LUC. The fact that ℓ^1 with the norm is not WUC is a consequence of the following interesting result of V. Zizler.

THEOREM 2.6.45 (V. Zizler, 1969, 1971) On the space ℓ^1 there does not exist an equivalent norm $x \to \|x\|^*$ such that $(\ell^1, \| \|^*)$ is WUC.

Proof: Refer to Zizler's papers.

EXAMPLE 2.4.46 We give now an example of a Banach space which has the Radon-Riesz property and is not strictly convex. Consider the space ℓ^p of all sequences $x = (x_i)$ with $\|x\|_p = (\Sigma_1^\infty |x_i|^p)^{1/p} < \infty$ and for each $x = (x_1, x_2, x_3, \ldots)$ define the elements $x_1(x), x_2(x)$ in ℓ^p as follows:

$$x_1(x) = (x_1, 0, 0, \ldots)$$
$$x_2(x) = (0, x_2, x_3, x_4, \ldots)$$

The new equivalent norm is defined by the formula

$$x \to \max \{\|x_1(x)\|_p, \|x_2(x)\|_p\} = \|x\|^*$$

We suppose that p is a fixed number in $(1,\infty)$. Obviously $(\ell^p, \| \|^*)$ is not strictly convex. We show that it has the Radon-Riesz property. If (x_n) and x are in ℓ^p and

$$\|x\|^* = 1 \qquad \lim \|x_n\|^* = 1$$
$$x_n \rightharpoonup x$$

then we show that $\lim (x_n - x) = 0$. Indeed, from the form of the norm $\| \|^*$ we conclude that for some sequence (n_k) we have the relation

$$\lim \|x_1(x_{n_k})\|_p = \|x_1(x)\|_p$$

or

$$\lim \|x_2(x_{n_k})\|_p = \|x_2(x)\|_p$$

Since the ℓ^p space has the Radon-Riesz property (see Sec. 2.7, Theorem 2.7.11 and Theorems 2.4.25 and 2.4.26), we conclude that

$$\lim x_1(x_{n_k}) = x_1(x)$$

and this implies that (reasoning in a similar way)

$$\lim x_2(x_{n_k}) = x_2(x)$$

Now we summarize the connections between the classes of Banach spaces considered above. We use the following notation:

UC = the class of uniformly convex spaces

LUC = the class of locally uniformly convex spaces

WUC = the class of weakly uniformly convex spaces

MLUC = the class of midpoint locally uniformly convex spaces

UCED = the class of uniformly convex spaces in every direction

R-R = the class of Banach spaces with the Radon-Riesz property [sometimes this class is denoted by (H)]

C = the class of strictly convex spaces

We mention now a class of space introduced by A. R. Lovaglia which is related to uniform convexity and to the class LUC.

DEFINITION 2.6.47 (A. R. Lovaglia, 1955) The Banach space X is said to be weakly locally uniformly convex (WLUC) if the following property holds: if x and (x_n) are elements of X with

$$\|x\| = 1 \qquad \lim \|x_n\| = \lim \|\tfrac{1}{2}(x + x_n)\| = 1$$

then

$$x_n \rightharpoonup x$$

Then the following implications hold, where the arrow $\rightarrow$ means that we have an inclusion (proper inclusion):

UC $\rightarrow$ LUC $\rightarrow$ R-R $\rightarrow$ C

UC $\rightarrow$ LUC $\rightarrow$ MLUC $\rightarrow$ C

UC $\rightarrow$ LUC $\rightarrow$ WLUC $\rightarrow$ C

UC $\rightarrow$ WUC $\rightarrow$ WLUC $\rightarrow$ C

UC $\rightarrow$ WUC $\rightarrow$ UCED $\rightarrow$ C

For other relations between classes of Banach spaces which extend (in some sense) the class of uniformly convex spaces the reader may consult the papers listed in the Bibliography.

2.7 MODULI OF CONVEXITY AND SMOOTHNESS OF A BANACH SPACE

The properties of a Banach space are determined largely by the unit ball, and thus it is an important object to study. One way to do so is to find certain numerical invariants. In what follows we are concerned with two such invariants, namely, the modulus of convexity and the modulus of smoothness of a Banach space. We give various properties of these (isometric) invariants and some applications, among them, to unconditionally convergent series and to series diverging for every choice of signs, due to M. I. Kadets and J. Lindenstrauss, respectively. Let us first define these invariants.

(J. A. Clarkson, 1936) 2.7.1 The modulus of convexity of a Banach space X (of dimension at least 2) is a function δ_X defined on [0,2] with values in [0,1] by the formula

$$t \to \delta_X(t) = \inf \{1 - \|\frac{x + y}{2}\| : \|x\| = \|y\| = 1, \|x - y\| = t\}$$

DEFINITION 2.7.2 The modulus of smoothness of a Banach space X (of dimension at least 2) is a function ρ_X defined on $[0,\infty)$ with values in $[0,\infty)$ by the formula

$$t \to \rho_X(t) = \frac{1}{2} \sup \{\|x + y\| + \|x - y\| - 2 : \|x\| = 1, \|y\| = t\}$$

As we already mentioned (2.4.15) the Banach space X is uniformly convex if and only if for each $t > 0$, $\delta_X(t) > 0$. The class of uniformly smooth spaces is introduced in the following.

DEFINITION 2.7.3 The Banach space X is said to be uniformly smooth if the following property holds:

$$\lim_{t \to 0} \frac{\rho_X(t)}{t} = 0$$

The following lemma gives some important properties of the modulus of smoothness.

LEMMA 2.7.4 For any Banach space X the function

$$t \to \rho_X(t)$$

is convex, increasing, and

$$\rho_X(t) \le t$$

Proof: Since

$$t \to \{\|x + y\| + \|x - y\| : \|x\| = 1, \|y\| = t\}$$

is obviously continuous and convex, the modulus of smoothness is obviously a convex function, being a sup of convex functions. Also, it is clear that $t - \rho_X(t)$ is increasing, and since for each t,

$$\|x + y\| + \|x - y\| - 2 \le t$$

for $\|x\| = 1$, $\|y\| = t$, the inequality

$$\rho_X(t) \le t$$

follows.

It is an interesting problem to study the relation between the functions δ_X and ρ_X for Banach spaces. The basic result was obtained by J. Lindenstrauss and is given in the following theorem.

THEOREM 2.7.5 (J. Lindenstrauss, 1963a) For every Banach space X the following equality holds:

$$\rho_{X^*}(t) = \sup_{0 \le s \le 2} \left[\frac{ts}{2} - \delta_X(s)\right]$$

Proof: First we prove the following inequality: for any t,s we have

$$\delta_X(t) + \rho_{X^*}(s) \ge \frac{st}{2} \qquad (*)$$

Indeed, let $x,y \in X$ with the properties that $\|x\| = \|y\| = 1$ and $\|x - y\| = t$. Then we find x^* and y^* in X^* such that:

$$\|x^*\| = \|y^*\| = 1$$

$$x^*(x + y) = \|x + y\| \qquad y^*(x - y) = \|x - y\|$$

Then

$$\begin{aligned} 2\rho_{X^*}(s) &\ge x^* + sy^* + x^* - sy^* - 2 \ge x^*(x) + sy^*(x) + x^*(y) - sy^*(y) - 2 \\ &= x^*(x + y) + sy^*(x - y) - 2 = \|x + y\| + s\|x - y\| - 2 \\ &= \|x + y\| + st - 2 \end{aligned}$$

or

$$2 - \|x + y\| \ge st - 2\rho_{X^*}(s)$$

which is the desired inequality. This inequality obviously implies that $\rho_{X^*}(s) \ge \sup_{0 \le t \le 2} [(ts/2) - \rho_X(s)]$. Let us consider now x^*,y^* in X^* such that $\|x^*\| = 1$ and $\|y^*\| = s$. Let $r > 0$. From the definition of the norm in X^* it follows that there exist x,y in X such that

$$\|x\| = \|y\| = 1$$
$$(x^* + y^*)(x) = \|x^* + y^*\| - r \qquad (x^* - y^*)(y) = \|x^* - y^*\| - r$$

Then we have

$$\begin{aligned}
\|x^* + y^*\| + \|x^* - y^*\| &\le x^*(x) + y^*(x) + x^*(y) - y^*(y) + 2r \\
&= x^*(x + y) + y^*(x - y) + 2r \\
&\le \|x + y\| + \|x - y\| + 2r \\
&\le 2 + 2 \sup_{0\le t\le 2} [ts - \delta_X(t)] + 2r
\end{aligned}$$

and since r is arbitrary we have

$$\rho_{X^*}(s) \le \sup_{0\le t\le 2} \left[\frac{ts}{2} - \delta_X(t)\right]$$

This inequality combined with the inequality obtained above gives the equality of the theorem.

In a similar way we can prove the following "dual" result.

(J. Lindenstrauss, 1963a) 2.7.6 For every Banach space X the following relation holds:

$$\delta_X(s) = \sup_{0\le t\le 2} \left[\frac{ts}{2} - \rho_{X^*}(t)\right]$$

It is important to know more about the connections between the moduli δ_X and ρ_{X^*} but at this time no simple relation is known. One candidate for a new connection between δ_X and ρ_{X^*} is the function defined as follows:

$$t \to \sup_{s=0} \left[\frac{ts}{2} - \delta_{X^*}(s)\right] = \bar{\delta}(t)$$

This is obviously convex, and from Lindenstrauss' results given above the following relation holds:

$$\bar{\delta}(t) \le \delta_X(t)$$

But the equality is not possible for all Banach spaces. Indeed, V. I. Liokumovich (Liokumovič) (1973) has found an example of a Banach space with the property that $t \to \delta_X(t)$ is not a convex function.

Now using the modulus of convexity we can, in some sense, compare Banach spaces.

DEFINITION 2.7.7 The Banach space X is said to be *more convex* than the Banach space Y if for all t,

$$\delta_X(t) \geq \delta_Y(t)$$

and the Banach space X is said to be *smoother* than the Banach space Y if

$$\rho_X(s) \leq \rho_Y(s)$$

for all s.

Suppose now that X is a Hilbert space H. Then according to the computation in Sec. 1.8, Proposition 1.8.9,

$$\delta_H(t) = 1 - \left[1 - \left(\frac{t}{2}\right)^2\right]^{1/2}$$

We prove now the following result of G. Nordlander (1960) which says that the Hilbert spaces are the most convex spaces in the class of uniformly convex spaces.

THEOREM 2.7.8 (G. Nordlander, 1960) If X is any uniformly convex space then the following inequality holds:

$$\delta_X(t) \leq \delta_H(t) = 1 - \left[1 - \left(\frac{t}{2}\right)^2\right]^{1/2}$$

Proof: We remark that it is sufficient to consider the case of two-dimensional Banach spaces, and thus without loss of generality we can take X as the plane. The unit ball of X, $\bar{S}_1(0) = \{x : \|x\| \leq 1\}$, is a closed convex set containing the origin in the interior. Let us denote by r the boundary curve of $\bar{S}_1(0)$, and set x = OA and y = OB where O is the origin and $\|x\| = 1 = \|y\|$. If x and y traverse r and $\|x - y\| = t$ then $(1/2)(x + y)$ traverses a simple closed curve r_t. Then if a_t denotes the area inside r_t and A* is the area of $\bar{S}_1(0)$ the following relation holds:

$$a_t = 1 - \frac{t^2}{4} A^*$$

To see that this assertion is true we use a rectangular coordinate system. If we set $A = (x_1, y_1)$, $B = (x_2, y_2)$ (where OA = x, OB = y) then $(1/2)(x + y)$ has the coordinates $((1/2)(x_1 + x_2), (1/2)(y_1 + y_2))$. In this case the area a_t is given by the formula

$$a_t = \frac{1}{4}\int_\alpha^\beta [y_1(t) + y_2(t)]\, d[x_1(t) + x_2(t)]$$

and for A* we have the formula

$$A^* = \int_\alpha^\beta y_1(t)\, dx_1(t) = \int_\alpha^\beta y_2(t)\, dx_2(t)$$

which implies that

$$A^* - a_t = \frac{1}{4}\left(\int_\alpha^\beta [y_1(t) - y_2(t)]\, d[x_1(t) - x_2(t)]\right)$$

But the integral represents the area of a domain with boundary traversed by the vectors $x - y$ ($\|x\| = \|y\| = 1$, $\|x - y\| = t$), and the boundary of this domain is a curve having the same shape as the curve r. Further we remark that it is in fact the curve tr. This clearly implies that

$$A^* - a_t = \frac{1}{4} t^2 A^*$$

Now let m be the midpoint of AB (suppose for a moment that A,B are fixed). If $p(t,\varphi)$ is the norm of mc, where c is the image of m under the radial projection of 0 to r, and φ is the angle made by mc with the x axis, then the modulus of convexity $\delta_X(t)$ is given by the formula $\delta_X(t) = \inf p(t,\varphi)$. But we have

$$\int_0^{2\pi} [1 - p(t,\varphi)^2 \ell^2(\varphi)\, d\varphi = \left(1 - \frac{t^2}{4}\right) \int_0^{2\pi} \ell^2(\varphi)\, d\varphi$$

where $\ell(\varphi)$ is the length of the vector 0c. Since $\ell(\varphi) > 0$ for all φ we find a φ_0 with the property that

$$[1 - p(t,\varphi_0)]^2 = 1 - \frac{t^2}{4}$$

which gives that

$$p(t,\varphi_0) = 1 - \left(1 - \frac{t^2}{4}\right)^{1/2}$$

and thus the modulus of convexity $\delta_X(t)$ is equal to

$$\inf p(t,\varphi) = 1 - \left(1 - \frac{t^2}{4}\right)^{1/2}$$

which is clearly equivalent with the assertion of the theorem.

From this result we can obtain a characterization of complete inner product spaces in terms of the behavior of the modulus of smoothness.

First we remark that from the above result, there follows:

COROLLARY 2.7.9 For any space X which is uniformly convex the following inequality holds:

$$\rho_X(s) \geq (1 + s^2)^{1/2} - 1$$

Proof: Since, by the above results,

$$\pi_X(s) = \sup_{0\leq t\leq 2} \left[\frac{st}{2} - \delta_{X^*}(t)\right] = \sup_t \left[\frac{ts}{2} - \delta_H(t)\right] = \rho_H(s) \geq (1 + s^2)^{1/2} - 1$$

the assertion is proved.

COROLLARY 2.7.10 If X is uniformly convex then X is a complete inner product space if and only if

$$\rho_X(s) = (1 + s^2)^{1/2} - 1$$

Proof: Indeed, if this relation holds then we get the following inequality for all x,y in X:

$$\|x + y\| + \|x - y\| \leq 2(\|x\|^2 + \|y\|^2)^{1/2}$$

and this inequality implies that X is an inner product space (M. M. Day, 1958; V. I. Istrăţescu, 1983b).

In what follows we give some examples of uniformly convex spaces, using the modulus of convexity. We give also an estimate of the modulus of convexity of the Banach spaces L^p for p in $(1,\infty)$. First we need another characterization of uniformly convex spaces, which is given in the following lemma.

LEMMA 2.7.11 The Banach space X is uniformly convex if and only if for each fixed p [in $(1,\infty)$] there exists a function

$$t - \delta_p(t)$$

such that if x,y are in X and $\|x - y\| \geq t$, then

$$\|x + y\|^p \leq 2^p[1 - \delta_p(t)] \frac{\|x\|^p + \|y\|^p}{2}$$

Proof: It is obvious that this condition is sufficient for X to be uniformly convex. Thus we prove the necessity only. First we remark that the following inequality holds:

$$(1 + t)^p \le 2^{p-1}(1 + t^p) \qquad t \in [0,1]$$

with strict inequality for t in [0,1). Indeed, if we consider the function

$$t \to h(t) = \frac{1 + t^p}{(1 + t)^p}$$

we see that it is continuous and has exactly one minimum for t in [0,1], namely, at t = 1. This clearly implies that

$$2^{p-1}(1 + t)^p \le 1 + t^p$$

which is equivalent to our inequality.

We remark that it is sufficient to prove the lemma only in the case $\|x\| = 1$, $\|y\| \le 1$. Indeed, suppose that $\|x\| \ge \|y\|$ (the same argument works if the inequality $\|x\| \le \|y\|$ holds), and thus

$$t \le \|x - y\| = \|x\|\left\|\frac{x}{\|x\|} - \frac{y}{\|x\|}\right\|$$

which implies that

$$\frac{t}{\|x\|} \leqq \left\|\frac{x}{\|x\|} - \frac{y}{\|x\|}\right\|$$

and since the inequality in the lemma is homogeneous the assertion follows. Thus we suppose without loss of generality that $\|x\| = 1$, $\|y\| \le 1$ and $\|x - y\| \ge t$.

Suppose that the lemma is false. Then for some $t_0 > 0$ and sequences $(x_n),(y_n)$ satisfying

$$\|x_n\| = 1 \qquad \|y_n\| \le 1 \qquad \|x_n - y_n\| \ge t_0$$

we have

$$\lim \frac{(\|x_n + y_n\|)^p/2^p}{(\|x_n\|^p + \|y_n\|^p)/2} = 1 \tag{*}$$

We may suppose that

$$\lim \|y_n\| = 1$$

since, in the contrary case, for some subsequence and some a < 1, $\|y_{n_k}\| < a$, and thus

$$\frac{\|x_{n_k} + y_{n_k}\|^p}{2^p} \le \frac{1}{2} (1 + \|y_{n_k}\|)^p \le \frac{b}{2} (1 + \|y_{n_k}\|^p) \le \frac{b}{2} (\|x_{n_k}\|^p + \|y_{n_k}\|^p)$$

with b in (0,1). But this contradicts (*), and thus $\lim \|y_n\| = 1$. We set

$$z = \frac{y_n}{\|y_n\|}$$

and thus $\|z_n\| = 1$ and

$$\|z_n - y_n\| = \left(\frac{1}{\|y_n\|} - 1\right)\|y_n\| \to 0$$

Thus for n large,

$$\|z_n - x_n\| \ge \frac{t_0}{2}$$

This together with (*) implies that

$$\lim \|x_n + y_n\| = 2$$

which contradicts the uniform convexity of X. The lemma is proved.

The following basic result was obtained by J. A. Clarkson and gives the important fact that the spaces L^p are uniformly convex [p in $(1,\infty)$].

THEOREM 2.7.12 (J. A. Clarkson, 1936) Let $(\Lambda,\mathcal{B},P)$ be a measure space. Then for each fixed p in $(1,\infty)$ the Banach spaces $L^p(\Lambda,\mathcal{B},P) = L^p$ are uniformly convex.

Proof: (E. J. McShane, 1950; G. Köthe's formulation, 1969) Let t be given and f,g be two elements in L^p with the properties that $\|f\| \le 1$, $\|g\| \le 1$, and $\|f - g\| \ge t$. We consider the following set in Λ:

$$M = \{s : t^p[|f(s)|^p + |g(s)|^p] \le 4|f(s) - g(s)|^p\}$$

and applying Lemma 2.7.11 (for δ corresponding to $4^{-1/p}t$) we obtain the following inequality: for all s in M,

$$\left|\frac{f(s) + g(s)}{2}\right|^p \le (1 - \delta) \frac{|f(s)|^p + |g(s)|^p}{2}$$

Now on the complement C_M of M we have

$$\int_{C_M} |f(s) - g(s)|^p \, d\mu \le \frac{t^p}{4} \int_\Lambda (|f(s)|^p + |g(s)|^p) \, d\mu \le \frac{t^p}{2}$$

which implies that

$$\int_{C_M} |f(s) - g(s)|^p \, d\mu \geq \frac{t^p}{2}$$

From these inequalities we have

$$\max\left\{\int_M |f(s)|^p \, d\mu, \int_M |g(s)|^p \, d\mu\right\} \geq \frac{t^p}{2^{p+1}}$$

and thus

$$\int_\Lambda \frac{|f(s)|^p + |g(s)|^p}{2} \, d\mu - \int_\Lambda \left|\frac{f(s) + g(s)}{2}\right|^p \, d\mu$$
$$\geq \int_M \frac{|f(s)|^p + |g(s)|^p}{2} \, d\mu - \int_M \left|\frac{f(s) + g(s)}{2}\right|^p \, d\mu$$
$$\geq \int_M \delta \, \frac{|f(s)|^p + |g(s)|^p}{2} \, d\mu \geq \delta \, \frac{t^p}{2^{p+2}}$$

Then

$$\int_\Lambda \left|\frac{f(s) + g(s)}{2}\right|^p \, d\mu \leq \int_\Lambda \frac{|f(s)|^p + |g(s)|^p}{2} \, d\mu - \frac{t^p}{2^{p+2}} \leq 1 - \frac{t^p}{2^{p+2}}$$

and this obviously implies the assertion of the theorem.

COROLLARY 2.7.13 The spaces L^p [p in $(1,\infty)$] are uniformly smooth.

Proof: This follows from Lindenstrauss' theorem and the fact that the dual space of L^p is L^q $(1/p + 1/q = 1)$.

We present now more precise results on the moduli of the spaces L^p. These were obtained by O. Hanner (1956). For the formulation, we need some notions which also are of interest in themselves.

DEFINITION 2.7.14 The Banach space is said to be p-uniformly convex $(p \geq 2)$ if the modulus of convexity of X satisfies the inequality

$$\delta_X(t) \geq ct^p$$

for some positive constant c and all t. The Banach space X is said to be p-uniformly smooth $(1 < p \leq 2)$ if the modulus of smoothness of X satisfies the inequality

$$\rho_X(s) \leq Cs^p$$

for some positive constant C and all s.

Using these notions Hanner's result is as follows.

THEOREM 2.7.15 Let $X = L^p$ with $1 < p < \infty$. Then the following assertions hold:

(1) For $1 < p < 2$, L^p is p-uniformly smooth.

(2) For $2 \le p < \infty$, L^p is p-uniformly convex.

Proof: In what follows we present the proof for the case $\Lambda = [0,1]$ and $d\mu = dx$, the Lebesgue measure, since the extension to general measure spaces is obvious. For this purpose we need two inequalities for real numbers, which are given below.

For any real numbers a and b the following inequalities hold: If $1 < p < 2$,

$$|a|^p + p|a|^{p-1}b \text{ sign } a + k_p|b|^2(|a|^{2-p} + |b|^{2-p}) \le |a + b|^p \tag{*}$$

and if $2 \le p < \infty$,

$$|a|^p + p|a|^{p-1}b \text{ sign } a + k_p|b|^p \le |a + b|^p \tag{**}$$

where k_p is a strictly positive constant depending upon p only. We give the proof for (*) only since the proof of (**) is similar. For this we consider the function

$$u \to f(u) = (|1 + u|^p - 1 - pb)(|b|^{-p} + |b|^{-2})$$

for $|u| > |b/2a|$. (We suppose that both a and b are nonzero since in the contrary case the assertion of the inequalities is obvious.) We remark that the function f is nonzero and continuous, with

$$\lim_{u\to\infty, u\to-\infty} f(u) = 1$$

and this implies that for some $k_p > 0$,

$$f(u) \ge k_p$$

Taking $u = b/a$ we obtain our inequality.

Now from these inequalities we have for any $g \in L^p$,

$$\|f\|_p^p + p\|f\|_p^{p-1} \int_0^1 g(u) \text{ sign } f(u)du + k_p \int_0^1 \frac{|f(u)|^2}{|f(u)|^{2-p} + |g(u)|^{2-p}} du$$
$$\le \|f + g\|_p^p \tag{***}$$

if $1 < p < 2$, and

$$\|f\|_p^p + p \int_0^1 |f(u)|^{p-1} g(u) \text{ sign } f(u)\, du + k_p \|g\|_p^p \le \|f + g\|_p^p \qquad (****)$$

Now if $p \ge 2$, using the Hölder inequality we have

$$\|g\|_p^p = \int_0^1 |g(u)|^p\, du = \int_0^1 |g(u)|^p \frac{[|f(u)|^{2-p} + |g(u)|^{2-p}]^{p/2}}{[|f(u)|^{2-p} + |g(u)|^{2-p}]^{p/2}}\, du$$

$$\le \int_0^1 \left|\frac{|g(u)|^2}{|f(u)|^{2-p} + |g(u)|^{2-p}}\right|^{p/2} [|f(u)|^{2-p} + |g(u)|^{2-p}]^{p/2}\, du$$

$$\le \left[\int_0^1 \frac{|g(u)|^2}{|f(u)|^{2-p} + |g(u)|^{2-p}}\, du\right]^{p/2}$$

$$\times \left\{\int_0^1 [|f(u)|^{2-p} + |g(u)|^{2-p}]^{p/(2-p)}\, du\right\}^{(2-p)/p} \qquad (*)_1$$

If p is in $(1,2)$ then $p/(2 - p) > 1$, which implies that

$$|f|^{2-p} + |g|^{2-p} \in L^{p/(2-p)}$$

$$\| |f|^{2-p} + |g|^{2-p} \|_{p/(2-p)} \le \|f\|_p^{2-p} + \|g\|_p^{2-p}$$

Using these inequalities in $(*)_1$ we get

$$\|g\|_p^p = \int_0^1 |g(u)|^p\, du = \left[\int_0^1 \frac{|g(u)|^2}{|f(u)|^{2-p} + |g(u)|^{2-p}}\, du\right]^{p/2}$$

$$\times \left(\|f\|_p^{2-p} + \|g\|_p^{2-p}\right)^{p/2}$$

and thus

$$\frac{\|g\|_p^2}{\|f\|_p^{2-p} + \|g\|_p^{2-p}} \le \int_0^1 \frac{|g(u)|^2}{|f(u)|^{2-p} + |g(u)|^{2-p}}\, du$$

Since

$$F_f = \frac{f^{p-1}}{\|f\|_p^{p-1}} \text{ sign } f \in L_q \qquad \frac{1}{p} + \frac{1}{q} = 1$$

the function F_f defines a continuous linear functional on L^p by the relation

$$x^*(h) = \int_0^1 F_f(u) h(u)\, du$$

and it is easy to verify that

$$\|x^*\| = 1$$
$$x^*(f) = \|f\|_p$$

Taking into account these relations, from the inequalities (*) and (**) we obtain the following inequalities:

$$\|f\|_p^p + p\|f\|_p^{p-1}x^*(g) + k_p\|g\|_p^2(\|f\|_p^{2-p} + \|g\|_p^{2-p})^{-1} \le \|f + g\|_p^p \qquad (****)$$

for p in (1,2), and

$$\|f\|_p^p + p\|g\|_p^{p-1}x^*(g) + k_p\|g\|_p^p \le \|f + g\|_p^p \qquad (*****)$$

if p is in $[2,\infty)$. We consider now the hyperplane

$$x^*(h) = 1 - \delta$$

for δ in (0,1). Let

$$M = \{h : h \text{ in } L^p, \|h\|_p \le 1, x^*(h) = 1 - \delta\}$$

and we note that f is in M. For any h in M the elements of the form $h - f = d$ clearly have the property that $\|d + f\| \le 1$ and

$$x^*(d) = x^*(h) - x^*(f) \ge 1 - \delta - 1 = -\delta$$

The inequality in (****) implies that

$$1 \ge 1 - p\delta + k_p \frac{\|g\|_p^2}{1 - \|g\|_p^{2-p}}$$

and thus

$$\|g\|_p^p \le \frac{p\delta}{k_p}(1 + \|g\|_p^{2-p})\|g\|_p^{p-2} = \frac{p\delta}{k_p}\frac{1}{\|g\|_p^{2-p} + 1} = 2\frac{p}{k_p}\delta$$

Now for any d_1, d_2 in M we obtain

$$\|d_1 - d_2\|_p = \|d_1 - f + f - d_2\|_p \le \|d_1 - f\|_p + \|d_2 - f\|_p$$

and thus

$$\sup\{\|d_1 - d_2\|_p : d_i \in M\} = H(\delta) \le 2\left(\frac{p\delta}{2k_p}\right)^{1/p}$$

which implies that

$$\delta(t) \geq t^p \frac{k_p}{2^p} 2p = ct^p$$

and thus (1) is proved. Using the inequalities obtained above for p in $[2,\infty)$, in a similar way we prove that (2) holds.

REMARK 2.7.16 From the above estimates for $\delta(t)$ we get that for p in (1,2) the following inequality holds:

$$\delta(t) \geq \bar{k}t^2$$

where $\bar{k}$ is a positive constant depending upon p only.

We present now two important and interesting applications of the results on the modulus of convexity and the modulus of smoothness obtained by M. I. Kadets and J. Lindenstrauss.

THEOREM 2.7.17 (M. I. Kadets, 1956) Let X be a uniformly convex Banach space and $\Sigma_{n=1}^{\infty} x_n$ be an unconditionally convergent series in X. Then the series $\Sigma_{n=1}^{\infty} \bar{\delta}_X(\|x_n\|)$ is convergent. Here $\bar{\delta}_X$ is the extended modulus of convexity of X defined as follows:

$$\bar{\delta}_X(t) = \begin{cases} \delta_X(t) & 0 \leq t \leq 2 \\ 1 & t > 2 \end{cases}$$

Proof: We consider the metric space $(-1,1)^\omega$ and we consider the map

$$\varphi : (-1,1)^\omega \to C$$

(C is the Cantor set) defined by the formula:

$$\varphi((t_k)) = \sum_{i=1}^{\infty} \frac{t_i + 1}{3^i}$$

From the form of φ we see that it is in fact a homeomorphism between C and $(-1,1)^\omega$. We consider now the function on C with values in X defined by the formula:

$$h(t) = \sum_{i=1}^{\infty} t_i x_i \qquad (t_i) = t$$

Since $\Sigma_{i=1}^{\infty} x_i$ is supposed to be unconditionally convergent, h is well defined and it is continuous. Then obviously $t \to \|h(t)\|$ is continuous on C, and C being compact, $\|h(t)\|$ achieves its maximum value for some t_0 in C. But $(-1)x_i = 1(-x_i)$ and thus we may assume without loss of generality that t_0 is the following element of C:

$$t_0 = \frac{2}{3} + \frac{2}{3^2} + \cdots + \frac{2}{3^n} + \cdots = 1$$

Then we have the inequality

$$\|h(t)\| \le \|h(1)\| = \left\| \sum_{n=1}^{\infty} x_n \right\|$$

If we consider the elements (t_m) of C,

$$t_m = \frac{2}{3} + \cdots + \frac{2}{3^{m-1}} + \frac{2}{3^{m+1}} + \cdots = 1 - \frac{2}{3^m}$$

we have

$$h(1 - t_m) = \sum_{i=1}^{\infty} x_i - 2x_m$$

and thus, using the above inequality,

$$\|h(1 - t_m)\| \le \|h(1)\| = \left\| \sum_{i=1}^{\infty} x_i \right\|$$

We can suppose without loss of generality that $\|h(1)\| \le 2$. We assume also that $x = \Sigma_{i=1}^{\infty} x_i \neq 0$. Let $x^* \in X^*$ such that $x^*(x) = \|x\|$, $\|x^*\| = 1$, and consider the hyperplane

$$\left\{ y : y \in X,\ x^*(y) = x^*\left(\frac{y - 2x_i}{\|x\|} \right) \right\} = H_i$$

If H_i has positive distance from 0, with

$$1 - \delta_i = x^*\left(\frac{x - 2x_i}{\|x\|} \right)$$

then

$$x^*\left(\frac{x}{\|x\|} \right) = 1 > 1 - \delta_i$$

$$x^*\left(\frac{x - 2x_i}{\|x\|} \right) = 1 - \delta_i$$

which implies that the set cut off from the unit sphere by H_i contains the elements

$$\frac{x}{\|x\|} \qquad \frac{x - 2x_i}{\|x\|}$$

Thus we have

$$\left\|\frac{2x_i}{\|x\|}\right\| = \left\|\frac{x}{\|x\|} - \frac{x - 2x_i}{\|x\|}\right\| = u(\delta_i)$$

where u(t) is the least upper bound of the diameters of sets cut off from the unit sphere of X by hyperplanes (determined by functionals in the unit sphere of X*), which are at the distance 1 - t from the origin. Then we have that u(t) is exactly the modulus of convexity $\delta_X(t)$ of X. [This interesting geometric fact was first noted by M. I. Kadets (1955).]

From the above inequality we have further that

$$\delta_X\left(\frac{2x_i}{\|x\|}\right) \le \delta_i \le x^*\left(\frac{2x_i}{\|x\|}\right)$$

and since

$$\sum_{i=1}^{\infty} \delta_X(\|x_i\|) \le \sum_{i=1}^{\infty} \delta_X\left(\frac{2\|x_i\|}{\|x\|}\right) \le \sum_{i=1}^{\infty} x^*\left(\frac{2x_i}{\|x\|}\right) = 2x^*\left(\sum_{i=1}^{\infty} \frac{\|x_i\|}{\|x\|}\right)$$
$$= 2x^*\left(\frac{x}{\|x\|}\right) = 2$$

the theorem is proved.

REMARK 2.7.18 The results of Kadets in the case of the L^p spaces ($1 < p < \infty$) are related to W. Orlicz's (1933a,b) results on unconditionally convergent series in L^p. As is well known, Orlicz's results are valid also for the case p = 1. (We mention that L^1 is not uniformly convex; this may be seen easily by observing that no point of $\{x : \|x\| = 1\}$ is an extreme point of $\{x : \|x\| \le 1\}$, and this implies in fact that L^1 is not even LUC.)

The second application, a result of J. Lindenstrauss, is in fact dual to Kadets' theorem.

THEOREM 2.7.19 (J. Lindenstrauss, 1963a) Let X be a Banach space which is uniformly smooth and (x_i) be a sequence of elements of X with the property that $\Sigma_{i=1}^{\infty} \pm x_i$ diverges for every choice of signs. Then

$$\sum_{i=1}^{\infty} \rho_X(\|x_i\|) = \infty$$

For the proof we need some results which are also of independent interest.

LEMMA 2.7.20 If X is a Banach space and

$$r_X = \limsup_{t\to 0} \frac{\rho_X(2t)}{\rho_X(t)}$$

then the following inequality holds:

$$2 \le r_X \le 4$$

Proof: Since ρ_X is a convex function we have

$$\rho_X(t) \le \frac{\rho_X(2t)}{2}$$

which implies that

$$2 \le r_X$$

To prove the inequality

$$r_X \le 4$$

we consider arbitrary elements x,y in X with $\|x\| = 1$ and $\|y\| = t < 1/2$. Let $a = \|x + y\|$ and $b = \|x - y\|$. Then we have

$$\|x + 2y\| - 1 \le \|x + y + y\| + \|x + y - y\| - 2 = a\left(\left\|\frac{x + y}{a} + \frac{y}{a}\right\| + \left\|\frac{x + y}{a} - \frac{y}{a}\right\| - 2\right) + 2a - 2$$

$$\le 2a\rho_X\left(\frac{t}{a}\right) + 2a - 2$$

Since $\|-y\| = t$ we obtain that

$$\|x - 2y\| - 1 \le 2b\rho_X\left(\frac{t}{b}\right) + 2b - 2$$

and thus

$$\frac{\|x + 2y\| + \|x - 2y\| - 2}{2} \le a\rho_X\left(\frac{t}{a}\right) + b\rho_X\left(\frac{t}{b}\right) + 2\rho_X(t) \qquad (*)$$

Since

$$\max\,\{a,b\} \le 1 + t$$

and

$$\max\left\{\frac{1}{a},\frac{1}{b}\right\} \le 1 + 2t$$

taking the supremum on the left-hand of (*) we obtain

$$\rho_X(2t) \le (4 + 2t)\rho_X(t(1 + 2t)) \le (4 + 2t)[(1 - 2t)\rho_X(t) + 2t\rho_X(2t)]$$

For $t \to 0$ we have

$$\rho_X(2t) \le [4 + O(t)]\rho_X(t)$$

and this clearly implies that

$$r_X \le 4$$

LEMMA 2.7.21 Let X be a Banach space and $(y_i)_1^n$ be a family of elements in X. Let a,b be positive numbers such that $b \ge \|y_i\|$, $i = 1, 2, 3, \ldots, n$. Then there exist signs $\varepsilon_i = \pm 1$, $i = 1, 2, 3, \ldots, n$ such that

$$\left\| \sum_{i=1}^{k} \varepsilon_i y_i \right\| \le (a^{-1} + b) \prod_{i=1}^{k} [1 + \rho_X(a\|y_i\|)] \tag{*}$$

for $k = 1, 2, 3, \ldots, n$.

Proof: First we remark that for $k = 1$ the assertion is obvious. Suppose now that we have ε_i, $i \le h$, such that (*) holds. Let

$$S_h = \varepsilon_1 y_1 + \cdots + \varepsilon_h y_h$$

and if $\|S_h\| \le a^{-1}$ we have

$$\|S_h + \varepsilon_{h+1} y_{h+1}\| \le a^{-1} + b$$

which shows that (*) holds for $k = h + 1$ for any ε_{h+1}. If $\|S_h\| > a^{-1}$ then

$$\|S_h\|S_h\|^{-1} + ay_{h+1}\| + \|S_h\|S_h\|^{-1} - ay_{h+1}\| \le 2[1 + \rho_X(a\|y_{h+1}\|)]$$

which implies that there exists a sign ε_{h+1} such that

$$\|S_h\|S_h\|^{-1} + \varepsilon_{h+1} a y_{h+1}\| \le 1 + \rho_X(a\|y_{h+1}\|)$$

But this clearly implies that

$$\|S_h + \varepsilon_{h+1} a\|S_h\| y_{h+1}\| \le \|S_h\|[1 + \rho_X(a\|y_{h+1}\|)]$$

Also, $S_h + \varepsilon_{h+1} y_{h+1}$ is in the segment $[S_h, S_h + \varepsilon_{h+1} a S_h y_{h+1}]$, which gives

$$\|S_{h+1}\| \le \|S_h + \varepsilon_{h+1} y_{h+1}\| \le \|S_h\ [1 + \rho_X(a\|y_{h+1}\|)]$$

and thus (*) holds for $k = h + 1$. Then it holds for all k.

Now we can prove Theorem 2.7.19.

Proof: It suffices to prove that if $\Sigma_{i=1}^{\infty} \rho_X(\|x_i\|) < \infty$ then there exists a choice of signs ε_i such that $\Sigma_{i=1}^{\infty} \rho_X(\|x_i\|)$ converges. Suppose thus that $\Sigma_{i=1}^{\infty} \rho_X(\|x_i\|) < \infty$. In this case clearly $\lim_{i\to\infty} \rho_X(\|x_i\|) = 0$. From Theorem 2.7.5 we get $\lim \|x_i\| = 0$. Since $r_X \le 4$ we obtain for each k, $\Sigma_{i=1}^{\infty} \rho_X(2^k\|x_i\|) < \infty$, and thus there exists an increasing sequence of integers (n_k) such that

$$\sum_{i=n_k}^{\infty} [1 + \rho_X(2^k\|x_i\|)] < 2$$

and $\|x_i\| \le 2^{-k}$ for $i \ge n_k$. By Lemma 2.7.21 we find $\varepsilon_i = \pm 1$ such that for every k and h with $n_k \le h \le h_{k+1}$,

$$\left\| \sum_{i=n_k}^{h} \varepsilon_i x_i \right\| \le 2^{2-k}$$

and this obviously implies that $\Sigma_{i=1}^{\infty} \varepsilon_i x_i$ converges. The theorem is proved.

The results of Kadets and Lindenstrauss suggest the consideration of certain indices associated with the convergence and a divergence of series in Banach spaces.

DEFINITION 2.7.22 (J. Lindenstrauss, 1963a) Let X be a Banach space. The numbers α_X and β_X associated with X are defined as follows:

$\alpha_X = \inf\{p :$ If $\Sigma \pm x_i$, $x_i \in X$, converges for every choice of signs then $\Sigma_{i=1}^{\infty} \|x_i\|^p < \infty\}$

$\beta_X = \sup\{p:$ If $\Sigma \pm x_i$, $x_i \in X$, diverges for every choice of signs then $\Sigma \|x_i\|^p = \infty\}$

If X is finite dimensional, $\alpha_X = 1$ and $\beta_X = \infty$, and for arbitrary Banach spaces the following inequality holds:

$$1 \le \alpha_X \le 2 \le \beta_X \le \infty$$

From the Kadets and Lindenstrauss theorems given above we have that for uniformly convex spaces and for uniformly smooth spaces, respectively, the following relations hold:

$$\alpha_X \le \limsup_{t\to\infty} \frac{\log \delta_X(t)}{\log t}$$

$$\beta_X \ge \liminf_{s\to\infty} \frac{\log \rho_X(s)}{\log s}$$

For more information and problems connected with α_X and β_X we refer to the excellent paper of Lindenstrauss (1963a).

We give now a characterization of p-uniformly smooth spaces obtained by J. Hoffman-Jørgensen (1974).

THEOREM 2.7.23 The Banach space X is p-uniformly smooth if and only if there exists a positive constant C such that for all x,y in X the following inequality holds:

$$\|x + y\|^p + \|x - y\|^p \le 2\|x\|^p + C\|y\|^p$$

Proof: First we remark that the Banach space X is p-uniformly smooth if and only if there exists a constant c > 0 such that for all x,y in X,

$$\left\|\frac{1}{2}(x + y)\right\| + \left\|\frac{1}{2}(x - y)\right\| \le \|x\| \; 1 + c\left(\frac{\|x\|}{\|y\|}\right)^p$$

Now suppose first that X is p-uniformly smooth. Then for some k > 0 we have

$$\frac{1}{2}\,[\|x + y\| - (\|x\| + \|y\|) + \|x - y\| - (\|x\| - \|y\|)] \le \|x\|\left(1 + k\,\frac{\|x\|^p}{\|y\|^p}\right)$$

If $\|y\| \le \|x\|$ then

$$\max\{\|x + y\|, \|x - y\|\} \le 2\|x\|$$

which implies that

$$\begin{aligned}\frac{1}{2}(\|x + y\|^p + \|x - y\|^p) &\le \frac{1}{2}\left|(\|x\|^p + \|y\|^p) + (\|x\| - \|y\|^p\right.\\ &\qquad \left. + \, p2\|x\|^{p-1}\|x\|k\,\frac{\|y\|^p}{\|x\|^p}\right|\\ &\le \|x\|^p + \|y\|^p + 2^{p-1}pk\|y\|^p\\ &= \|x\|^p + (1 + 2^{p-1}pk)\|y\|^p\end{aligned}$$

by use of inequalities

$$a^p - b^p \le pa^{p-1}(a - b) \qquad a,b \ge 0,\ p \ge 1$$

$$|a + b|^p + |a - b|^p \le 2(a^p + b^p) \qquad a,b \in \mathbb{R},\ p \in [1,2]$$

If we set

$$C = \max\{2^p, 1 + p2^{p-1}k\}$$

we obtain the required inequality.

Suppose now that X is a Banach space satisfying the inequality of the theorem for all x,y in X. Now, using the definition of the modulus of smoothness, we have

$$\rho_X(s) = \sup\{\frac{1}{2}\|x + y\| + \frac{1}{2}\|x - y\| - 1 : \|x\| = 1, \|y\| = s\}$$

$$\le \sup\left\{\left(\frac{\|x + y\|}{2} + \frac{\|x - y\|}{2}\right)^p - 1 : \|x\| = 1, \|y\| = s\right\}$$

$$\le \sup\{\frac{1}{2}(\|x + y\|^p + \|x - y\|^p) - 1 : \|x\| = 1, \|y\| = s\} \le Cs^p$$

and the theorem is proved.

We leave to the reader the proof of the following characterization of p-uniformly convex spaces.

THEOREM 2.7.24 The Banach space X is p-uniformly convex if and only if for all x,y in X and for some positive constant C the following inequality holds:

$$\|x + y\|^p + \|x - y\|^p \ge 2\|x\|^p + C\|y\|^p$$

For more information concerning the modulus of convexity we refer to the paper of T. Figiel (1976). We consider in what follows some notions which can be regarded as localizations of the modulus of convexity and the modulus of smoothness.

DEFINITION 2.7.25 (A. R. Lovaglia, 1955) Let X be a Banach space and x in X with $\|x\| = 1$. The local modulus of convexity at x of X is the function defined on [0,2] with values in [0,1] by the formula:

$$t \to \delta_X(x,t) = \inf\{1 - \frac{1}{2}\|x + y\| : \|y\| = 1, \|x - y\| \ge t\}$$

The Banach space is said to be uniformly convex at x, $\|x\| = 1$, if for each $t > 0$, $\delta_X(x,t) > 0$.

Obviously we have the following result.

THEOREM 2.7.26 The Banach space X is LUC if for each x, $\|x\| = 1$, and $t > 0$.

$$\delta_X(x,t) > 0$$

DEFINITION 2.7.27 (V. I. Istrăţescu, 1981a) The local modulus of smoothness of the Banach space X at x, $\|x\| = 1$, is a function defined on $[0,\infty)$ with values in $[0,\infty)$ by the formula

$$\rho_X(x,s) = \frac{1}{2} \sup \{\|x + y\| + \|x - y\| - 2 : \|y\| = s\}$$

DEFINITION 2.7.28 (V. I. Istrăţescu, 1981a) The Banach space X is said to be uniformly smooth at x if

$$\lim_{s\to 0} \frac{\rho_X(x,s)}{s} = 0$$

DEFINITION 2.7.29 (V. I. Istrăţescu, 1981a) The Banach space X is said to be hilbertian at x if

$$\rho_X(x,t) = \delta_H(t)$$

(the modulus of convexity of Hilbert spaces).

DEFINITION 2.7.30 (V. I. Istrăţescu, 1981a) The Banach space X is said to be locally p-uniformly convex at x if there exists a strictly positive constant k(p,x) such that

$$\delta_X(x,t) \geq k(p,x)t^p$$

DEFINITION 2.7.31 (V. I. Istrăţescu, 1981a) The Banach space X is said to be locally p-uniformly smooth at x if there exists a strictly positive constant c(p,x) such that

$$\rho_X(x,s) \leq c(p,x)s^p$$

The following characterization of locally p-uniformly smooth spaces may be proved similarly to Theorem 2.7.23.

THEOREM 2.7.32 The Banach space X is locally p-uniformly smooth at x if there exists a positive constant C(x,p) such that for all y in X the following inequality holds:

$$\|x + y\|^p + \|x - y\|^p \leq 2\|x\|^p + C(x,p)\|y\|^p$$

We mention also the following interesting result of G. Pisier (1974a, b): Every uniformly convex space X admits an equivalent norm $\| \ \|^*$ such that $(X,\| \ \|^*)$ is p-uniformly convex (p-uniformly smooth) for some $p > 1$ $(p < \infty)$, respectively. We mention also that there exist other functions like the moduli defined above, and for information about these we refer to the interesting papers of V. D. Milman (1967, 1971), Bui-Min-Chi and V. I. Gurarii (1969), V. I. Gurarii (1965, 1966, 1967), V. I. Liokumovič (1977), as well as to the literature quoted there.

Below, following R. McGuigan, we show how we can use a function defined by V. D. Milman to describe several geometric properties of Banach spaces, as well as to obtain new characterizations of uniformly convex spaces and of the uniformly nonsquare Banach spaces of R. C. James (1964b). We begin with the following notions.

DEFINITION 2.7.33 Let X be a Banach space and C be a convex subset of X. Then $d(x,C) = \inf \{\|x - y\| : y \in C\}$. We write this as $d(x,C) = \|x - C\|$.

DEFINITION 2.7.34 For any x and $a > 0$ we set

$$d_C(x,a) = \inf \{\max\{\|x + ay) - C\|, \|(x - ay) - C\|\} : \|y\| = 1\}$$

DEFINITION 2.7.35 A point $x \in C$ is called a strongly extreme point if and only if for all $a > 0$, $d_C(x,a) > 0$.

If $C = \{x : \|x\| = 1\}$ then we use the notation

$$d_C(x,a) = d_X(x,a)$$

This is equal to

$$\inf \{\max\{\|x + ay\|, \|x - ay\| : \|y\| = 1\} - 1\}$$

The function

$$a \to d_X(x,a)$$

is called the modulus of extreme convexity of X at x.

DEFINITION 2.7.36 The Banach space X is called strongly strictly convex if any point x, $\|x\| = 1$, is a strongly extreme point of $\{x : \|x\| = 1\}$.

It is not difficult to see that when C is a compact set any extreme point is strongly extreme, but in general, the class of strongly extreme points is a proper subset of the set of extreme points.

In what follows we show first how we can characterize the class of uniformly nonsquare Banach spaces of James and the class of uniformly convex spaces. For this we need a result about the function $d_X(x,a)$, given in the following lemma.

LEMMA 2.7.37 For each x, the function

$$a \to d_X(x,a)$$

is an increasing function.

Proof: It is clear that it suffices to prove the following assertion: if $s > 0$ then $d_X(x,a) < d_X(x, a + s)$. Indeed, let y be an arbitrary point in X with $\|y\| = 1$; we note that the point $x + ay$ lies on the segment with endpoints x and $x + (a + s)y$, and similarly, $x - ay$ lies on the segment with endpoints x and $x - (a + s)y$. From the convexity of the norm we obtain

$$\max \|x \pm ay\| \leq \max \|x \pm (a + s)y\|$$

and this obviously implies the assertion of the lemma.

THEOREM 2.7.38 The Banach space X is uniformly nonsquare if and only if

$$d_X(1) > 0$$

where

$$d_X(a) = \inf \{d_X(x,a) : \|x\| = 1\}$$

Proof: Suppose that the assertion is false. Thus $d_X(1) = 0$ for a uniformly nonsquare Banach space X. Then from the definition of $d_X(1)$, it follows that there exist two sequences $(x_n),(y_n)$ such that

$$\|x_n\| = \|y_n\| = 1$$
$$a_n \to 1$$

where $a_n = \max \|x_n \pm ay_n\|$. Let $r > 0$, arbitrary but fixed. Let

$$b_n = \left\| \frac{1}{2} \left[\frac{x_n + y_n}{a_n} \pm \frac{x_n - y_n}{a_n} \right] \right\| = \frac{1}{a_n}$$

and since $\lim a_n = 1$, for some N, $b_N > 1 - r$. But this clearly implies that X is uniformly nonsquare. This contradiction shows that we must have $d_X(1) > 0$.

Conversely, suppose that X is a Banach space with the property that $d_X(1) > 0$. We must show that X is uniformly nonsquare. Suppose that this is false. Then there exist for each n elements z_n and w_n such that

$$\|z_n\| \leq 1 \qquad \|w_n\| \leq 1$$

$$\left\|\frac{1}{2} (z_n + w_n)\right\| > 1 - \frac{1}{n + 1}$$

Let

$$x_n = \frac{1}{2} (z_n + w_n) \qquad y_n = \frac{1}{2} (z_n - w_n)$$

and thus

$$1 \geq \|x_n\| \geq 1 - \frac{1}{n + 1} \qquad 1 \geq \|y_n\| > 1 - \frac{1}{n + 1}$$

and

$$\|x_n \pm y_n\| \leq 1$$

Since

$$\left\|\frac{x_n}{\|x_n\|} \pm \frac{y_n}{\|y_n\|}\right\| = \frac{1}{\|x_n\|\|y_n\|}\|\|y_n\|x_n \pm \|x_n\|y_n\|$$

we have

$$\begin{aligned}\|\|y_n\|x_n \pm \|x_n\|y_n\| &\leq \|\|y_n\|x_n - x_n\| + \|x_n \pm y_n\| + \|\mp y_n \pm \|x_n\|y_n\| \\ &< (1 - \|y_n\|)\|x_n\| + \|x_n \pm y_n\| + \|y_n\|(1 - \|x_n\|) \\ &\leq 1 + 2\left(1 - \frac{1}{n + 1}\right) = 1 + \frac{2}{n + 1}\end{aligned}$$

and since

$$\frac{1}{\|x_n\|\|y_n\|} < \left(\frac{n}{n + 1}\right)^2$$

we have the inequality

$$\left\|\frac{x_n}{\|x_n\|} \quad \frac{y_n}{\|y_n\|}\right\| < \left(\frac{n}{n + 1}\right)^2 1 + \frac{2}{n + 1}$$

which for $n \to \infty$ tends to 1. This implies that $d_X(1) = 0$, and this contradiction proves the theorem.

Now the characterization of uniformly convex spaces is as follows.

THEOREM 2.7.39 The Banach space X is uniformly convex if and only if

$$d_X(a) > 0 \qquad a > 0$$

For the proof we need the following important property of the modulus of convexity of a Banach space.

THEOREM 2.7.40 For any Banach space X,

$$t \to \delta_X(t)$$

is a nondecreasing function.

Proof: First we note that it is sufficient to consider only the case of Banach spaces of finite dimension since it is clear that the following formula holds:

$$\delta_X(t) = \inf \{\delta_{X_1}(t) : X_1 \subset X, \dim X_1 < \infty\}$$

For the proof in the case of two-dimensional Banach spaces we need the following inequality, due to J. Lindenstrauss and L. Tzafriri (1977): for any x,y in X, $\|x\| \le 1$, $\|y\| \le 1$,

$$\|x + y\| \le 2[1 - \delta_X(\|x - y\|)]$$

Indeed, let u,v be in $\{x : \|x\| \le 1\}$ such that u + v is maximal subject to $\|u - v\| = t$. If t = 0 then the assertion of the inequality is valid. Suppose now that t > 0 and consider x* in X* such that $\|x^*\| = 1$ and $x^*(u + v) = \|u + v\|$ (which exists). The inequality is proved provided we show that if $\|v\| < 1$ then $x^*(v - u) = \|u - v\| = t$ and $\|u\| < 1$. Similarly we get that $x^*(u - v) = t$ and thus t = -t, which is a contradiction. Let M be the set

$$\{x : x \in X, \|x - u\| = t\}$$

Then for x in M with $\|x\| \le 1$ we have

$$x^*(x + u) \le \|x + u\| \le \|u + v\| = x^*(u + v)$$

and thus for $\|v\| < 1$,

$$\|u\| \le \frac{1}{2} (\|u + v\| + \|u - v\|) = \frac{1}{2} [x^*(u + v) + x^*(u - v)] = x^*(v) < 1$$

and the assertion is proved.

Now we prove that $\delta_X(t)$ is a nondecreasing function. Let us consider t,s in [0,2] with t < s. For x,y with the properties that $\|x\| = \|y\| = 1$, $\|x - y\| = s$, $\|x + y\| \le 2[1 - \delta_X(s)]$, we have, applying the above inequality for the elements

$$x_1 = x + c(y - x) \qquad y_1 = y - c(y - x) \qquad c = \frac{s - t}{2s}$$

that the following holds:

$$\delta_X(t) \le 1 - \frac{1}{2} \|x_1 + y_1\| = 1 - \frac{1}{2} \|x + y\| = \delta_X(s)$$

Now the proof of Theorem 2.7.39 is as follows.

Let us first suppose that X is a Banach space with the property that $d_X(a) > 0$ for all a > 0. We show that X is uniformly convex. Let t be in (0,2) and for each n, let x_n and y_n be elements of X with the properties that

$$\|x_n\| = \|y_n\| = 1 \qquad \|x_n - y_n\| = t$$
$$\delta_X(t) \le 1 - \frac{1}{2}\|x_n + y_n\| < \delta_X(t) + \frac{1}{n}$$

Then clearly,

$$2[1 - \delta_X(t)] \ge \|x_n + y_n\| > 2\left[1 - \frac{1}{n} - \delta_X(t)\right]$$

If we consider the segment with endpoints $2x_n/\|x_n + y_n\|$, $2y_n/\|x_n + y_n\|$, then the midpoint is $(x_n + y_n)/\|x_n + y_n\|$, and we have the inequality

$$\left\|\frac{2x_n}{\|x_n + y_n\|} - \frac{x_n + y_n}{\|x_n + y_n\|}\right\| \le \frac{t}{\|x_n + y_n\|}$$

We know that $d_X(a)$ is an increasing function of a, and thus

$$d_X\left(\frac{t}{2[1 - \delta_X(t)]}\right) + 1 \le \frac{2}{\|x_n + y_n\|}\,\frac{1}{1 - \delta_X(t) - 1/n}$$

Since the left-hand side of this inequality does not depend on n, we have

$$1 + d_X\left(\frac{t}{2[1 - \delta_X(t)]}\right) \le \frac{1}{1 - \delta_X(t)}$$

or

$$1 - \frac{1}{d_X(t/[2 - 2\delta_X(t)] + 1} \le \delta_X(t)$$

Since $d_X(a) > 0$ we see that the term on the left-hand side of the above inequality is strictly positive, so that $\delta_X(t)$ is strictly positive, and $t \to \delta_X(t)$ being an increasing function, X is uniformly convex.

Now we show that if X is uniformly convex then $d_X(a) > 0$ for all $a > 0$. We consider the sequence (y_n) in X with the property that

$$\lim\,(\max \|x \pm ay_n\|) = d_X(x,a) + 1$$

In this case we have also,

$$\left\|\frac{x \pm ay_n}{a_n}\right\| \le 1 \qquad an = \max \|x \pm ay_n\|$$
$$\left\|\frac{x + ay_n}{a_n} + \frac{x - ay_n}{a_n}\right\| = \frac{2}{a_n}$$

and $t \to \delta_X(t)$ being an increasing function, we have

$$1 - \frac{1}{a_n} \geq \delta_X\left(\frac{2a}{a_n}\right) \geq \delta_X\left(\frac{2a}{1 + a}\right) > 0$$

for a in (0,1]. For $n \to \infty$ this implies that

$$1 - \frac{1}{d_X(x,a) + 1} \geq \delta_X\left(\frac{2a}{1 + a}\right) > 0$$

Since $d_X(a)$ is an increasing function we have then $d_X(a) > 0$ and the theorem is proved.

REMARK 2.7.41 The above characterizations of uniformly nonsquare Banach spaces and uniformly convex Banach spaces using the function $d_X(a)$ suggest the following question: Is it possible to find a characterization of other classes of Banach spaces related to uniformly Banach spaces using functions derived from the functions $d_C(x,a)$?

2.8 DIFFERENTIABILITY PROPERTIES OF THE NORM AND CLASSES OF BANACH SPACES

The connection between differentiability properties of the norm of a Banach space and various types of convexity and smoothness was first studied by V. L. Smulian in a series of well-known papers published between 1938 and 1941. This subject has attracted the attention of many mathematicians, among whom we mention only a few (see the References): L. Alaoglu, G. Birkhoff, D. F. Cudia, A. R. Lovaglia, M. I. Kadets, V. Klee, K. Sundaresan, V. Zizler, and E. Asplund. To state precisely the results obtained we first give some definitions.

DEFINITION 2.8.1 Let X be a Banach space and $S(X) = \{x : x \in X, \|x\| = 1\}$, $B_1(X) = \{x : x \in X, \|x\| \leq 1\}$. The point x in S(X) is a point of weak smoothness or is Gateaux smooth if the norm of X is Gateaux differentiable at x, i.e., if

$$\lim_{t \to 0} \frac{\|x + th\| - \|x\|}{t} = D(x,h) \qquad (*)$$

exists for every $h \in X$. The Banach space X is weakly smooth or Gateaux smooth (G-smooth) if every point of S(X) is a point of weak smoothness.

DEFINITION 2.8.2 The Banach space X is uniformly Gateaux smooth (or uniformly weakly smooth) if the norm is uniformly Gateaux differentiable for

$x \in S(X)$, i.e., the limit in (*) is uniform with respect to x in S(X). We denote this class of Banach spaces as the class (UG).

DEFINITION 2.8.3 The point $x \in S(X)$ of the Banach space X is said to be a point of strong smoothness of S(X) (or Fréchet smoothness) if the limit in (*) is uniform in h over S(X). The Banach space X is said to be strongly smooth (or Fréchet smooth) if every point of S(X) is a point of Fréchet smoothness. We denote the class of Banach spaces which are Fréchet smooth by (F).

DEFINITION 2.8.4 We say that the Banach space X is uniformly strongly smooth (or uniformly Fréchet smooth) if the limit in (*) is uniform with respect to x and h in S(X). We denote this class of Banach spaces by (UF).

We define now the higher derivatives for norms in Banach spaces.

DEFINITION 2.8.5 We say that the norm of the Banach space X,

$$x \to \|x\|$$

is twice directionally differentiable at $x \neq 0$ if there exists a symmetric bounded bilinear mapping B_s such that for all $y \in S(X)$,

$$\|x + ty\| = \|x\| + tT(x)y + t^2 B_x(y,y) + \theta(t,x,y)$$

with

$$\lim_{t \to 0} \frac{\theta(t,x,y)}{t^2} = 0$$

where $T \in L(X)$. We say that the Banach space X is twice directionally differentiable if the norm is twice directionally differentiable at each $x \neq 0$.

DEFINITION 2.8.6 We say that the Banach space is twice smooth at $x \neq 0$ (or x is a point of twice weak smoothness) if the mapping $T : X \setminus 0 \to X^*$ [T(x) is the derivative of the norm at x] is differentiable at x, i.e., there exists a linear operator G_x such that

$$\lim_{\|h\| \to 0} \frac{1}{\|h\|} \|T(x + h) - T(x) - G_x(h)\| = 0$$

We say that the Banach space X has a twice differentiable norm if all $x \neq 0$ are twice smooth points.

DEFINITION 2.8.7 We say that the point $x \neq 0$ is a point of twofold Fréchet differentiability of the norm $x \to \|x\|$ if there exists a symmetric bounded bilinear mapping B_x such that

$$\|x + y\| = \|x\| + G(x)y + B_x(y,y) + \theta(x,h)$$

and

$$\lim_{h \to 0} \frac{\theta(x,h)}{\|h\|^2} = 0$$

We say that the Banach space X has a twice Fréchet differentiable norm if all points $x \neq 0$ are points of twofold differentiability of the norm.

DEFINITION 2.8.8 Let X be a Banach space. A mapping $x \to f_x$ defined on $X \setminus 0$ with values in $X^* \setminus 0$ is called a support mapping if the following conditions are satisfied:

1. If $\|x\| = 1$, then $\|f_x\| = 1 = f_x(x)$.
2. If $s \geq 0$ then $f_{sx} = sf_x$.

DEFINITION 2.8.9 We say that the norm of the Banach space X is weakly differentiable in the direction $y \in S(X)$ if y is a point of weak smoothness of X.

DEFINITION 2.8.10 We say that the norm of the Banach space X is uniformly weakly differentiable if the norm is uniformly Gateaux differentiable.

If X is a Banach space with dual X* for each $x \in S(X)$, we set

$$D(x) = \{x^* : x^*(x) = \|x^*\| = 1\}$$

From the extension theorems we know that this set is nonempty. This implies that the set of all support mappings is not empty.

We mention now some results on the differentiability of the norm in Banach space; our account does not pretend to be in any way complete. Apparently the first theorem on the Fréchet differentiability of the norm of a Banach space was obtained by S. Banach (1932), who proved that the norm of the Banach space $C_{[0,1]}$ is differentiable at $f \neq 0$ if and only if f attains its supremum at only one point in [0,1]. A similar result was obtained by S. Mazur (1933) for the case of $L^p_{[0,1]}(dx)$ (dx is the Lebesgue measure on [0,1]). V. L. Smulian (1939a,b,c; 1940; 1941) found interesting necessary and sufficient conditions for Gateaux and Fréchet differentiability of the norm, as well as a duality between strict convexity and Gateaux differentiability. He also studied the differentiability in some concrete Banach spaces. L. Alaoglu and G. Birkhoff (1940) proved that if X^* is G-smooth then X is strictly convex. Results concerning the existence of equivalent norms on Banach spaces with the properties G-smoothness (with respect to this new norm), (F), and (UG) were obtained by V. L. Klee, D. F. Cudia, J. Lindenstrauss, E. Asplund, J. Kurzweil, K. Sundaresan, S. L. Troyanski, M. I. Kadets, K. John, V. Zizler, J. Gilles, G. Restrepo, and J. Schaefer.

To formulate some further notions of differentiability of the norm we need the following definitions.

DEFINITION 2.8.11 We say that the Banach space X is weakly uniformly rotund (or weakly uniformly strictly convex) in the direction $y^* \in S(X^*)$ if for all $\varepsilon > 0$ there exists $\delta(y,\varepsilon) > 0$ such that for all x^*,y^* in $S(X^*)$ with $|y^*(x - y)| < \varepsilon$,

$$\|x + y\| > 2 - \delta(y^*,\varepsilon)$$

DEFINITION 2.8.12 We say that the Banach space X^* (the dual of the Banach space X) is weakly uniformly rotund (or weakly uniformly strictly convex) in the direction $y \in S(X)$ if for every $\varepsilon > 0$ there exists $\delta(y,\varepsilon) > 0$ such that for all x^*,y^* in $S(X^*)$ with $|(x^* - y^*)(y)| < \varepsilon$,

$$\|x^* + y^*\| > 2 - \delta(y,\varepsilon)$$

In what follows we give some lemmas which will enable us to prove "duality" relations between strict convexity (sometimes called *rotundity*) and smoothness, and between differentiability and smoothness, as well as some useful properties related to support mappings.

LEMMA 2.8.13 For any Banach space X the following assertions are equivalent:

1. The norm of X is weakly differentiable (i.e., Gateaux differentiable) at $x \in S(X)$ if there exists a support mapping $x \to f_x$ such that

$$\lim_{t\to 0} \operatorname{Re}\left(\frac{f_{x+ty}(y)}{\|x + ty\|} - \frac{f_x(y)}{\|x\|}\right) = 0$$

for all $y \in S(X)$.

2. The point x is a point of strong smoothness.
3. The space is uniformly strongly smooth.

Proof: Let $x \to f_x$ be a support mapping. Then for each x,y in S(X) and $t > 0$, $s > 0$, the following inequalities hold:

$$\frac{\|x + ty\| - \|x\|}{t} \geq \frac{|f_x(x + ty)| - \|x\|^2}{t\|x\|} \geq \text{Re}\ \frac{f_x(x + ty) - \|x\|^2}{t\|x\|}$$

$$\geq \text{Re}\ \frac{f_x(y)}{\|x\|} \qquad (*)$$

$$\frac{\|x + sy\| - \|x\|}{s} \leq \frac{\|x + sy\|^2 - f_{x+sy}(x)}{s\|x + sy\|}$$

$$\leq \text{Re}\ \frac{f_{x+sy}(x) + sf_{x+sy}(y) - |f_{x+sy}(x)|}{s\|x + sy\|}$$

$$\leq \text{Re}\ \frac{f_{x+sy}(y)}{s\|x + sy\|} \qquad (**)$$

If $t < 0$ and $s < 0$ then the inequalities in (*) and (**) are reversed. From these inequalities it is clear that if the support mapping $x \to f_x$ satisfies one of the properties (1), (2), or (3), then the other two are satisfied.

LEMMA 2.8.14 If the norm of the Banach space X is weakly differentiable at $x \in S(X)$ then $D(x)$ reduces to a single point. (In some papers, in this case x is called a point of smoothness or simply a smooth point.)

Proof: From the inequality (*) given above we have for all x,y in $S(X)$ and x^* in $D(x)$ the following inequalities:

$$\frac{\|x + ty\| - \|x\|}{t} \leq \text{Re}\ x^*(y) \qquad t > 0$$

$$\frac{\|x + ty\| - \|x\|}{t} \geq \text{Re}\ x^*(y) \qquad t < 0$$

Now since

$$\lim_{t\to 0} \frac{\|x + ty\| - \|x\|}{t}$$

exists, Re $x^*(y)$ is constant for all $x^* \in D(x)$, and since for each f in X^*,

$$f(u) = \text{Re}\ f(u) - i\ \text{Re}\ f(iu)$$

we obtain that $D(x)$ is a singleton.

LEMMA 2.8.15 Let X be a Banach space and suppose that for $x \in S(X)$, $D(x)$ is a singleton. Then the support mapping $x \to f_x$ is continuous at x with respect to the norm topology on X and the weak* topology on X^*.

Proof: Suppose that the assertion is false, i.e., there exists a support mapping $x \to f_x$ which is not continuous at x with the above topologies on X and X^*. In this case we find a weak* neighborhood U of f_x such that for every n there exists y_n, $\|y_n\| = 1$, $f_{y_n} \notin U$, and $\|y_n - x\| < 1/n$. By Alaoglu's theorem, $B_1(X)$ is weak* compact and thus we may suppose that (f_{y_n}) is weak* convergent to f [choosing a subsequence and renumbering if (f_{y_n}) is not weak* convergent to f]. Then we have

$$|f(x) - 1| \le |f(x) - f_{y_n}(x)| + |f_{y_n}(x) - f_{y_n}(y_n)|$$

$$\le |f(x) - f_{y_n}(x)| + \|x - y_n\|$$

and this implies that $|f(x) - 1|$ is arbitrarily small, which implies that $f(x) = 1$. Thus f is in D(x). Since D(x) is a singleton we must have $f_x = f$. But U is a weak* neighborhood of f, and thus for n sufficiently large we have that $f_x = f$ is in U_n. But this is a contradiction, and the lemma is proved.

LEMMA 2.8.16 Let X be a Banach space and suppose that the norm is strongly smooth (i.e., Frêchet differentiable) at $x \in S(X)$. Then if (f_n) is a sequence in $S(X^*)$ with the property that $\lim f_n(x) = \|x\|$, then $\lim \|f_n - f_x\| = 0$.

Proof: Since the norm is supposed to be strongly smooth at x this is the case if and only if

$$\lim_{y\to 0} \frac{\|x + y\| - \|x\| - \operatorname{Re} f_x(y)}{\|y\|} = 0$$

We suppose that the lemma is false. Thus for some $\varepsilon_0 > 0$ and a sequence (n_k) we have

$$\operatorname{Re} [f_{n_k}(y_k) - f_x(y_k)] \ge 2\varepsilon_0$$

where $y_n \in S(X)$. If $(x_n) \subset S(X)$ then we have

$$\|x + x_k\| - \|x\| - \operatorname{Re} f_x(x_k) \ge |f_{n_k}(x + x_k)| - \operatorname{Re} f_x(x + x_k)$$

$$\ge \operatorname{Re} (f_{n_k} - f_x)(x + x_k)$$

If we choose

$$\|x_k\| \le \varepsilon_0^{-1} \operatorname{Re} (f_x - f_{n_k})(x)$$

we have

$$\operatorname{Re}\,(f_{n_k} - f_x)(x + x_k) \geq \operatorname{Re}\,(f_x - f_{n_k})(x)$$

and this gives

$$\frac{\|x + x_k\| - \|x\| - \operatorname{Re} f_x(x_k)}{\|x_k\|} \geq \varepsilon_0$$

This is a contradiction and the lemma is proved.

LEMMA 2.8.17 For a Banach space X the following assertions are equivalent:

1. There exists a support mapping $x \to f_x$ which is uniformly continuous on $S(X)$ when $S(X)$ has the topology induced from the norm topology of X, and X^* has the norm topology.
2. X is strongly smooth.
3. X is uniformly smooth, i.e., for given $\varepsilon > 0$ there exists $\delta > 0$ such that for all $x \in S(X)$,

$$\|x + y\| + \|x - y\| \leq 2 + \varepsilon\|y\|$$

for all $\|y\| \leq \delta$.

4. X^* is uniformly convex.

Proof: From Lemma 2.8.13(3) we obtain that (1) → (2). We show now that (2) → (3). Indeed, if (2) holds then for each $\varepsilon > 0$ there exists $\delta > 0$ such that

$$\frac{\|x + y\| - \|x\| - \operatorname{Re} f_x(y)}{\|y\|} < \frac{\varepsilon}{2}$$

$$\frac{\|x - y\| - \|x\| + \operatorname{Re} f_x(y)}{\|y\|} < \frac{\varepsilon}{2}$$

for all $y \in S(X)$ and $\|y\| < \delta$. Then we have the inequality

$$\|x + y\| + \|x - y\| \leq 2 + \varepsilon\|y\|$$

which is (3).

Suppose now that (3) holds. We show that (4) is true. Indeed, from (3), for $\varepsilon > 0$ given we find $\delta > 0$ such that

$$\|x + y\| + \|x - y\| \leq 2 + \varepsilon\,\frac{\|y\|}{4}$$

for all y, $\|y\| \leq \delta$, and all $x \in S(X)$. Let x^*, y^* be elements in $S(X^*)$ with $\|x^* - y^*\| \geq \varepsilon$. From the definition of the norm in X^* we find y in X such that $\|y\| = \delta/2$ and $(x^* - y^*)(y) \geq \varepsilon\delta/4$. Then we have

$$\|x^* + y^*\| = \sup\{(x^* + y^*)(x) : \|x\| = 1\}$$
$$= \sup\{x^*(x + y) + y^*(x - y) - (x^* - y^*)(y) : \|x\| = 1\}$$
$$\sup\{\|x + y\| + \|x - y\| - \frac{1}{4}\varepsilon\delta : \|x\| = 1\} < 2 - \frac{\varepsilon\delta}{8}$$

Thus X^* is uniformly convex.

To complete the proof of the lemma it remains to show that (4) → (1). Let $x \to f_x$ be a support mapping. Since

$$2(\|x\|^2 + \|y\|^2) = (f_x + f_y)(x + y) + (f_x - f_y)(x - y)$$

(like a parallelogram law!) we obtain

$$4 \le \|f_x + f_y\|\|x + y\| + \|f_x - f_y\|\|x - y\|$$

Suppose now that X^* is uniformly convex. Thus for $\varepsilon > 0$ given there exists $\delta > 0$ such that for all x^*, y^* in $S(X^*)$, $\|x^* - y^*\| < \varepsilon$ when $\|x^* + y^*\| > 2 - \delta$. Now if we suppose that $\|x - y\| \le \delta$, then from the above inequality we have

$$2 \le \|f_x + f_y\| + \|x - y\| \le \|f_x + f_y\| + \delta$$

i.e.,

$$\|f_x + f_y\| > 2 - \delta$$

Then

$$\|f_x - f_y\| \le \varepsilon$$

which implies that the support mapping $x \to f_x$ is uniformly continuous. Thus (4) → (1).

We now prove a result which connects the above notions.

THEOREM 2.8.18 Let X be a Banach space. Then the following assertions hold:

1. The space is Gateaux smooth at x if and only if there exists a support mapping $x \to f_x$ which is continuous from X (with the norm topology) to X^* (with the weak* topology).
2. The point $x \in S(X)$ is strongly smooth if and only if there exists a support mapping $x \to f_x$ which is continuous from X (with the norm topology) to X^* (with the norm topology).
3. X is uniformly strongly smooth if and only if there exists a support mapping $x \to f_x$ which is uniformly continuous from X (with the norm topology) to X^* (with the norm topology).

Proof: (1) The necessity follows from Lemma 2.8.14 and the sufficiency follows from Lemma 2.8.15.

(2) If the norm is strongly differentiable at $x \in S(X)$ then it is weakly differentiable, and thus according to (1) there exists a support mapping $x \to f_x$ with the property that (f_{x_n}) is weak* convergent to f_x if (x_n) has the property that $\lim f_{x_n}(x) = \|x\|$. Then Lemma 2.8.16 asserts that $\lim \|f_{x_n} - f_x\| = 0$. This implies that the condition is necessary. The sufficiency part follows from Lemmas 2.8.15, 2.8.16, and 2.8.17.

(3) The assertion follows easily from Lemma 2.8.17.

Below we show an important duality relation between uniformly strongly smooth and uniformly convex spaces. To formulate it we need a result concerning subreflexivity.

DEFINITION 2.8.19 (E. Bishop and R. Phelps, 1961) The Banach space X is called subreflexive if the set

$$\{x^* : x^* \in X^*, x^* \text{ attains its norm on } S(X)\}$$

is dense in X^*.

The following result shows that this is an important property of Banach spaces.

THEOREM 2.8.20 (E. Bishop and R. Phelps, 1961) Every Banach space is subreflexive.

Proof: Refer to the paper of Bishop and Phelps (1961).

THEOREM 2.8.21 (Gilles-Smulian Theorem: J. R. Giles, 1971) Let X be a Banach space with the property that the norm of X^* is strongly smooth [i.e., every point of $S(X^*)$ is a point of strong smoothness]. Then $D(S) = \cup \{D(x) : x \in S(X)\} = S(X^*)$ and X is a reflexive Banach space.

Proof: Since X is subreflexive, for each x^* in $S(X^*)$ there exists a sequence (x_n^*) in $D(S)$ which converges to x^*. But the norm of X^* is strongly differentiable, and thus there exists a unique support mapping $x^* \to f_{x^*}$ (of X^* into X^{**}) and $(\hat{x}_n) = (f_{x_n^*})$ is norm convergent to f_{x^*}. Since the image of X in X^{**} $(\hat{X})$ is norm (weakly) closed we obtain that f_{x^*} is in $\hat{X}$. Thus x^* is in $D(S)$ and $D(S) = S(X^*)$. From the smoothness of X^* and the fact that the set of x^* which attain their norms on $S'(X^*)$ is $\hat{X}$, subreflexivity implies that $\hat{X}$ is dense in X^{**}. But $\hat{X}$ is closed, thus $\hat{X} = X^{**}$, and the reflexivity is proved.

The duality result announced above is the following theorem.

THEOREM 2.8.22 For a Banach space X the following assertions are equivalent:

(1) The space is (UF) if and only if X* is uniformly convex.

(2) The space is uniformly convex if and only if X* is (UF).

Proof: We remark that (1) is contained in Lemma 2.8.17 [(2) → (4)]. To prove assertion (2) we need two preliminary results given as lemmas.

LEMMA 2.8.23 Let X be a Banach space and suppose that there exists a support mapping $x^* \to f_{x^*}$ of X* into X** such that on D(S*), $f_{x^*} = \hat{x}$ with $\hat{x} \in D(x^*) \cap \{\hat{x} : \|\hat{x}\| = 1\}$, and for f_{x^*} is continuous on D(X*) when X* has the norm topology and X** has the norm topology. Then D(S) = S(X*).

Proof: Since X* is subreflexive, for any x* in S(X*) there exists a sequence (x_n^*) in D(S) converging to x*. But $\hat{x}_n = f_{x_n^*}$ and $(\hat{x}_n)$ is a Cauchy sequence which converges to an element $\hat{x} \in \hat{X}$. ($\hat{x} = m_x$ where m is the canonical embedding of X into X**.) Then we have that

$$|x^*(x) - 1| \leq |x^*(x - x_n)| + |x^*(x_n) - x_n^*(x_n)| \leq \|x - x_n\| + \|x^* - x_n^*\|$$

and thus we must have x*(x) = 1. Thus x* is in D(S) and the lemma is proved.

LEMMA 2.8.24 Suppose that X is a uniformly convex space. Then X* is uniformly strongly smooth.

Proof: Consider the support mapping

$$x^* \to f_{x^*}$$

For any u*,v* in X* we have the identity

$$2(\|u^*\|^2 + \|v^*\|^2) = (u^* + v^*)(x_u + x_v) + (u^* - v^*)(x_u - x_v)$$

This implies that

$$4 \leq \|u^* + v^*\|\|u + v\| + \|u^* - v^*\|\|u - v\|$$

and the uniform convexity of X implies that

$$x^* \to f_{x^*}$$

is uniformly continuous on D(S). Then the above lemma and Theorem 2.8.18(3) give the assertion.

Now the second assertion of Theorem 2.8.22 follows from Lemma 2.8.24 and Theorem 2.8.18(1). Since X* has a strongly differentiable norm, X** is then uniformly convex and from the fact that X is isometrically isomorphic to a closed subspace of X^{**} (the space X) the assertion of uniform convexity of X is clear.

EXAMPLE 2.8.25 We present now an example of a Banach space X which is Gateaux smooth and X^* is not strictly convex. The example was constructed by S. L. Troyanski (1970). We consider the Banach space ℓ^1 of all sequences $x = (x_i)$ with $|x| = \Sigma_1^\infty |x_i| < \infty$. We know that the dual space of ℓ^1 is m, the space of all bounded sequences $y = (y_i)$ with the norm

$$\|y\| = \sup_i |y_i|$$

We define a new equivalent norm on m. For this purpose we consider a real-valued function on $(-1,1)$ with the following properties:

$$h(t) = h(-t) \qquad h(t) < h(s) \qquad t < s$$

$$h(pt + qs) < ph(t) + qh(s) \qquad \text{all } t,s \text{ and } p,q > 0,\ p + q = 1$$

An example of such a function is $h(t) = t^2/(1 - t^2)$. Now for each $y = (y_i)$ in m and a in $\mathbb{R}$ we set

$$u(y,a) = \sum_{i=1}^{\infty} \frac{1}{2^i} h\left(\frac{y_i}{a}\right)$$

$$I(y) = \{a : a \in \mathbb{R},\ a \geq \|y\|,\ u(y,a) \leq 1\}$$

It is obvious that $I(y)$ is a closed subset of $\mathbb{R}$. On the space m we define the function

$$y \to \|y\|_1 = \min \{a : a \in I(y)\}$$

It is clear that this function satisfies the inequality $\|y\|_1 \geq \|y\|$ for all y in m. Also, it is clear from the definition of the function $y \to \|y\|_1$ that it defines a norm on m. We show that this norm is equivalent to the norm $\|\ \|$ on m. Since the inequality $\|y\|_1 \geq \|y\|$ holds for all y it remains to prove that for some fixed constant $K > 0$,

$$\|y\|_1 \leq K\|y\|$$

Let y be in m and suppose that $u(y,a) = 1$. From the definition of the function u we get that there exists an integer j such that

$$h\left(\frac{y_j}{\|y\|_1}\right) = 1$$

and thus if K is the root of the equation $h(1/t) = 1$ we have the inequality

$$\|y\|_1 \leq K\|y\|$$

Now if y is in m and u(y,a) < 1 then $\|y\|_1 = \|y\|$, and this obviously implies that the norms $\| \ \|_1$ and $\| \ \|$ are equivalent.

Now we consider the following norm on ℓ^1.

$$x \to \|x\|_1 = \sup_{y \neq 0} \frac{\Sigma_{i=1}^{\infty} y_i x_i}{\|y\|_1}$$

where $y = (y_i) \in m$. We prove now that $(m, \| \ \|_1)$ is the dual space of $(\ell^1, \| \ \|_1)$. We know that every bounded linear functional on ℓ^1 is of the form

$$y^*(x) = \sum_{i=1}^{\infty} x_i a_i$$

with $(a_i) \in m$. To prove the assertion of duality we must show that

$$\|y^*\| = \|(a_i)\|_1$$

For each n we consider the following functional on ℓ^1:

$$y_n^*(x) = y^*((x_1, \ldots, x_n, 0, 0, \ldots))$$

This is exactly

$$y_n^*(x) = x_1 a_1 + \cdots + x_n a_n$$

and the norm of y_n^* is equal to

$$\|(a_1, \ldots, a_n, 0, 0, \ldots)\|_1$$

(in m). From the fact that $\lim y_n^* = y^*$ the desired equality follows, since

$$\lim \|(a_1, \ldots, a_n, 0, 0, \ldots)\|_1 = \|(a_1, a_2, \ldots)\|_1$$

If x is in ℓ^1 with $\|x\|_1 = 1$ and $x = (x_i)$, and $y = (y_i)$ is in m with $\|y\|_1 = 1$, and

$$\Sigma\, x_i y_i = 1$$

then a(y,1) = 1. Indeed, in the contrary case, a(y,1) < 1, and since $\|y\|_1 = 1$ we have $|y_i| < 1$, $\lim\sup y_i = 1$. Then there exists an index j such that $x_j > 0$ and

$$a(y,1) + \frac{1}{2^j}[h(t) - h(x_k)] = 1 \qquad |x_k| < t < 1$$

We consider the element $y' = (y_i')$ where $y_i' = y_i$ for all $i \neq j$ and $y_j' = t$ sign y_j. Then $\|y'\|_1 = 1$, and since $x_j y_j < x_j y_j'$ we obtain

$$\sum_{i=1}^{\infty} x_i y_i' > 1$$

This contradiction proves the assertion.

To prove that $(\ell^1, \| \|_1)$ is smooth it suffices to show that we can choose a support mapping that is continuous on $S(\ell^1)$ (with the weak* topology) to m. But this follows from the definition of the norm $\| \|_1$ and the properties of h.

To show that $(m, \| \|_1)$ is not strictly convex, we consider the following elements in m:

$$y = ((-1)^i y_i) \qquad y^* = (y_i)$$

where y_i is the root of the equation

$$h(t) = \frac{i}{3}$$

Then we have

$$\|y\| = \|y^*\| = 1$$

and since h is monotone on [0,1] we obtain that for $t \geq 1$,

$$a(y,t) = a(y^*,t) = \frac{2}{3}$$

which implies that

$$\|y_1\| = \|y^*\|_1 = 1$$

and since

$$\|y + y^*\| \geq \|y + y^*\| = 2$$

we get that

$$\|y + y^*\| = 2$$

and the non-strict convexity of m (with respect to the norm $\| \|_1$) is proved. We mention that there exists a Banach space X which is strictly convex and X* is not smooth.

We now present a result which, in some sense, complements Theorem 2.8.22.

THEOREM 2.8.26 (J. Giles, 1971) For a Banach space X the following assertions are equivalent:

1. The norm of X is weakly differentiable in the direction of $y \in S(X)$ if and only if X^* is weak* uniformly rotund in the direction y.
2. The norm of X is uniformly weakly differentiable in the direction g $S(X^*)$ if and only if X is weakly uniformly rotund in the direction g.

Proof: As we have remarked above, for any duality mapping $x \to f_x$ the following identity holds:

$$2(\|x\|^2 + \|y\|^2) = (f_x + f_y)(x + y) + (f_x - f_y)(x - y)$$

and thus

$$2(\|x\|^2 + \|y\|^2) \leq \|f_x + f_y\|\|x + y\| + \|f_x - f_y\|\|x - y\|$$

Suppose now that X^* is weak* uniformly rotund in the direction $y \in S(X)$. Thus for each $\varepsilon > 0$ there exists $\delta(y,\varepsilon) > 0$ such that f_x, f_y are in $S(X^*)$ with $|(f_x - f_y)(y)| < \varepsilon$ when $\|f_x + f_y\| > 2 - \delta$, and if $\|x - y\| < \delta$ then (by the above inequality),

$$2 \leq \|f_x + f_y\| + \|x - y\| < \|f_x + f_y\| + \delta$$

and thus

$$|(f_x - f_y)(y)| < \varepsilon$$

Thus we have that X is uniformly weakly differentiable in the direction y. But if the norm of X is uniformly weakly differentiable in the direction y, $\|y\| = 1$, then for each $\varepsilon > 0$ there exists $\delta(y,\varepsilon) > 0$ such that

$$\left|\frac{\|x + ty\| - \|x\| - f_x(y)}{t}\right| < \varepsilon \qquad t > 0$$

$$\left|\frac{\|x + ty\| - \|x\| + f_x(y)}{t}\right| < \varepsilon \qquad t < 0$$

for every $x \in S(X)$ and $|t| < \delta$. This gives

$$\|x + ty\| + \|x - ty\| < 2 + 2\varepsilon t$$

for all $x \in S(X)$ and $|t| < \delta$. Thus, if x^*, y^* are in $S(X^*)$ we obtain

$$\begin{aligned}\|x^* + y^*\| &= \sup\{(x^* + y^*)(x) : \|x\| = 1\}\\ &= \sup\{x^*(x + ty) + y^*(x - ty) - (x^* - y^*)(y) : \|x\| = 1)\}\\ &\le \sup\{\|x + ty\| + \|x - ty\| - 3\varepsilon\delta : \|x\| = 1\} < 2 - \varepsilon\delta\end{aligned}$$

for all $0 < |t| < \delta$, where $|(x^* - y^*)(y)| \ge 3\varepsilon$. Thus

$$|(x^* - y^*)(y)| < \varepsilon$$

if

$$\|x^* + y^*\| > 2 - \varepsilon\delta$$

This means that X^* is weak* uniformly rotund in the direction y. In a similar way we prove the second part of the theorem.

We give now an example of a Banach space which is (G) but not (F), and the space is reflexive. The first example of such a space was constructed by V. Klee (1969); the example below seems to be simpler.

EXAMPLE 2.8.27 (V. I. Istrăţescu, 1982a) We consider the Hilbert space H with orthonormal basis $(e_i)_0^\infty$. If $x \in H$, $x = x_0e_0 + \cdots + x_ne_n + \cdots$, we set

$$\|x\|_1^2 = \max\{|x_0^2| + |x_2^2| + \cdots + |x_{2n}^2| + \cdots, |x_1^2| + \cdots + |x_{2n+1}^2| + \cdots\}$$

and it is obvious that this defines on H an equivalent norm. Further, we set

$$\|x\|^* = \left(\|x\|_1^2 + \sum_{i=1}^{\infty} \frac{1}{2^i}|x_i^2|\right)^{1/2}$$

and clearly this is again a norm on H equivalent to the original one. From the form of $\|\ \|^*$ we see that $(H, \|\ \|^*)$ is a strictly convex space. Now we show that there exists a sequence (x_n) of elements in H and an element u with the following properties:

1. $x \rightharpoonup u$.
2. $\lim \|x_n\|^* = \|u\|^*$.
3. (x_n) does not converge to u.

Clearly the existence of such a sequence implies that $(H, \|\ \|^*)$ is not (F). We take

$$x_n = e_0 + e_{2n+1} \qquad u = e_0$$

and it is easy to see that this satisfies properties 2 and 3. Since (e_n) converges weakly to zero, property 1 is also satisfied (since weak convergence does not depend on the equivalent norm used).

We present now some results on convexity and smoothness of the higher duals.

THEOREM 2.8.28 (F. Sullivan, 1977) If X is a Banach space with the property that for each x^{***} in X^{***}, $D(x^{***})$ is a singleton, then X is reflexive.

Proof: For the proof we use the very interesting characterization of reflexivity obtained by R. C. James: X is nonreflexive if and only if there exists x^* which does not attain its norm on $S(X)$.

Suppose that the theorem is false. Thus there exists x^* in X^* which does not attain its norm on $S(X)$. Since, by the Bishop-Phelps theorem, every Banach space is subreflexive, we find a sequence (x_n^*) in $S(X^*)$ (we may suppose without loss of generality that $\|x^*\| = 1$) such that $x_n^* \to x^*$ and each x_n^* attains its norm on $S(X)$. The sequence (x_n) (as a sequence in X^{**}) does not converge weakly. Indeed, in the contrary case, let $F = \lim x_n = X$, which is an element of $\hat{X}$. If $F(g) = g(x)$ for all $g \in X^*$ we have

$$
\begin{aligned}
|1 - x^*(x)| = |1 - F(x^*)| &= |x_n^*(x_n) - F(x^*)| \\
&\le |x_n^*(x_n) - x^*(x_n)| + |x^*(x_n) - F(x^*)| \\
&\le \|x_n^* - x^*\| + |x^*(x_n) - F(x^*)|
\end{aligned}
$$

which implies that $x^*(x) = 1$. Since x as an element of X^{**} is F, $\|F\| = 1$ and we have a contradiction, since by hypothesis x^* does not attain its norm. Thus $\lim x_n = F$ is not true. We remark that F attains its norm at x^*, and since $x_n^* \to x^*$, the fact that $D(x^{***})$ is a singleton implies that $x_n \to F$ in the weak* topology of X^{***}. But we noted above that this is not true. Thus we have a contradiction, and x^* attains its norm on $S(X)$, and X is then, by James' theorem, a reflexive Banach space.

COROLLARY 2.8.29 (J. Dixmier, 1948) If a Banach space has the property that X^{****} is strictly convex then X is reflexive.

We mention the following result of M. Smith.

THEOREM 2.8.30 (M. A. Smith, 1976a) There exists a Banach space X (nonreflexive) with X^{***} strictly convex.

In fact this Banach space is the famous James space with an appropriate norm. We refer to Smith's paper for details. J. R. Giles (1974) and J. Rainwater (1969) have shown that for a nonreflexive Banach space X, X^{***} is not smooth. For other results we refer to M. A. Smith (1976a,b), V. Zizler (1969), and F. Sullivan (1977).

2.9 STRICTLY CONVEXIFIABLE SPACES. THE RENORMING PROBLEM

For some problems in the theory of Banach spaces it is useful to know if, given a norm on the Banach space X, say $x \to \|x\|$, there exists an equivalent norm, say $x \to \|x\|_1$, such that $(X, \|\ \|_1)$ is a space having some special properties (for example, strict convexity on the Radon-Riesz property). The usefulness of equivalent norms with special properties may be seen from the fact that, for example, N. Aronszajn and K. T. Smith (1954) in proving their famous result that on separable Banach spaces every bounded linear and compact operator has a nontrivial invariant subspace (the von Neumann-Aronszajn-Smith theorem) used the existence of an equivalent norm with the strict convexity property. Also, equivalent norms with special properties were used (and constructed) by M. I. Kadets (1967) in the famous problem of the homeomorphism of all separable Banach spaces. In what follows we give some results concerning the existence (and the nonexistence) of equivalent norms with special properties. We mention that the first existence proof (in fact a construction) of a norm on separable Banach spaces with the strict convexity property was given by J. A. Clarkson (1936). Further interesting and important constructions were proposed by M. I. Kadets, V. Klee, E. Asplund, J. Lindenstrauss, J. Rainwater, and M. Smith, to mention only a few. First we give the following definition.

DEFINITION 2.9.1 The Banach space X, with the norm $x \to \|x\|$, is said to be convexifiable if there exists a new equivalent norm $x \to \|\ \|$ such that $(X, \|\ \|_1)$ is a strictly convex space.

The following general result is useful in proving the existence of convexifiable spaces.

THEOREM 2.9.2 (V. Klee, 1953) The Banach space X is strictly convexifiable if and only if there exists a strictly convex Banach space Y and $T \in L(X,Y)$ which is injective.

Proof: The condition is obviously necessary, since we can take $(X, \|\ \|_1)$ where the norm $\|\ \|_1$ is strictly convex and $T = I$ (the identity operator). Conversely, if there exists such Y and T then we define a norm on X by the formula

$$\|x\|_1 = \|x\| + \|Tx\|$$

which is obviously equivalent to the original norm of X. We show now that $(X, \|\ \|_1)$ is strictly convex. Indeed, if x,y are two elements in X such that

$$\|x\|_1 = \|y\|_1 \qquad \|x + y\|_1 = \|x\|_1 + \|y\|_1$$

then we obtain $\|Tx + Ty\| = \|Tx\| + \|Ty\|$ and from the strict convexity of Y, for some $c > 0$, $Tx = cTy$ or $T(x - cy) = 0$. The injectivity of T implies that $x = cy$ and the theorem is proved.

COROLLARY 2.9.3 (J. A. Clarkson, 1936) Every separable Banach space is strictly convexifiable.

Proof: Since X is separable we can choose a sequence (x_i^*) in X^* which is dense in X and (we may suppose without loss of generality that $\|x_i^*\| = 1$) we define the following norm on X:

$$\|x\|_1 = \|x\| + \|Tx\|$$

where

$$Tx = \left(\frac{x_i^*(x)}{2^i}\right)$$

Since obviously $T : X \to \ell^2$ the assertion follows from the above theorem.

REMARK 2.9.4 Another proof of this corollary may be given using the famous result of S. Banach and S. Mazur (S. Banach, 1932) which asserts that $C_{[0,1]}$ is universal for the class of separable Banach spaces [the notion of universal space for a class of spaces was introduced by the Soviet mathematician P. Urysohn (1922)], i.e., for any separable Banach space there exists a closed linear subspace in $C_{[0,1]}$ to which it is isomorphic and isometric. Now the assertion follows from the following result.

THEOREM 2.9.5 (J. A. Clarkson, 1936) The space $C_{[0,1]}$ is strictly convexifiable.

Proof: We choose a sequence of points (t_i) dense in [0,1] and we define the following function on $C_{[0,1]}$:

$$f \to \left[\|f\|^2 + \sum_{i=1}^{\infty} \frac{1}{2^i} f(t_i)^2\right]^{1/2} \qquad \|f\| = \sup \{|f(t)| : t \in [0,1]\} \quad (*)$$

Using Minkowski's inequality we obtain easily that in fact this is a norm. Since (t_i) is dense, the Hölder inequality implies that this norm is strictly convex. If we take on any separable Banach space the norm induced from the norm (*) it is clear that this norm is strictly convex.

From Klee's theorem we have the following very simple result.

PROPOSITION 2.9.6 If a Banach space X is isomorphic to a strictly convexifiable Banach space then X is strict convexifiable.

We note now the very interesting result that the class of Banach spaces which are weakly compactly generated (i.e., spaces in which there exists a weakly compact set K such that the linear span of K is dense in X) is contained in the class of strictly convexifiable spaces. The dual space of any weakly compactly generated (WCG) space is also strictly convexifiable. For a proof we refer to J. Diestel (1975). It is worth mentioning that the WCG class contains the family of reflexive Banach spaces and the family of Banach spaces which are separable [here K may be, for example, the set (x_n/n) where (x_i) is a dense subset in X].

We give now, following M. M. Day (1955), a Banach space that does not admit an equivalent strictly convex norm. For this we consider an index set I and we denote by m(I) the space of all real-valued bounded functions x on I with the norm

$$x \to \sup_{t \in I} |x(t)|$$

Clearly this is a Banach space. Further, we consider the subspace $m_0(I)$ which consists of all x in m(I) which vanish except on a countable subset of I. Clearly this is a Banach space, and if I is countable these spaces are m and c_0, respectively. In what follows we suppose that I is uncountable. Then we have the following result.

THEOREM 2.9.7 The spaces m(I) and $m_0(I)$ for I uncountable are not strictly convexifiable.

Proof: Suppose that the theorem is false and there exists on $m_0(I)$ an equivalent norm $x \to |x|$ such that $(m_0, |\ |)$ is a strictly convex space. We may suppose without loss of generality that

$$\|x\| \le |x| \le k\|x\|$$

for some constant $k \ge 1$. Let S be the unit sphere of $m_0(I)$ with respect to the norm $|\ |$. Let x have norm 1 and define the set F_x as follows:

$$F_x = \{y : y \in S,\ y(i) = x(i) \text{ at every point at which } x(i) \ne 0\}$$

We call F_x the facet determined by x. Let

$$M_x = \sup\{|y| : y \in F_x\} \qquad m_x = \inf\{|y| : y \in F_x\}$$

We show that the following inequality holds:

$$m_x + M_x \geq 2|x|$$

Let $\varepsilon > 0$ be arbitrary and take y in F_x such that $|y| < m_x + \varepsilon$. Then we have

$$|x \pm (y - x)| = 1$$

and thus

$$x \pm (y - x) \in F_x$$

But

$$\begin{aligned} 2|x| = |2x| &= |x + (y - x) + x - (y - x)| \\ &\leq |x + (y - x)| + |x - (y - x)| < m_x + \varepsilon + M_x \end{aligned}$$

and since $\varepsilon > 0$ is arbitrary the above inequality is proved.

Let x_1 be an element such that $|x_1| \geq (3k + 1)/4$. Then

$$m_{x_1} \geq 2\,\frac{3k + 1}{4} - k = \frac{k + 1}{2}$$

which implies that

$$M_{x_1} - m_{x_1} \leq \frac{k - 1}{2}$$

Further, we consider x_2 in F_{x_1} with the property that $|x_2| \geq (3M_{x_1} + |x_1|)/4$. Then

$$m_{x_2} \geq \frac{M_{x_1} + |x_1|}{2}$$

which gives

$$M_{x_2} - m_{x_2} \leq \frac{M_{x_1} - |x_1|}{2} \leq \frac{M_{x_1} - m_{x_1}}{2} \leq \frac{k - 1}{2^2}$$

Suppose now that $x_1, \ldots, x_n$ are given, and we choose x_{n+1} in the same manner as x_2 was constructed using x_1. Then we have a sequence (x_n) in $m_0(I)$ with the following properties:

$$x_{n+1} \in F_{x_n} \qquad n = 1, 2, 3, \ldots$$

$$M_{x_n} - m_{x_n} \leq \frac{k - 1}{2^n} \qquad n = 1, 2, 3, \ldots$$

From these properties we see that (M_{x_n}) and (m_{x_n}) have the same limit, say a. Thus for all $y \in F_{x_n}$ $(n = 1, 2, 3, \ldots)$,

$$m_{x_n} \le |y| \le M_{x_n}$$

which implies that $|y| = a$. We define a point $x = (x(i))$ in $m_0(I)$ as follows:

$$x(i) = \begin{cases} x_n(i) & \text{if } x_n(i) \ne 0 \text{ for some } n \\ 0 & \text{if all } x_n(i) = 0 \end{cases}$$

Clearly this element is in $m_0(I)$, and y is in F_x if and only if $y \in F_{x_n}$ for all n. Thus we must have $|x| = |y| = a$ for all y in F_x. But F_x contains many segments of length 2, and then the sphere of $(m_0(I), |\ |)$ of radius 1 contains many segments. This implies that $(m_0(I), |\ |)$ is not strictly convex. Since this obviously implies that $(m(I), |\ |)$ is not strictly convex, the theorem is proved.

REMARK 2.9.8 The fact that m(I) is not strictly convexifiable was first noted by R. S. Phillips (1943, unpublished).

Let X be a Banach space and consider the set F_X of all norms on X equivalent to a given norm. There exist simple examples of a Banach space which has a property when it is considered with a norm, say $\|\ \|_1$, and may not have that property when it is considered with another norm, say $\|\ \|_2$; of course, we suppose that both $\|\ \|_1$ and $\|\ \|_2$ are in F_X. For example, this is the case for the properties of strict convexity and uniform convexity. It seems quite natural to try to characterize those Banach spaces that are isomorphic with a uniformly convex or a strictly convex space. The solution for uniform convexity is given in the following theorem of R. James (1972) and P. Enflo (1974). To formulate the theorem we need the notion of *finite tree property*.

DEFINITION 2.9.9 Let X be a Banach space and $\varepsilon > 0$. We say that the points $(x_1, x_2) \in X$ are a $(1,\varepsilon)$ part of a tree if $\|x_1 - x_2\| \ge \varepsilon$. Suppose now that the $(n - 1, \varepsilon)$ part of a tree is defined. We say that the 2^n-tuple $(x_1, \ldots, x_{2^n})$ is an (n,ε) part of a tree if

$$\|x_{2j-1} - x_{2j}\| \ge \varepsilon \qquad j = 1, 2, \ldots, 2^{n-1}$$

and

$$\left\{\frac{1}{2}(x_1 + x_2), \frac{1}{2}(x_3 + x_4), \ldots, \frac{1}{2}(x_{2^n-1} + x_{2^n})\right\}$$

is an $(n - 1, \varepsilon)$ part of a tree. The Banach space X is said to have the finite tree property if there exists $\varepsilon > 0$ such that for all n there exists an (n,ε) part of a tree in the unit ball of X.

We say that

$$\left\{\frac{1}{2}(x_1 + x_2), \frac{1}{2}(x_2 + x_3), \ldots, \frac{1}{2}(x_{2^n-1} + x_{2^n})\right\}$$

if the first simplification of the (n,ε) part of the tree $(x_1,\ldots,x_{2^n})$;

$$\left\{\frac{1}{4}(x_1 + x_2 + x_3 + x_4), \ldots, \frac{1}{4}(x_{2^n-3} + x_{2^n-2} + x_{2^n-1} + x_{2^n})\right\}$$

is the second simplification of the (n,ε) part of the tree $(x_1,\ldots,x_{2^n})$, etc.

Now the James-Enflo result is as follows.

THEOREM 2.9.10 A Banach space X is isomorphic with a uniformly convex space iff X does not have the finite tree property.

Proof: Necessity (R. C. James, 1972d): Suppose that the contrary assertion is true. Thus for each $\varepsilon > 0$ there exist (n,ε) parts of trees in the unit ball of X. Of course, we can suppose that there exists a norm $\|\ \|^*$ in F_X such that $(X,\|\ \|^*)$ is a uniformly convex space. Thus there exist a and b such that

$$a\|x\| \leq \|x\|^* \leq b\|x\|$$

for all $x \in X$. For $\varepsilon > 0$ there exists $\delta(\varepsilon) > 0$ such that if $\|x\|^* = 1$, $\|y\|^* = 1$, and $\|x - y\|^* \geq (a/b)\varepsilon$, then $\|(1/2)(x + y)\|^* \leq 1 - \delta(\varepsilon)$, since $(X,\|\ \|^*)$ is a uniformly convex space. Thus if $\|x\|^* = 1$, $\|y\|^* = 1$, and $\|x - y\|^* \geq a\varepsilon$ then $\|x + y\| \leq 2b(1 - \delta)$. Let n be sufficiently large such that

$$b(1 - \delta)^{n-1} < \frac{a\varepsilon}{2}$$

[This clearly is possible since $1 - \delta(\varepsilon) < 1$.] Now if $(x_1,\ldots,x_{2^n})$ is the (n,ε) part of a tree we have that all x_i are in the unit ball of X with respect to the norm $\|\ \|$, and thus $\|x_i\|^* = b$. But we have

$$\|x_{2j} - x_{2j-1}\| \geq \varepsilon \qquad j = 1, 2, 3, \ldots, 2^{n-1}$$

Thus we have

$$\|x_{2j} - x_{2j-1}\| \geq a\varepsilon$$

which implies that

$$\|x_{2j} + x_{2j-1}\| = 2b(1 - \delta)$$

Let y_1 and y_2 be the points which are obtained by simplification of the (n,ε) part of the tree $n - 1$ times, i.e.,

$$y_1 = \frac{x_1 + \cdots + x_{2^{n-1}-1}}{2^n} \qquad y_2 = \frac{x_{2^{n-1}+1} + \cdots + x_{2^n}}{2^n}$$

Then (y_1,y_2) form a $(1,\varepsilon)$ part of a tree. Also, $\|y_1 - y_2\| \geq a\varepsilon$. Since

$$\|y_1\|^* \leq b(1 - \delta)^{n-1} \qquad \|y_2\|^* \leq b(1 - \delta)^{n-1}$$

we have that

$$\|y_1 - y_2\|^* \leq 2b(1 - \delta)^{n-1} < 2a\,\frac{\varepsilon}{2} = a\varepsilon$$

This is a contradiction and the necessity is proved.

Sufficiency (P. Enflo, 1974): This part of the proof is longer. We need some results that are of interest in themselves.

Suppose that X does not have finite tree property. This means that there exists $\varepsilon > 0$ and an integer n such that an (n,ε) part of a tree is not in the unit ball of X, and similarly, if $\eta > 0$, then there exists n such that no (n,ε) part of a tree is in $(1 + \eta)B(X)$ [$B(X)$ denotes the unit ball of X].

DEFINITION 2.9.11 Let z be an arbitrary fixed point of X and (x_1,x_2) be a pair of points in X. This pair is said to be a $(1,\varepsilon)$ partition of z if the following conditions hold:

$$\|x_1\| = \|x_2\| = 1 \qquad \left\|\frac{x_1}{\|x_1\|} - \frac{x_2}{\|x_2\|}\right\| \geq \varepsilon \qquad z = x_1 + x_2$$

If an $(n - 1, \varepsilon)$ partition of a z is defined, then we say that $(x_1,\ldots, x_{2^n})$ is an (n,ε) partition of z if the following conditions hold:

(1) $\|x_{2j}\| = \|x_{2j-1}\|$ $\qquad j = 1, 2, 3, \ldots, 2^{n-1}$

(2) $\left\| \frac{x_{2j}}{\|x_{2j}\|} - \frac{x_{2j-1}}{\|x_{2j-1}\|} \right\| \geq \varepsilon$

(3) $(x_1 + x_2, x_3 + x_4, \ldots, x_{2^n-1} + x_{2^n})$ is an $(n - 1, \varepsilon)$ partition of z.

Of course from (3) we get that $z = x_1 + \cdots + x_{2^n}$. If $(x_1,\ldots,x_{2^n})$ is an (n,ε) partition of z then we say that the (k,ε) partition of z,

$$(x_1 + \cdots x_{2^{n-k}}, x_{2^{n-k}} + \cdots + x_{2^{n-k+1}}, \ldots, x_{2^n-2^{n-k+1}} + \cdots + x_{2^n})$$

is the kth division of the (n,ε) partition $(x_1,\ldots,x_{2^n})$ of z.

Now we have:

LEMMA 2.9.12 Suppose that X is a Banach space not having the finite tree property. Then for all $\varepsilon > 0$ there exist n and $\delta > 0$ such that for all z and for all (n,ε) partitions $(x_1,\ldots,x_{2^n})$ of z, the following inequality holds:

$$\sum_{i=1}^{2^n} \|x_i\| \geq (1 + \delta)\|z\|$$

Proof: We may suppose without loss of generality that $\|z\| = 1$. Let $\varepsilon > 0$ and $(x_1,\ldots,x_{2^n})$ be a (n,ε) partition of z. Further, let

$$(x_1 + \cdots + x_{2^{n-1}}, x_{2^{n-1}+1} + \cdots + x_{2^n})$$

have the property that

$$\|x_1 + \cdots + x_{2^{n-1}}\| = \|x_{2^{n-1}+1} + \cdots + x_{2^n}\|$$

Since $\|z\| = 1$ we obtain

$$\|x_1 + \cdots + x_{2^{n-1}}\| = \|x_{2^{n-1}+1} + \cdots + x_{2^n}\| = \frac{1}{2}$$

and since

$$(x_1 + \cdots + x_{2^{n-1}}, x_{2^{n-1}+1} + \cdots + x_{2^n})$$

is a $(1,\varepsilon)$ partition of z we have

$$\left\| \frac{x_1 + \cdots + x_{2^{n-1}}}{\|x_1 + \cdots + x_{2^{n-1}}\|} - \frac{x_{2^{n-1}+1} + \cdots + x_{2^n}}{\|x_{2^{n-1}+1} + \cdots + x_{2^n}\|} \right\| \geq \varepsilon$$

Thus

$$\|x_1 + \cdots + x_{2^{n-1}} - x_{2^{n-1}+1} - \cdots - x_{2^n}\| \geq \frac{\varepsilon}{2}$$

This implies that the elements $2(x_1 + \cdots + x_{2^{n-1}})$ and $2(x_{2^{n-1}+1} + \cdots + x_{2^n})$ form a $(1,\varepsilon)$ part of a tree and each element has norm at least 1. We prove by induction that these results extend to the kth division of z. First we note that the assertion is true for $k = 1$. Suppose now that the assertion is true for all the divisions through $k - 1$ and consider the kth division of the partition $(x_1,\ldots,x_{2^n})$, which has the form

$$(x_1 + \cdots + x_{2^{n-k}},\ x_{2^{n-k}+1} + \cdots + x_{2^{n-k+1}},\ \ldots,\ x_{2^n - 2^{n-k}+1} + \cdots + x_{2^n})$$

Since

$$\|x_1 + \cdots + x_{2^{n-k}}\| = \|x_{2^{n-k}+1} + \cdots + x_{2^{n-k+1}}\|$$

$$\|x_1 + \cdots + x_{2^{n-k}} + x_{2^{n-k}+1} + \cdots + x_{2^{n-k+1}}\| \geq \frac{1}{2^{k-1}}$$

we obtain as above that

$$\|x_1 + \cdots + x_{2^{n-k}}\| \geq \frac{1}{2^k} \qquad \|x_{2^{n-k}+1} + \cdots + x_{2^n}\| \geq \frac{1}{2^k}$$

$$\|x_1 + \cdots + x_{2^{n-k}} - x_{2^{n-k}+1} - \cdots - x_{2^n}\| \geq \frac{\varepsilon}{2^k}$$

Similarly we prove that all the elements in the kth division have norm greater than $1/2^k$. In this case the elements obtained by multiplying by 2^k form an $(n - 1, \varepsilon)$ part and have norm at least 1. This proves the assertion. But there exists n and $\eta > 0$ such that no (n,ε) part is in $(1 + \eta)B(X)$, which implies that each (n,ε) part, obtained in the way described above, has at least one element of norm $1 + \eta$. Setting $\eta = 2^n\delta$, we get

$$2^n \sum_1^{2^n} \|x_i\| \geq 2^n + 2^n\delta = 2^n(1 + \delta)$$

and the lemma is proved.

We consider now the notion of an *écart*.

DEFINITION 2.9.13 A function $f : X \to \mathbb{R}$ is called an écart on X if the following conditions are satisfied:

1. $f(x) \geq 0$ and $f(x) = 0$ iff $x = 0$.
2. $f(ax) = |a|f(x)$ for all $a \in \mathbb{R}$ and $x \in X$.

Concerning écarts we prove the following.

THEOREM 2.9.14 If X is a Banach space which does not have the finite tree property, then for each $\varepsilon > 0$ and each n and δ as in Lemma 2.9.12, if $0 < \delta < \varepsilon < 1/8$ there exists an écart f on X such that for some $\delta_1 > 0$ the following inequalities hold:

1. $(1 - \delta)\|x\| \leq f(x) \leq (1 - \delta/3)\|x\|$.
2. If $1 = \|x\| = \|y\|$ and $\|x - y\| \geq \varepsilon$ then $f(x + y) < f(x) + f(y) - \delta_1$.

Proof: Consider an arbitrary element $x \in X$ and define

$$f(x) = \inf\left\{\sum_{i=1}^{2^m} \frac{\|x_i\|}{1 + (\delta/2)(1 + 1/4 + \cdots + 1/4^m)} : 0 \leq m \leq n,\right.$$
$$\left.(x_1, \ldots, x_{2^m}) \text{ is an } (m,\varepsilon) \text{ partition of } x\right\}$$

We show that

$$x \to f(x)$$

is an écart satisfying the lemma. First we note that it is obvious that f is an écart. We remark that if $m = 0$ then from the definition of f it follows that

$$f(x) \leq \frac{\|x\|}{1 + \delta/2} < \left(1 - \frac{\delta}{3}\right)\|x\|$$

Further, if $(x_1, \ldots, x_{2^m})$ is an (m,ε) partition of x, since $x = \Sigma_1^{2^m} x_i$, we get

$$\sum_{i=1}^{2^m} \frac{\|x_i\|}{1 + (\delta/2)(1 + 1/4 + \cdots + 1/4^m)} \geq \frac{\|x\|}{1 + (\delta/2)(1 + 1/4 + \cdots + 1/4^m)}$$
$$\geq \frac{\|x\|}{1 + \delta} > (1 - \delta)\|x\|$$

which implies that property (1) in the lemma is satisfied. We prove now that the second property is also satisfied. Let x,y be elements as in (2). Suppose that $(x_1, \ldots, x_{2^k})$ and $(y_1, \ldots, y_{2^k})$ are (k,ε) partitions of x and y, respectively. Then

$$(x_1, \ldots, x_{2^k}, y_1, \ldots, y_{2^k})$$

is a $(k + 1, \varepsilon)$ partition of $x + y$. Now let $r \in (0, \delta/4^{2n})$. From the definition of $f(x)$ and $f(y)$ there exists a (k,ε) partition of x, say $(x_1, \ldots, x_{2^k})$ and an (ℓ,ε) partition of y, say $(y_1, \ldots, y_{2^\ell})$, such that

$$f(x) > \sum_{i=1}^{2^k} \frac{\|x_i\|}{1 + (\delta/2)(1 + 1/4 + \cdots + 1/4^k)} - r$$

$$f(y) > \sum_{i=1}^{2^\ell} \frac{\|y_i\|}{1 + (\delta/2)(1 + 1/4 + \cdots + 1/4^\ell)} - r$$

According to Lemma 2.9.12, for $m \leq n$ and $(x_1, \ldots, x_{2^n})$ an (n,ε) partition of x, we obtain

$$\sum_{i=1}^{2^n} \|x_i\| \geq (1 + \delta)\|x\|$$

and thus

$$\sum_{i=1}^{2^n} \frac{\|x_i\|}{1 + (\delta/2)(1 + 1/4 + \cdots + 1/4^m)}$$
$$\geq (1 + \delta) \frac{\|x\|}{1 + (\delta/2)(1 + 1/4 + \cdots + 1/4^m)} \geq \frac{\|x\|}{1 + \delta/2}$$

which is the value for $m = 0$. Then we get that the infimum cannot be attained for $m \leq n$. Suppose that $0 \leq \ell, k \leq n - 1$ and by symmetry we may assume that $k \leq \ell$. Let $(w_{k,1}, \ldots, w_{k,2^k})$ be the kth partition of the $(1,\varepsilon)$ partition of y and set

$$A_k = 1 + \frac{\delta}{2}\left(1 + \frac{1}{4} + \cdots + \frac{1}{4^k}\right)$$

Since

$$\|x\| = 1 \leq \sum_{i=1}^{2^k} \|x_i\| \qquad \|y\| = 1 \leq \sum_{j=1}^{2^k} \|w_{k,j}\|$$

and

$$\sum_{i=1}^{2^k} \frac{\|x_i\|}{A_k} \leq f(x) + r \leq \left(1 - \frac{\delta}{3}\right) + r$$

we obtain that

$$\sum_{i=1}^{2^k} \|x_i\| \le \left(1 - \frac{\delta}{3} + r\right) A_k$$

and

$$\sum_{i=1}^{2^k} \|w_{k,j}\| \le 1 + \delta$$

From the partitions $(x_1, \ldots, x_{2^k})$ and $(w_{k,1}, \ldots, w_{k,2^k})$ of x and y, respectively, we get a $(k + 1, \varepsilon)$ partition of $x + y$,

$$(x_1, \ldots, x_{2^k}), w_{k,1}, \ldots, w_{k,2^k})$$

Again using Lemma 2.9.12 we have

$$f(x + y) \le \sum_1^{2^k} \frac{\|x_j\| + \|w_{k,j}\|}{A_k}$$

Then

$$f(y) > \sum_{j=1}^{2^\ell} \frac{\|w_{\ell,j}\|}{A_\ell} - r \ge \sum_{j=1}^{2^k} \frac{\|w_{k,j}\|}{A_\ell} - r$$

$$\ge \frac{\Sigma_{j=1}^{2^k} \|w_{k,j}\|}{1 + (\delta/2)[1 + 1/4 + \cdots + 1/4^{k+1} + 1/(3 \times 4^{k+1})]} - r$$

and thus

$$f(x) + f(y) - f(x + y) \ge \frac{\Sigma_{j=1}^{2^k} (\|x_j\| + \|w_{k,j}\|)}{1 + (\delta/2)(1 + 1/4 + \cdots + 1/4^{k+1})} - 2r$$

From the inequalities just obtained for f(x), f(y), and f(x + y) we get

$$f(x) + f(y) - f(x + y) \ge \sum_{j=1}^{2^k} \|x_j\| \left(\frac{1}{1 + \frac{\delta}{2}\left(1 + \frac{1}{4} + \cdots + \frac{1}{4^k}\right)} - \frac{1}{1 + \frac{\delta}{2}\left(1 + \frac{1}{4} + \cdots + \frac{1}{4^{k+1}}\right)} \right)$$

$$- \sum_{j=1}^{2^k} \|w_{k,j}\| \left(\frac{1}{1 + \frac{\delta}{2}\left(1 + \frac{1}{4} + \cdots + \frac{1}{4^{k+1}}\right)} - \frac{1}{1 + \frac{\delta}{2}\left(1 + \frac{1}{4} + \cdots + \frac{1}{4^{k+1}} + \frac{1}{3 \times 4^{k+1}}\right)} \right) - 2r$$

and using the fact that $\Sigma_{j=1}^{2^k} \|x_j\|$ and $\Sigma_{j=1}^{2^k} \|w_{k,j}\|$ are in the interval $[1, 1 + \delta]$ we get further

$$f(x) + f(y) - f(x + y) \geq \frac{\delta}{2} \times \frac{1}{4^{k+1}}$$

$$\times \frac{1}{\left(1 + \frac{\delta}{2}\left(1 + \frac{1}{4} + \cdots + \frac{1}{4^k}\right)\right)\left(1 + \frac{\delta}{2}\left(1 + \frac{1}{4} + \cdots + \frac{1}{4^{k+1}}\right)\right)} - (1 + \delta)\frac{\delta}{2}\frac{1}{3 \times 4^{k+1}}$$

$$\times \frac{1}{\left(1 + \frac{\delta}{2}\left(1 + \frac{1}{4} + \cdots + \frac{1}{4^k}\right)\right)\left(1 + \frac{\delta}{2}\left(1 + \frac{1}{4} + \cdots + \frac{1}{4^{k+1}} + \frac{1}{3 \times 4^{k+1}}\right)\right)} - 2r$$

But $r < \delta/4^{2n}$ and thus

$$f(x) + f(y) - f(x + y) \geq \delta\left(\frac{1}{2 \times 4^{k+1}} - \frac{2}{4^{2n}}\right) \geq \frac{\delta}{2 \times 4^{k+1}}\left(1 - \frac{4}{4^n}\right)$$

$$\geq \frac{\delta}{2 \times 4^{k+3}}$$

and finally (since $0 < \delta < 1/8$),

$$f(x) + f(y) - f(x + y) \geq \frac{\delta}{2 \times 4^{n+2}}$$

and the lemma is proved.

LEMMA 2.9.15 Let X be a Banach space without the finite tree property. If $\varepsilon > 0$ and δ are as in Lemma 2.9.14 and if $0 < \delta < \varepsilon/(1 + \varepsilon) < 1/8$ then there exists a norm $\| \ \|_{5\varepsilon}$ on X such that:

(a) $(1 - \delta)\|x\| \leq \|x\|_{5\varepsilon} \leq (1 - \delta/3)\|x\|$.

(b) If $\|x\| = \|y\| = 1$ and $\|x - y\| \geq 5\varepsilon$ then

$$\|x + y\|_{5\varepsilon} \leq \|x\|_{5\varepsilon} + \|y\|_{5\varepsilon} - \varepsilon\delta_1$$

where δ_1 is as in Lemma 2.9.14.

Proof: Let f be the écart constructed above and define the norm $\| \ \|_{5\varepsilon}$ as follows: for each $x \in X$,

$$\|x\|_{5\varepsilon} = \inf \left\{ \sum_1^n f(x_i) \; : \; x = \sum_1^n x_i \right\}$$

Clearly this is a gauge for $\{z : f(z) \le 1\}$. Indeed, if $0 < \|x\|_{5\varepsilon} < 1$ then x is of the form $x = \Sigma_1^m x_i$, with $0 < f(x_i) < 1$. Then $x = \Sigma_1^m r_i y_i$ where

$$r_i = \frac{f(x_i)}{\Sigma_1^m f(x_i)} \qquad y_i = \frac{x_i}{r_i}$$

and obviously $r_i \ge 0$, $\Sigma r_i = 1$, $y_i \in \text{conv}\{z: f(z) \le 1\}$, where conv denotes convex hull. Thus x is in $\text{conv}\{z : f(z) \le 1\}$ and

$$\|x\|_{5\varepsilon} \le \sum_1^m f(r_i y_i) \le 1$$

Now from the relation for the écart f we obtain further that

$$\frac{1}{1 - \delta/3}\{z : \|z\| \le 1\} \subseteq \{z : f(z) \le 1\} \subseteq \frac{1}{1 - \delta}\{z : \|z\| \le 1\}$$

and thus property (a) is satisfied.

For the second property we consider x and y as in the lemma and let $a > 0$ such that $0 < a < \min\{\delta(1/3 - \delta), \delta_1\varepsilon/4\}$. By the definition of the norm $\| \ \|_{5\varepsilon}$ there exists a decomposition

$$x = \sum_1^n x_i \qquad y = \sum_1^m y_j$$

such that

$$\|x\|_{5\varepsilon} > \Sigma f(x_i) - a \qquad \|y\|_{5\varepsilon} > \Sigma f(y_j) - a$$

We may assume without loss of generality that $m \le n$ and that $\|x_i\| = \|y_i\|$ for $i = 1, 2, 3, \ldots, m$. Indeed, if we have $\|x_1\| < \|y_1\|$ then we can write $y_1 = sy_1 + (1 - s)y_1$ with $s = \|x_1\|/\|y_1\|$ and then $\|x_1\| = \|sy_1\|$. But $f(y_1) = f(sy_1) + f((1 - s)y_1)$ and in the decomposition of y we may replace y_1 by two elements, sy_1 and $(1 - s)y_1$. Of course, we can proceed in this manner for all $i = 1, 2, 3, \ldots, m$, and renumber the elements.

We note that we have

$$1 \le \sum_1^n \|x_i\| < 1 + \delta \qquad 1 = \sum \|y_j\| < 1 + \delta$$

Indeed,

$$1 \le \sum \|x_i\| \le \frac{1}{1 - \delta} \sum f(x_i) \le \frac{1}{1 - \delta} (\|x\|_{5\varepsilon} + a)$$
$$\le \frac{1}{1 - \delta}\left[\left(1 - \frac{\delta}{3}\right)\|x\| + a\right] < \frac{1}{1 - \delta}\left(1 - \frac{\delta}{3} + \frac{\delta}{3} - \delta^2\right) = 1 + \delta$$

Similarly we see that the relation on $\sum \|y_j\|$ is true. Since $\|x_i\| = \|y_i\|$, $i = 1, 2, 3, \ldots, m$, we obtain that

$$\sum_{m+1}^n \|x_i\| < \delta \tag{*}$$

Now we consider the sets

$$\{1,2,3,\ldots,m\} = \{i : 1 \le i \le m, \|x_i - y_i\| < \varepsilon\|x_i\|\}$$
$$\cup \{i : 1 \le i \le m, \|x_i - y_i\| > \varepsilon\|x_i\|\} = J_1 \cup J_2$$

and (*) implies that

$$5\varepsilon \le \|x - y\| = \sum_{i\in J_1} \|x_i - y_i\| + \sum_{i\in J_2} \|x_i - y_i\| + \sum_{m+1}^n \|x_i\|$$
$$\le \varepsilon \sum_{i\in J_1} \|x_i\| + \sum_{i\in J_2} \|x_i - y_i\| + \delta$$

But $\delta < \varepsilon/(1 + \varepsilon)$ and thus

$$\sum_{i\in J_2} \|x_i - y_i\| \ge 5\varepsilon - \varepsilon \sum_1^n \|x_i\| - \delta > 5\varepsilon - \varepsilon(1 + \delta) - \delta > 3\varepsilon$$

Further,

$$x + y = \sum_{i\in J_1} (x_i + y_i) + \sum_{i\notin J_2} x_i + \sum_{i\notin J_2} y_i$$

and then

$$\|x + y\|_{5\varepsilon} \le \sum_{i\in J_1} f(x_i + y_i) + \sum_{i\notin J_2} f(x_i) + \sum_{i\notin J_2} f(y_i)$$

Thus we have

$$\|x\|_{5\varepsilon} + \|y\|_{5\varepsilon} - \|x + y\|_{5\varepsilon} \geq \sum_1^m f(x_i) - a + \sum_1^m f(y_i) - a$$

$$- \sum_{i \in J_2} f(x_i + y_i) - \sum_{i \notin J_2} f(x_i) - \sum_{i \notin J_2} f(y_i)$$

$$= \sum_{i \in J_2} [f(x_i) + f(y_i) - f(x_i + y_i)] - 2a$$

and since $\|x_i\| = \|y_i\|$ $(i = 1, 2, 3, \ldots, m)$, for i in J_2 we have that

$$\left\| \frac{x_i}{\|x_i\|} - \frac{y_i}{\|y_i\|} \right\| = \frac{\varepsilon}{\|x_i\|} \geq \varepsilon$$

which implies that for the écart f we have the relations

$$f\left(\frac{x_i}{\|x_i\|} - \frac{y_i}{\|y_i\|}\right) \leq f\left(\frac{x_i}{\|x_i\|}\right) + f\left(\frac{y_i}{\|y_i\|}\right) - \delta_1$$

Then we get

$$f(x_i) + f(y_i) - f(x_i + y_i) > \delta_1 \|x_i\|$$

for all $i \in J_2$. This implies that

$$\|x\|_{5\varepsilon} + \|y\|_{5\varepsilon} - \|x + y\|_{5\varepsilon} \geq \delta_1 \sum_{i \in J_2} \|x_i\| - 2a$$

From the inequality proved above we obtain

$$3\varepsilon \leq \sum_{i \in J_2} \|x_i - y_i\| \leq \sum_{i \in J_2} (\|x_i\| + \|y_i\|) \leq 2 \sum_{i \in J_2} \|x_i\|$$

and thus

$$\sum_{i \in J_2} \|x_i\| \geq \frac{3}{2} \varepsilon$$

But $a < \delta_1 \varepsilon / 4$ and thus

$$\|x\|_{5\varepsilon} + \|y\|_{5\varepsilon} - \|x + y\|_{5\varepsilon} > \frac{3}{2} \varepsilon \delta_1 - 2a > \delta_1 \varepsilon$$

and the lemma is proved.

LEMMA 2.9.16 Let X be a Banach space with the norm $\| \ \|$ and suppose that there exists a new norm $\| \ \|_1$ on X such that:

1. $\|x\| \le \|x\| \le M\|x\|_1$

2. If $\|x\| = \|y\| = 1$ and $\|x - y\| \ge \varepsilon$, then $\|x + y\|_1 \le \|x\|_1 + \|y\|_1 - \delta(\varepsilon)$ for a function $\delta(\varepsilon) > 0$, $\varepsilon > 0$.

Then $(X, \| \|_1)$ is uniformly convex.

Proof: Let $a \in [0, 1 - \varepsilon/2]$ or $a \ge 1 + \varepsilon/2$. Then we have

$$\|ax - y\|_1 = |\|ax\|_1 - \|y\|_1| \ge \varepsilon - \frac{\varepsilon}{2} = \frac{\varepsilon}{2}$$

If $|a - 1| \le \varepsilon/2$ then

$$\|ax - y\|_1 = |\|x - y\|_1 - |a - 1|\|x\|_1| \ge \varepsilon - \frac{\varepsilon}{2} = \frac{\varepsilon}{2}$$

Finally, if $a < 0$,

$$a \to \|ax - y\| = h(a)$$

is obviously a convex function with $h(0) = 1$ and $h(1) < 1$. Now if $\|x\|_1 = 1$, $\|y\|_1 = 1$, and $1 \ge \|x - y\|_1 \ge \varepsilon$, there exist real numbers α and β such that:

$$\frac{1}{M} \le \alpha \le 1 \qquad \frac{1}{M} \le \beta \le 1$$

$$\|\alpha x\| = 1 \qquad \|\beta y\| = 1$$

Then we have

$$\|ax - \beta y\| \ge \|ax - \beta y\|_1 = |\beta|\|\frac{a}{\beta} x - y\|_1 \ge \frac{1}{M}\frac{\varepsilon}{2}$$

and according to the hypothesis,

$$\|ax + \beta y\|_1 \le a + \beta - \delta \frac{\varepsilon}{2M}$$

Thus

$$\|x + y\| \le \|\alpha x + \beta y\|_1 + \|(1 - \alpha)x + (1 - \beta)y\|_1 \le \alpha + \beta - \delta \frac{\varepsilon}{2M}$$
$$+ 1 - \alpha + 1 - \beta$$
$$= 2 - \delta \frac{\varepsilon}{2M}$$

and the lemma is proved.

We are ready to complete the proof of Theorem 2.9.10. We consider the following norm on X:

$$x \to |||x||| = \sum_1^\infty \frac{1}{2^n} \|x\| \frac{\varepsilon}{2^{n-1}}$$

This is obviously equivalent to the original norm of X. Suppose now that $\|x\| = \|y\| = 1$ and $\|x - y\| \geq \varepsilon_1$. Then $\varepsilon_1 > \varepsilon/2^k$ for large k and

$$|||x + y||| = \sum_{i=0}^{\infty} \frac{1}{2^{i+1}} \|x + y\| \frac{\varepsilon}{2^i} = \sum_{i \neq k} \frac{1}{2^{i+1}} \|x + y\| \frac{\varepsilon}{2^i}$$

$$+ \frac{1}{2^{k+1}} \|x + y\| \frac{\varepsilon}{2^k}$$

$$\leq \sum_{i \neq k} \frac{1}{2^{i+1}} \left(\|x\| \frac{\varepsilon}{2^i} + \|y\| \frac{\varepsilon}{2^i}\right) + \frac{1}{2^{k+1}}\left(\|x\| \frac{\varepsilon}{2^k} + \|y\| \frac{\varepsilon}{2^k}\right)$$

$$- \frac{\varepsilon}{2^k} + \delta_1'$$

where δ_1' is the δ given by Lemma 2.9.14 for $\varepsilon/2^k$. Thus we have

$$|||x + y||| \leq \sum_i \frac{1}{2^i}\left(\|x\| \frac{\varepsilon}{2^i} + \|y\| \frac{\varepsilon}{2^i}\right) - \frac{\varepsilon}{2^{2k+1}} \delta_1'$$

or

$$|||x + y||| \leq |||x||| + |||y||| - \frac{\varepsilon}{2^{2k+1}} \delta_1'$$

and according to Lemma 2.9.16, $(X, ||| \; |||)$ is uniformly convex, and the proof is complete.

We now consider the class of Banach spaces with basis. We suppose that this basis has some special properties and we prove that the Banach space is isomorphic with a Banach space with is LUC or WLUC (i.e., weakly locally uniformly convex).

A basis $(x_i)_1^\infty$ in a Banach space X is a sequence of elements with the property that for each $x \in X$ there exists a unique sequence of scalars $(a_i)_1$ such that

$$x = \lim_n \sum_{i=1}^{n} a_i x_i$$

DEFINITION 2.9.17 (A. R. Lovaglia, 1955) The basis (x_i) is said to have property (A) if for each number $c > 0$ there exists a number $r_c > 0$ such that

(A) If $\|\Sigma_{i=1}^n a_i x_i\| = 1$ and $\|\Sigma_{i=n+1}^\infty a_i x_i\| = c$ then $\|\Sigma_{i=1}^\infty a_i x_i\| \geq 1 + r_c$.

EXAMPLE 2.9.18 (A. R. Lovaglia, 1955) The space ℓ^1 has the (A) property.

Let X be a Banach space with the basis (x_i) and for each n let X_n be the closed linear span of $(x_{n+1}, x_{n+2}, \ldots)$.

DEFINITION 2.9.19 A Banach space X with basis (x_i) is said to have property $(\bar{A})$ if for any $x^* \in X^*$,

$$\lim \|x^*\|/X_n = 0$$

where

$$\|x^*\|/X_n = \sup_{\substack{\|x\|=1 \\ x \in X_n}} |x^*(x)|$$

EXAMPLE 2.9.20 (A. R. Lovaglia, 1955) The space c_0 which we consider is (e_i), where

$$e_i = (0,0,\ldots,0,1,0,0,\ldots) \qquad 1 \text{ in the ith place}$$

It is well known that the dual of c_0 is ℓ^1. Let $x^* \in c_0^*$, i.e.,

$$x^* = (b_1, b_2, b_3, \ldots)$$

with

$$\Sigma\, |b_i| < \infty$$

Now, since

$$\|x^*\|/X_n = \sum_{i=n+1}^{\infty} |b_i|$$

it is clear that property $(\bar{A})$ holds.

THEOREM 2.9.21 (A. R. Lovaglia, 1955) If X is a Banach space with basis $(x_i)_1^\infty$ having the property (A) then there exists an equivalent norm $\|\ \|^*$ on X such that $(X, \|\ \|^*)$ is a locally uniformly convex space.

Proof: Let $x \in X$, $x = \Sigma_1^\infty a_i x_i$ and set

$$x \to \|x\|^* = \left\{\|x\|^2 + \sum \frac{1}{2^{2i}} |a_i|^2\right\}^{1/2}$$

Then it is obvious that this is an equivalent norm on X. Now we prove that $(X, \|\ \|^*)$ is a locally uniformly convex space. Indeed, consider an arbitrary point $x \in X$, $\|x\|^* = 1$, and a sequence $(y_m)_1^\infty$ of points in X with the following properties:

$$\lim \|y_m\|^* = 1$$

$$\lim \|y_m + x\|^* = 2$$

We prove that $\lim (y_m - x) = 0$. Suppose that this is not true and we show that this leads to a contradiction. Since (y_m) does not converge to x there exists $\varepsilon > 0$ and a subsequence (m_k) such that

$$\|y_{m_k} - x\|^* \geq \varepsilon$$

It is not difficult to see that we can suppose that the sequence (m_k) [or a subsequence of (m_k), which we may suppose to be (m_k)] satisfies the following properties:

$$\lim \|y_{m_k}\|^* = \|x^*\|$$

$$\lim b_i^{m_k} = a_i \qquad i = 1, 2, 3, \ldots$$

where

$$y_m = \sum_1^\infty b_i^m x_i$$

Since $\|\ \|$ and $\|\ \|^*$ are equivalent norms we have that

$$\|y_{m_k} - x\| \geq \varepsilon_1 > 0$$

for all k. In this case for a subsequence of (m_k), say (m_k^*), we have that

$$\varepsilon_1 < \|x - y_{m_k^*}\| = \left\|\sum_1^n (a_i - b^{m_k^*})x_i + \sum_{n+1}^\infty a_i x_i - \sum_{n+1}^\infty b_i^{m_k^*} x_i\right\|$$

$$\leq \left\|\sum_1^n (a_i - b^{m_k^*})x_i\right\| + \left\|\sum_{n+1}^\infty a_i x_i\right\| + \left\|\sum_{n+1}^\infty b^{m_k^*} x_i\right\|$$

and thus

$$\left\|\sum_{n+1}^\infty b_i^{m_k^*} x_i\right\| > \varepsilon_1 - \left[\left\|\sum_1^n (a_i - b_i^{m_k^*})x_i\right\| + \left\|\sum_{n+1}^\infty a_i x_i\right\|\right]$$

which implies that

$$\liminf_{m_k^*} \left\|\sum_{n+1}^\infty b_i^{m_k^*} x_i\right\| \geq \varepsilon - \left\|\sum_{i=n+1}^\infty a_i x_i\right\|$$

If N_0 is such that for all $n \geq N_0$, $\varepsilon_1 - \|\Sigma_{n+1}^\infty a_i x_i\| \geq u > 0$ we obtain that

$$\liminf_{m_k^*} \|b_i^{m_k^*} x_i\| \geq u > 0$$

Suppose now that $\|x\| < 1$ and thus we find m_0 such that

$$\left\|\sum_1^{m_0} a_i x_i\right\| < 1$$

Hence for all $n \geq \max N_0, m_0$ we have the inequality

$$\liminf_{m_k^*} \frac{\|b_i^{m_k^*} x_i\|}{\|\Sigma_1^n a_i x_i\|} \geq \frac{u}{\|\Sigma_1^n a_i x_i\|} \geq u > 0$$

Thus for fixed n there exists a subsequence of (m_k^*), say (m_k^{**}), with the property that

$$\left\|\frac{\Sigma_{n+1}^{\infty} b_i^{m_k^{**}} x_i}{\|\Sigma_1^n a_i x_i\|}\right\| \geq u \qquad \left\|\frac{\Sigma_1^n b_i^{m_k^{**}} x_i}{\|\Sigma_1^n b_i^{m_k^{**}}\|}\right\| = 1$$

Since the basis (x_i) has property (A) we find $r_u > 0$ such that

$$\left\|\frac{\Sigma_1^n b_i^{m_k^{**}} x_i}{\|\Sigma_1^n b_i^{m_k^{**}} x_i\|} + \frac{\Sigma_{n+1}^{\infty} b_i^{m_k^{**}} x_i}{\|\Sigma_1^n a_i x_i\|}\right\| = 1 + r_u$$

This implies that

$$\liminf_{m_k^{**}} \frac{\Sigma_1^n b_i^{m_k^{**}} x_i}{\|\Sigma_1^n b_i^{m_k^{**}} x_i\|} + \frac{\Sigma_{n+1}^{\infty} b_i^{m_k^{**}} x_i}{\|\Sigma_1^n a_i x_i\|} = \frac{1}{\|\Sigma_1^n a_i x_i\|} \liminf_{m_k^{**}} \left\|\sum_1^{\infty} b_i^{m_k^{**}} x_i\right\|$$

$$\geq 1 + r_u$$

or that

$$\liminf_{m_k^{**}} \|y_{m_k^{**}}\| \geq \left\|\sum_1^n a_i x_i\right\| (1 + r_u)$$

But

$$\liminf_{m_k^{**}} \|y_{m_k^{**}}\| = \lim \|y_{m_k^{**}}\| = \|x\|$$

and thus

$$\|x\| \geq \left\|\sum_1^n a_i x_i\right\|(1 + r_u)$$

For $n \to \infty$ we get the inequality

$$\|x\| \geq \|x\|(1 + r_u)$$

which is clearly a contradiction, and the theorem is proved.

Now we give a result concerning the class of Banach spaces with basis with the property $(\bar{A})$.

DEFINITION 2.9.22 (A. R. Lovaglia, 1955) A Banach space X is said to be weakly locally uniformly convex (WLUC) if whenever $\|x\| = 1$ and $\lim \|x_n + x\| = 2$ for $\|x_n\| = 1$, then for any x^* in X^*, $\lim x^*(x_n) = x^*(x)$.

REMARK 2.9.23 It is obvious that any locally uniformly convex space is WLUC.

THEOREM 2.9.24 (A. R. Lovaglia, 1955) If X is a Banach space with basis $(x_i)_1^\infty$ having the property $(\bar{A})$ then there exists an equivalent norm $\| \ \|^*$ on X such that $(X, \| \ \|^*)$ is WLUC.

Proof: If $x \in X$ is of the form

$$x = \sum_1^\infty a_i x_i$$

we set

$$\|x\|^* = \left(\|x\|^2 + \frac{\Sigma_1^\infty |a_i|^2}{2^{2i}}\right)^{1/2}$$

It is clear that this is an equivalent norm on X and we show that $\| \ \|^*$ satisfies the theorem. Let $\|x\|^* = 1$, $\|(y_n)\|^* = 1$, and $\lim \|x + y_n\|^* = 2$. We show now that for each $x^* \in X^*$, $\lim x^*(y_n) = x^*(x)$. Suppose that the assertion is false. Indeed, there exists then a subsequence (m_k) and $\varepsilon > 0$ such that

$$|x^*(y_{m_k}) - x^*(x)| \geq \varepsilon > 0$$

We can suppose without loss of generality that for a subsequence (m_k^*) we have the relations:

$$\lim \|y_{m_k^*}\| = \|x\|$$

$$\lim b_i^{m_k^*} = a_i \qquad i = 1, 2, 3, \ldots$$

This implies that

$$\left\|\sum_{i=n+1}^{\infty} b_i^{m_k^*} x_i\right\| \le \|y_{m_k^*}\| + \left\|\sum_{i=1}^{n} b_i^{m_k^*} x_i\right\|$$

and thus

$$\limsup_{m_k^*} \left\|\sum_{i=n+1}^{\infty} b_k^{m_k^*} x_i\right\| \le \|x\| + \left\|\sum_{i=1}^{n} a_i x_i\right\|$$

But

$$f(y_{m_k^*}) = f\left(\sum_1^n b_i^{m_k^*} x_i\right) + f\left(\sum_{n+1}^{\infty} b_i^{m_k^*} x_i\right)$$

and thus

$$\left|f\left(\sum_{n+1}^{\infty} b_i^{m_k^*} x_i\right)\right| \le \|f\|_{n+1}\left\|\sum_{n+1}^{\infty} b_i^{m_k^*} x_i\right\|$$

where $\|f\|_m = \sup\{\|f(x)\| : \|x\| = 1,\ x \in \operatorname{span}(x_1, x_2, \ldots, x_m)\}$. Then we have

$$\limsup_{m_k^*} \left|f\left(\sum_{n+1}^{\infty} b_i^{m_k^*} x_i\right)\right| \le \|f\|_{n+1}\left(\|x\| + \left\|\sum_1^n a_i x_i\right\|\right)$$

and

$$\begin{aligned} |f(y_{m_k^*}) - f(x)| = |f(y_{m_k^*} - x)| &= \left|f\left(\sum_1^n (b_i^{m_k^*} - a_i)\right)x_i \right. \\ &\quad \left. + f\left(\sum_{n+1}^{\infty} b_i^{m_k^*} x_i\right) - f\left(\sum_{n+1}^{\infty} a_i x_i\right)\right| \\ &\le \left|f\left(\sum_1^n (b_i^{m_k^*} - a_i) x_i\right)\right| + \left|f\left(\sum_{n+1}^{\infty} b_i^{m_k^*} x_i\right)\right| + \left|f\left(\sum_{n+1}^{\infty} a_i x_i\right)\right| \end{aligned}$$

which implies that

$$\limsup_{m_k^*} |f(y_{m_k^*} - x)| = \|f\|_{n+1}\ \|x\| + \left\|\sum_1^n a_i x_i\right\| + \|f\|_{n+1}\left\|\sum_{n+1}^{\infty} a_i x_i\right\|$$

Since n is arbitrary and $\lim \|\Sigma_1^n a_i x_i\| = \|x\|$, $\lim \|\Sigma_{n+1}^{\infty} a_i x_i\| = 0$, $\|f\|_{n+1} \to 0$, we obtain

$$\lim |f(y_{m_k^*} - x)| = 0$$

which is a contradiction, and the theorem is proved.

COROLLARY 2.9.25 The space c_0 is isomorphic to a weakly locally uniformly convex space.

REMARK 2.9.26 The space c_0 with the sup norm is not weakly locally uniformly convex.

Proof: Indeed, let $x = (1,1,0,0,0,\ldots)$, $\|x\| = 1$, and $y = (0,1,0,0,0,\ldots)$. If f is in $c_0^* = \ell^1$, $f = (1,0,0,0,\ldots)$, then we have that $\|f\| = 1 = f(x)$ and $f(y) = 0$ while $\|x + y\| = 2$. This shows that c_0 is not weakly locally uniformly convex.

Let Γ be an arbitrary set and denote by $\ell^{\infty}(\Gamma) = m(\Gamma)$ the set of all bounded functions defined on Γ. The subspace $c_0(\Gamma)$ consists of all those functions in $\ell^{\infty}(\Gamma)$ for which $\{i : |x(i)| \geq \varepsilon\}$ is finite for all $\varepsilon > 0$. If $\Gamma = \mathbb{N}$ then we write simply ℓ^{∞} or m and c_0, respectively. In what follows we give an important result about the possible equivalent norms on $c_0(\Gamma)$, namely, that there exists an equivalent norm $\|\ \|^*$ such that $(c_0(\Gamma), \|\ \|^*)$ is locally uniformly convex.

THEOREM 2.9.27 (J. Rainwater, 1969) For any set Γ there exists an equivalent norm $\|\ \|^*$ on $c_0(\Gamma)$ such that $(c_0(\Gamma), \|\ \|^*)$ is locally uniformly convex.

Proof: The norm $\|\ \|^*$ which we define now was first introduced by M. M. Day (1955) and uses a nondecreasing rearrangement of terms x(i). Let $x = (x(i))_{i \in \Gamma}$ be an element in $c_0(\Gamma)$ and set supp $x = \{i : x(i) \neq 0\}$. We can enumerate the support supp x so that

$$|x(i_k)| \geq |x(i_{k+1})|$$

for all k.

Day's norm $\|x\|^*$ of x is defined as follows. For the element x let $Dx = (u(r))$ with

$$u(r) = \begin{cases} 2^{-s} x(i_s) & \text{if } r = i_s \in \text{supp } x \\ 0 & \text{otherwise} \end{cases}$$

and

$$\|x\|^* = \|Dx\|_{\ell^2(\Gamma)}$$

We have clearly that $\|x\|^* = 0$ iff $x = 0$ and $\|ax\|^* = |a|\|x\|^*$. We prove now that

$$x \to \|x\|^*$$

is in fact an equivalent norm on $c_0(\Gamma)$. Since $|x(i_1)| = \|x\|_{c_0(\Gamma)}$ we have that

$$\frac{\|x\|_{c_0(\Gamma)}}{2} \le \|x\|^* = \frac{\|x\|_{c_0(\Gamma)}}{3^{1/2}}$$

It remains only to show that the triangle inequality is satisfied, i.e., for any x,y in $c_0(\Gamma)$,

$$\|x + y\|^* \le \|x\|^* + \|y\|^*$$

For this we need the following identity: if π is a permutation of the set of positive integers, then for any nonincreasing sequences of nonnegative numbers (a_n) and (b_n), we have

$$\Sigma\, a_i b_i - \Sigma\, a_i b_{\pi(i)} = \Sigma\, (a_i - a_{i+1})[b_1 + \cdots + b_i - b_{\pi(1)} - \cdots - b_{\pi(i)}]$$

From this we have the following lemmas.

LEMMA 2.9.28 The following inequality holds:

$$\Sigma\, a_i b_i \le \Sigma\, a_i b_{\pi(i)}$$

where the sequences $(a_n),(b_n)$ are as above.

Proof: Since for each m,

$$\sum_1^m b_{\pi(i)} \le \sum_1^m b_i$$

(b_n) being nonincreasing, we get

$$0 \le \sum (a_i - a_{i+1})\left[\sum_1^m b_i - \sum_1^m b_{\pi(i)}\right] = \sum_i a_i b_i - \sum_i a_i b_{\pi(i)}$$

and the assertion follows.

LEMMA 2.9.29 We have either

$$(a_m - a_{m+1})(b_m - b_{m+1}) = \sum_{i=1}^m a_i b_i - \sum_{i=1}^m b_{\pi(i)}$$

for each m = 1, 2, 3, ..., or π permutes $\{1,2,3,\ldots,i\}$ onto itself for some i.

Proof: Since $\{\pi(1),\ldots,\pi(i)\}$ and $\{1,2,3,\ldots,i\}$ are different for all i, we obtain that

$$\sum_1^m b_{\pi(i)} \le b_1 + b_2 + \cdots + b_{m-1} + b_{m+1}$$

and since

$$b_1 + \cdots + b_{m-1} + b_m - b_1 - \cdots - b_{m-1} - b_{m+1} \le \sum_1^m b_i - \sum_1^m b_{\pi(i)}$$

we get

$$\begin{aligned}(a_m - a_{m+1})(b_m - b_{m+1}) &= (a_m - a_{m+1})(b_1 + \cdots + b_{m-1} + b_m \\ &\quad - b_1 - \cdots - b_{m-1} - b_{m+1}) \\ &\le (a_m - a_{m+1})\left[\sum_1^m b_i - \sum_1^m b_{\pi(i)}\right] \\ &\le \sum_i (a_i - a_{i+1})\left[\sum_1^m b_j - \sum b_{\pi(j)}\right] \\ &< \sum_i a_i b_i - \sum_i a_i b_{\pi(i)}\end{aligned}$$

We can now complete the proof of the theorem. Consider now $x \in c_0(\Gamma)$ and define

$$a_i = 4^{-i} \qquad b_i = |x(\alpha_i)|^2$$

where the α_i are in the support of x, supp x. Then we have

$$\Sigma_i\ 4^{-i}|x(\alpha_i)|^2 = \Sigma_i\ [2^{-i}|x(\alpha_i)|]^2 = \|Dx\|_{\ell^2(\Gamma)}^2 = \|x\|^{*2}$$

Let x, y be in $c_0(\Gamma)$ and supp x = (α_i), supp y = (β_i), supp (x + y) = (γ_i). Then we have

$$\begin{aligned}\|x + y\|^* = [\Sigma\ 4^{-i}|(x + y)(\gamma_i)|^2]^{1/2} &\le [\Sigma\ 4^{-i}|x(\gamma_i)|^2 + \Sigma\ 4^{-i}|y(\gamma_i)| \\ &\quad + \Sigma\ 4^{-i}|y(\gamma_i)|^2]^{1/2} \\ &\le [\Sigma\ 4^{-i}|x(\alpha_i)|^2]^{1/2} + [\Sigma\ 4^{-i}|y(\beta_i)|^2]^{1/2} = \|x\|^* + \|y\|^*\end{aligned}$$

and thus * is a norm equivalent with the original norm of $c_0(\Gamma)$.

Now we prove that $(c_0(\Gamma), \|\ \|^*)$ is a locally uniformly convex space. Suppose that $x \in c_0(\Gamma)$ and $\|x\|^* = 1$. Suppose further that we have a sequence (x_n) such that $\|x_n\|^* = 1$ and $\|x + x_n\|^* \to 2$. Let supp $x = (\alpha_i)$, supp $x_n = (\alpha_i^n)$, and supp $(x + x_n) = (\beta_i^n)$. Thus we have

$$0 \le \Sigma_i\ 4^{-i}|x(\beta_i^n) - x_n(\beta_i^n)|^2 = \Sigma_i\ 4^{-i}\{2x(\beta_i^n)^2 + 2x_n(\beta_i^n)^2 - [(x + x_n)(\beta_i^n)^2]\}$$

$$\le \Sigma\ 4^{-i}\{2x(\alpha_i)^2 + 2x_n(\alpha_i^n)^2 - [(x + x_n)(\beta_i^n)^2]$$

$$= 2\|x\|^{*2} + 2\|x_n\|^{*2} - \|x + x_n\|^{*2} \to 0$$

This implies that for each i,

$$\lim\ [x(\beta_i^n) - x_n(\beta_i^n)] = 0$$

Suppose now that the assertion about the local uniform convexity is not true. Then there exists a subsequence (n_k) and $\varepsilon > 0$ such that

$$\|x_{n_k} - x\| \ge \varepsilon$$

If i_1 is the largest integer with the property $|x(\alpha_i)| > \varepsilon/16$, then

$$|x(\alpha_{i+1})| < \frac{\varepsilon}{16} \le |x(\alpha_i)|$$

Thus we have

$$0 < \delta = 2(4^{-i_1} - 4^{-i_1-1})|x(\alpha_{i_1})^2 - x(\alpha_{i_1+1})^2|$$

and if n is large enough so that

$$2(\|x\|^{*2} + \|x_n\|^{*2}) - \|x + x_n\|^{*2} < \delta$$

we have

$$\Sigma_i\ 4^{-i}[2x(\alpha_i)^2 - 2x(\beta_i^n)^2] < \delta$$

Applying the inequality obtained for the sequences (d_i) and (β_i^n) we get by Lemma 2.9.28 that $\{\alpha_1, \dots, \alpha_{i_1}\} = \{\beta_1^n, \dots, \beta_{i_1}^n\}$ if n is sufficiently large and thus, considering eventually a subsequence, we may suppose from (*)

that for all n and for $j = 1, 2, \ldots, i_1$, $\beta_j = \beta_j^n$. Thus we have

$$\lim x_n(\alpha_j) = x(\alpha_j) \qquad j = 1, 2, \ldots, i_1$$

Thus $\lim x_n(s) = x(s)$ for each s in $\{\alpha_1, \ldots, \alpha_{i_1}\}$. Now given j we take γ_j in Γ with the property that

$$|(x - x_j)(\gamma_j)| = \|x - x_j\|_{c_0(\Gamma)} \geq \varepsilon$$

From the above result we have that for N sufficiently large and for all $n \geq N$,

(a) $|(x - x_n)(\alpha)| < \varepsilon$, $\alpha \in \{\alpha_1, \ldots, \alpha_{i_1}\}$

(b) $|x(\alpha)^2 - x_n(\alpha)^2| < \varepsilon^2/4^{i_1+4}$, $\alpha \in \{\alpha_1, \ldots, \alpha_{i_1}\}$

(c) $\|x_n\|^* - \|x\|^* < \varepsilon^2/4^{i_1+4}$

If $\gamma_n \in \{\alpha_1, \ldots, \alpha_{i_1}\}$, replacing supp x_n by a sequence starting with $\alpha_1, \ldots, \alpha_{i_1}, \gamma_n$, we have

$$\sum_{j=1}^{i_1} \frac{x_n(\alpha_j)^2}{4^j} + \frac{x_n(\gamma_n)^2}{4^{i_1+1}} \leq \|x_n\|^{*2}$$

But $|x(\alpha)| \leq \varepsilon 4^{-2}$ for $\alpha \in \{\alpha_1, \ldots, \alpha_{i_1}\}$ and since $\Sigma_{j=i_1+1}^{\infty} 4^{-j} \leq (1/3)4^{i_1}$ we get

$$\|x\|^{*2} < \sum_{j=1}^{i_1} 4^{-j} x(\alpha_j)^2 + \varepsilon^2 4^{-4} \frac{1}{3} 4^{i_1}$$

From (a), (b), and (c) we obtain further that

$$\frac{x_n(\gamma_n)}{4^{i_1+1}} < \frac{\varepsilon^2}{4^{i_1+1}}$$

for all $n \geq N$. But for all $n \geq N$ and γ in $\{\alpha_1, \ldots, \alpha_{i_1}\}$,

$$|(x - x_n)(\gamma)| < \varepsilon$$

and thus

$$|(x - x_n)(\gamma_n)| \leq |x(\gamma_n)| + |x_n(\gamma_n)| < \frac{\varepsilon}{4^2} + \frac{\varepsilon}{4} < \varepsilon$$

if $\gamma_n \notin (\alpha_1,\ldots,\alpha_{i_1})$. These contradict our assumption and thus $x_n \to x$ in $c_0(\Gamma)$, and the theorem is proved.

COROLLARY 2.9.30 (M. M. Day, 1955) If X is a Banach space and there exists a one-to-one linear mapping $T : X \to c_0(\Gamma)$ for some set Γ then X is strictly convexifiable.

COROLLARY 2.9.31 The space $\ell^\infty = m$ has an equivalent strictly convex norm.

Proof: We define the mapping T as follows:

$$Tx = \left\{x_1, \frac{x_2}{2}, \frac{x_3}{3}, \ldots, \frac{x_n}{n}, \ldots\right\}$$

and this satisfies the corollary.

By modifying Day's proof that if Γ is uncountable then $\ell^\infty(\Gamma)$ has no equivalent strictly convex norm (Theorem 2.9.7), J. Lindenstrauss proved the following interesting result [for a proof we refer to the paper of Lindenstrauss (1972)].

THEOREM 2.9.32 The space $\ell^\infty = m$ has no equivalent norm $\|\ \|^*$ such that $(m,\|\ \|^*)$ is a WLUC space.

For the problem of renorming Banach spaces it is of interest to construct new equivalent norms which combine properties of some given norms. We ask, more precisely, if $\|\ \|_1$ and $\|\ \|_2$ are two equivalent norms on X having the properties (P_1) and (P_2), respectively, is it possible to construct a new equivalent norm on X having both the properties (P_1) and (P_2)? In what follows we give a method (the so-called averaging method of Asplund) which presents a satisfactory answer to the above problem for the case when (P_1) is uniform convexity and (P_2) is uniform smoothness. For this we need some preliminary results.

LEMMA 2.9.33 Let X be a Banach space with the norm $\|\ \|$. Then the function

$$f(x) = \frac{1}{2}\|x\|^2$$

is convex and homogeneous of order 2.

Proof: We recall that a function F is called homogeneous of order 2 if

$$F(tx) = t^2F(x)$$

for all $x \in X$ and $t \in \mathbb{R}$.

Since the homogeneity of order 2 is obvious we prove only that f is a convex function. Let x,y be in X and $s \in (0,1)$. Then we have

$$\begin{aligned} f(sx + (1 - s)y) &= \frac{1}{2} (\|sx + (1 - s)y\|^2) \le \frac{1}{2} (s\|x\| + (1 - s)\|y\|)^2 \\ &\le \frac{1}{2} s^2\|x\|^2 + \frac{1}{2} (1 - s)^2\|y\|^2 + s(1 - s)\|x\|\|y\| \\ &\le \frac{1}{2} s^2\|x\|^2 + \frac{1}{2} (1 - s)^2\|y\|^2 + \frac{1}{2} s(1 - s)(\|x\|^2 + \|y\|^2) \\ &= \frac{1}{2} s\|x\|^2 + \frac{1}{2} (1 - s)\|y\|^2 = sf(x) + (1 - s)f(y) \end{aligned}$$

and the assertion is proved.

LEMMA 2.9.34 Let X be a Banach space and $f : X \to \mathbb{R}$ be a convex function, homogeneous of order 2, and $f(x) = 0$ iff $x = 0$. Then

$$x \to [2f(x)]^{1/2} = g(x)$$

is a norm on X.

Proof: We have to prove the following properties of g:

$g(x) = 0$ iff $x = 0$

$g(tx) = |t|g(x)$

$g(x + y) = g(x) + g(y)$

Since the first two properties of g are obvious we prove only the third. We have, for x,y arbitrary in X,

$$\begin{aligned} g(x + y)^2 &= 2f(x + y) = 2f\left(\frac{1}{1 + s} (1 + s)x + \frac{s}{1 + s} \frac{1 + s}{s} y\right) \\ &\le \frac{2}{1 + s} f((1 + s)x) + \frac{2s}{1 + s} f\left(\frac{1 + s}{s} y\right) \\ &\le 2(1 + s)f(x) + 2 \frac{1 + s}{s} f(y) \\ &\le 2f(x) + 2f(y) + 4[f(x)f(y)]^{1/2} \\ &= \{[2f(x)]^{1/2} + [2f(y)]^{1/2}\}^2 = [g(x) + g(y)]^2 \end{aligned}$$

where we have let $s = [f(y)/f(x)]^{1/2}$.

LEMMA 2.9.35 Let X be a Banach space and X* be its dual, and set

$$f(x) = \frac{1}{2} \|x\|^2$$

$$f^*(x^*) = \frac{1}{2} \|x^*\|^2$$

for all $x \in X$ and $x^* \in X^*$. Then for all $x^* \in X^*$, the Fenchel-Young conjugate of f is f*,

$$f^*(x^*) = \sup_{x \in X} [x^*(x) - f(x)]$$

Proof: Since

$$x^*(x) \leq \|x\|\|x^*\| \leq \frac{1}{2} (\|x\|^2 + \|x^*\|^2)$$

we obtain that

$$\frac{1}{2} x^{*2} \geq x^*(x) - \frac{1}{2} \|x\|^2 = x^*(x) - f(x)$$

and thus

$$f^*(x^*) \geq \sup \{x^*(x) - f(x) : x \in X\}$$

We prove now that in fact we have an equality. Let $x^* \in X^*$; we can suppose without loss of generality that $x^* \neq 0$. Then for $\varepsilon > 0$ given, there exists $x \in X$, $\|x\| = 1$, such that

$$x^*(x) + \varepsilon > \|x^*\|$$

and thus (letting $x^- = \|x^*\|x)$,

$$\begin{aligned} x^*(\|x^*\|x) = x^*(x^-) &= \|x^*\|x^*(x) \geq \|x^*\|(\|x^*\| - \varepsilon) \geq (\|x^*\| - \varepsilon)(\|x^-\| - \varepsilon) \\ &= \frac{1}{2} (\|x^*\| - \varepsilon)^2 + \frac{1}{2} (\|x^-\| - \varepsilon)^2 \end{aligned}$$

Then for $\varepsilon \to 0$ we obtain

$$\sup [x^*(x) - \frac{1}{2} \|x\|^2] \geq \frac{1}{2} \|x^*\|^2$$

and the lemma is proved.

DEFINITION 2.9.36 Let X be a Banach space and $f : X \to \mathbb{R}$ be a convex function homogeneous of order 2, and $f(x) = 0$ iff $x = 0$. We say that f is uniformly convex iff for any $2 \geq \varepsilon > 0$,

$$0 < \inf \left\{ f(x) + f(y) - 2f\left(\frac{x + y}{2}\right) : \|x\| \leq 1, \|y\| \leq 1, \|x - y\| \geq \varepsilon \right\}$$

We have the following result, which relates the uniform convexity of a Banach space X with the norm $\| \ \|$ to the uniform convexity of the function

$$f(x) = \frac{1}{2} \|x\|^2$$

LEMMA 2.9.37 The Banach space X is uniformly convex iff f is uniformly convex, where $f(x) = (1/2)\|x\|^2$.

Proof: Suppose that X is uniformly convex. Then for each $\varepsilon \in (0,2]$ there exists $\delta(\varepsilon) > 0$ such that if $\|x\| \le 1$, $\|y\| \le 1$, and $\|x - y\| \ge \varepsilon$, then

$$\|x + y\| \le 2[1 - \delta(\varepsilon)]$$

Then

$$f(x) + f(y) - 2f\left(\frac{x+y}{2}\right) = \frac{1}{2}\|x\|^2 + \frac{1}{2}\|y\|^2 - \left\|\frac{x+y}{2}\right\|^2$$

and we have to consider the following cases:

Case 1. $|\|x\| - \|y\|| \ge \varepsilon/2$. In this case we get

$$\begin{aligned} f(x) + f(y) - 2f\left(\frac{x+y}{2}\right) &\ge \frac{1}{2}\|x\|^2 + \frac{1}{2}\|y\|^2 - \frac{1}{4}\|x + y\|^2 \\ &= \frac{1}{4}(\|x\| - \|y\|)^2 \ge \frac{\varepsilon^2}{16} \end{aligned}$$

Case 2. $0 \le \|x\| - \|y\| \le \varepsilon/2$. Since

$$\|y\|\left[\frac{x}{\|x\|} + \frac{y}{\|y\|} + \left(\frac{\|x\|}{\|y\|} - 1\right)\frac{x}{\|x\|}\right] = x + y$$

and

$$\frac{1 + \|x\|^2/\|y\|^2}{2} \ge \frac{\|x\|/\|y\| + 1}{2}$$

we have

$$\begin{aligned} &\frac{1}{2}\|x\|^2 + \frac{1}{2}\|y\|^2 - \left\|\frac{x+y}{2}\right\|^2 \\ &= \frac{1}{2}\|y\|^2\frac{\|x\|^2}{\|y\|^2} - \frac{1}{4}\|y\|^2\left\|\frac{x}{\|x\|} + \frac{y}{\|y\|} + \left(\frac{\|x\|}{\|y\|} - 1\right)\frac{x}{\|x\|}\right\|^2 + \frac{1}{2}\|y\|^2 \\ &\ge \|y\|^2\left\{\frac{1}{2}\frac{\|x\|^2}{\|y\|^2} - \left[\left\|\frac{1}{2}\left(\frac{x}{\|x\|} + \frac{y}{\|y\|}\right)\right\| + \left(\frac{\|x\|/\|y\| - 1}{2}\right)\right]^2 + \frac{1}{2}\right\} \\ &\ge \|y\|^2\left\{\frac{1}{2}\left(1 + \frac{\|x\|^2}{\|y\|^2}\right) - \left[\frac{1}{2}\left\|\frac{x}{\|x\|} + \frac{y}{\|y\|}\right\| + \left(\frac{\|x\|/\|y\| - 1}{2}\right)\right]^2\right\} \\ &\ge \|y\|^2\left\{\frac{1}{2}\left(1 + \frac{\|x\|^2}{\|y\|^2}\right) + \frac{1}{2}\left(\frac{\|x\|^2}{\|y\|^2} - 1\right)\right. \\ &\quad \left. + \frac{1}{2}\left\|\frac{x}{\|x\|} + \frac{y}{\|y\|}\right\|\left[\frac{1}{4}\left(1 + \frac{\|x\|^2}{\|y\|^2}\right)^2 - \frac{1}{2}\left(\frac{\|x\|}{\|y\|} - 1\right) - \left\|\frac{x}{\|x\|} + \frac{y}{\|y\|}\right\|\right]\right\} \\ &\ge \|y\|^2\frac{\|x\|}{\|y\|}\left(1 - \frac{1}{2}\left\|\frac{x}{\|x\|} + \frac{y}{\|y\|}\right\|\right) = \|x\|\|y\|\left(1 - \frac{1}{2}\left\|\frac{x}{\|x\|} + \frac{y}{\|y\|}\right\|\right) \end{aligned}$$

But

$$\left\|\frac{x}{\|x\|} - \frac{y}{\|y\|}\right\| = \frac{1}{\|y\|}\left\|x - y - \frac{\|x\| - \|y\|}{\|x\|} x\right\| \geq \frac{1}{\|y\|}(\|x - y\| - |\|x\| - \|y\||)$$
$$\geq \frac{\varepsilon}{2\|y\|} \geq \frac{\varepsilon}{2}$$

because

$$\|y\| \geq \|x - y\| - \|x\| \geq \varepsilon - \|x\| \geq \varepsilon - (\|y\| + \frac{\varepsilon}{2})$$

and thus

$$\|y\| \geq \frac{\varepsilon}{4}$$
$$\|x\| \geq \frac{\varepsilon}{2} + \frac{\varepsilon}{4} \geq \frac{\varepsilon}{4}$$

Then

$$\left\|\frac{x}{\|x\|} + \frac{y}{\|y\|}\right\| \leq 2\left[1 - \delta\left(\frac{\varepsilon}{2}\right)\right]$$

and thus

$$\frac{1}{2}\|x\|^2 + \frac{1}{2}\|y\|^2 - \left\|\frac{x + y}{2}\right\|^2 \geq \frac{\varepsilon^2}{16}$$

Case 3. $0 < \|y\| - \|x\| \leq \varepsilon/2$. This may be treated similarly to case 2 by interchanging x with y, and the same estimates hold. Thus f is uniformly convex.

For the converse, if f is uniformly convex then obviously, if $\|x\| \leq 1$, $\|y\| \leq 1$, and $\|x - y\| \geq \varepsilon$, then from the form of f we have

$$\|x + y\| \leq 2[1 - \tilde{\delta}(\varepsilon)]$$

with $\tilde{\delta}(\varepsilon) > 0$.

DEFINITION 2.9.38 We say that a convex function f, homogeneous of order 2 and with $f(x) = 0$ iff $x = 0$, is locally uniformly convex iff for $\varepsilon \in [0,2]$ and $x \in X$, $\|x\| = 1$, we have

$$0 < \inf\left\{f(x) + f(y) - 2f\left(\frac{x + y}{2}\right) : \|y\| \leq 1, \|x - y\| \geq \varepsilon\right\}$$

In the same manner as for Lemma 2.9.37 we can prove the following result.

LEMMA 2.9.39 A Banach space X is locally uniformly convex iff the function $f(x) = (1/2)x^2$ is locally uniformly convex.

We present now the averaging procedure of Asplund. Let f_0 and g_0 be two functions defined on X such that:

1. f_0, g_0 are convex and homogeneous of order 2.
2. $f_0(x) = 0$ and $g_0(y) = 0$ iff $x = y = 0$.
3. There is a positive constant c such that

$$g_0(x) \leq f_0(x) \leq (1 + c)g_0(x) \qquad x \in X$$

[or $g_0 \leq f_0 \leq (1 + c)g_0$]. We define

$$f_1(x) = \frac{1}{2}(f_0 + g_0)$$
$$g_1(x) = \inf\{\frac{1}{2} f_0(x + y) + g_0(x - y) : y \in X\} \qquad x \in X$$

Then we show that:

1. f_1, g_1 are convex and homogeneous of order 2.
2. $f_1(x) = 0$, $g_1(y) = 0$ iff $x = y = 0$.
3. $f_1 \leq (1 + 2^{-1}c)g_1$.

Indeed, since $g_0 \leq f_0$ and g_0 is a convex function, we have

$$g_1(x) = \inf\left\{\frac{f_0(x + y) + g_0(x - y)}{2} : y \in X\right\}$$
$$= \inf\left\{\frac{g_0(x + y) + g_0(x - y)}{2} : y \in X\right\} \geq g_0(x)$$

and since $f_0 = (1 + c)g_0$ we have

$$f_1 \leq \frac{f_0 + g_0}{2} = \frac{2 + c}{2} g_0 \leq (1 + 2^{-1}c)g_0$$

Since f_0 and g_0 are homogeneous of order 2 we obtain obviously that f_1 and g_1 are homogeneous of order 2 and that f_1 is convex. We show now that g_1 is convex too.

Let $s \in (0,1)$ and x_1, x_2 be arbitrary points in X. Then we have that for $\varepsilon > 0$ there exist y_1 and y_2 in X such that

$$\frac{1}{2} f_0(x_1 + y_1) + \frac{1}{2} g_0(x_1 - y_1) \leq g_1(x_1) + \varepsilon$$
$$\frac{1}{2} f_0(x_2 + y_2) + \frac{1}{2} g_0(x_2 - y_2) \leq g_1(x_2) + \varepsilon$$

and since f_0, g_0 are convex functions,

$$
\begin{aligned}
sg_1(x_1) + (1 - s)g_1(x_2) &\geq s\left[\frac{1}{2} f_0(x_1 + y_1) + \frac{1}{2} g_0(x_1 - y_1)\right] \\
&\quad + (1 - s)\left[\frac{1}{2} f_0(x_2 + y_2) + \frac{1}{2} g_0(x_2 - y_2)\right] - \varepsilon \\
&\geq \frac{1}{2} f_0((x_1 + y_1) + (1 - s)(x_2 + y_2)) \\
&\quad + \frac{1}{2} g_0(s(x_1 - y_1) + (1 - s)(x_2 - y_2)) - \varepsilon \\
&\geq \frac{1}{2} f_0([sx_1 + (1 - s)x_2] + [sy_1 + (1 - s)y_2]) \\
&\quad + \frac{1}{2} g_0([sx_1 + (1 - s)x_2] - [sy_1 + (1 - s)y_2]) \\
&\quad - \varepsilon \\
&\geq g_1(sx_1 + (1 - s)x_2) - \varepsilon
\end{aligned}
$$

Since $\varepsilon > 0$ is arbitrary, g_1 is a convex function.

Now we define two sequences of functions (f_n) and (g_n) by the relations:

$$f_{n+1} = \frac{1}{2}(f_n + g_n) \qquad n = 1, 2, 3, \ldots$$

$$g_{n+1}(x) = \inf\left\{\frac{1}{2} f_n(x + y) + \frac{1}{2} g_n(x - y) : y \in X\right\}$$

and in the same manner as above we can prove the following assertions:

1. f_n, g_n are convex and homogeneous of order 2.
2. $f_n(x) = 0$, $g_n(y) = 0$ iff $x = y = 0$.
3. $f_n \leq (1 + 2^{-n}c)g_n$.

From these relations we obtain that $(f_n),(g_n)$ are convergent sequences and converge to a common function, say h. Then we have clearly,

$$(1 + 2^{-n}c)^{-1}h \leq g_n \leq h \leq f_n \leq (1 + 2^{-n}c)h$$

Now we prove the inequalities

$$g_n \leq f_n \leq (1 + 4^{-n}c)g_n$$

$$(1 + 4^{-n}c)^{-1}h \leq g_n \leq h \leq f_n \leq (1 + 4^{-n}c)h$$

Indeed, for $n = 1$ these relations are true. Suppose that they are true for all $i \leq n$ and we prove them for $i = n + 1$. Since

$$f_{n+1} \leq f_n \qquad f_{n+1} = \frac{f_n + g_n}{2}$$

we obtain that

$$f_{n+1} \le (1 + 4^{-n}c)g_n$$

and since f_n, g_n are homogeneous of order 2,

$$\frac{1}{2}\left[f_n(x + y) + g_n(x - y)\right] \ge \frac{1}{2}\left[\frac{1}{(1 + 4^{-n}c)^2} f_{n+1}((1 + 4^{-n}c)(x + y)) + \frac{1}{1 + 4^{-n}c} f_{n+1}(x - y)\right]$$

$$= \frac{1}{2}\frac{1 + 1 + 4^{-n}c}{(1 + 4^{-n}c)^2}\left[\frac{f_{n+1}((1 + 4^{-n}c)(x + y))}{1 + 1 + 4^{-n}c} + \frac{1 + 4^{-n}c}{1 + 1 + 4^{-n}c} f_{n+1}(x - y)\right]$$

$$\ge \frac{1 + 1 + 4^{-n}c}{2}(1 + 4^{-n}c)^2 f_{n+1}\left(\frac{2(1 + 4^{-n}c)}{1 + 1 + 4^{-n}c}\, x\right)$$

$$= \frac{2}{1 + 4^{-n}c} f_{n+1}(x)$$

and clearly this implies that

$$f_{n+1} \le (1 + 4^{-n}c)g_{n+1}$$

and thus the relations are established.

LEMMA 2.9.40 If f_0 is uniformly convex then h is also uniformly convex.

Proof: Since we can write

$$f_n = \frac{f_0}{2^n} + h_n \qquad h_n = \sum_0^{n-1} \frac{g_k}{2^{n-k}}$$

by the above relations we have

$$0 \le f_n - h \le f_n - g_n \le 4^{-n}cg_n \le 4^{-n}cf_0$$

and thus

$$\left(\frac{1}{2^n} - \frac{c}{4^n}\right) f_0 + h_n \le h \le \frac{1}{2^n} f_0 + h_n$$

Now, h_n being convex, for all x,y in X we have

$$h(x) + h(y) - 2h\left(\frac{x + y}{2}\right)$$

$$\geq \left[\frac{1}{2^n} f_0(x) - \frac{c}{4^n} f_0(x) + h_n(x)\right] - 2\left[\frac{1}{2^n} f_0\left(\frac{x + y}{2}\right) + h_n\left(\frac{x + y}{2}\right)\right]$$

$$+ \left[\frac{1}{2^n} f_0(y) - \frac{c}{4^n} f_0(y) + h_n(y)\right]$$

$$\geq \frac{1}{2^n}\left\{f_0(x) + f_0(y) - 2f_0\left(\frac{x + y}{2}\right) - \frac{c}{4^n}[f_0(x) + f_0(y)]\right\}$$

Thus if $\varepsilon > 0$, $\|x\| \leq 1$, $\|y\| \leq 1$, and $\|x - y\| \geq \varepsilon$ then for n sufficiently large, we obtain that h is uniformly convex.

In the same way we can prove the following assertion.

LEMMA 2.9.41 If f_0 is locally uniformly convex then h is locally uniformly convex.

We give now some relations involving the dual functions (the so-called Fenchel-Young conjugate functions). We recall that if $F : X \to \mathbb{R}$ is a function on X then the dual function associated with F is the function $F^* : X^* \to \mathbb{R}$ defined by the relation:

$$F^*(x^*) = \sup \{x^*(x) - F(x) : x \in X\}$$

Thus associated with the functions f_n, h, g_n are the dual functions f_n^*, h^*, and g_n^*, respectively. Since we have the relations

$$(1 + 4^{-n}c)h \leq g_n \leq h \leq f_n \leq (1 + 4^{-n}c)h \qquad n = 1, 2, 3, \ldots$$

then from the definition of the dual function we have the relations

$$(1 + 4^{-n}c)h^* \leq g_n^* \leq h^* \leq f_n^* \leq (1 + 4^{-n}c)h^*$$

and thus the sequences (f_n^*), (g_n^*) converge to the same function, namely h^*. A more precise relation between these functions is given in the following lemma.

LEMMA 2.9.42 The functions (f_n^*), (g_n^*) satisfy the relations:

$$f_{n+1}^*(x^*) = \inf\left\{\frac{1}{2}[f_n^*(x^* + y^*) + g_n^*(x^* - y^*)] : y^* \in X^*\right\}$$

$$g_{n+1}^* = \frac{1}{2}(f_n^* + g_n^*)$$

for all n = 1, 2, 3, ...

Proof: Since

$$f_{n+1} = \frac{1}{2}(f_n + g_n)$$

and if $\| \|_1^n$ and $\| \|_2^n$ are the norms defined by f_n and g_n then the norm $\| \|_1^{n+1}$ corresponding to f_{n+1} satisfies the relation

$$\frac{1}{2}\|x^*\|_1^{n+1} = \frac{1}{4}[(\|x^*\|_1^n)^2 + (\|x^*\|_2^n)^2]$$

Then f_{n+1}^* is given by the relation

$$\frac{1}{2}\|x^*\|_1^{n+1} = \inf_{y^*}\left\{\frac{1}{4}(\|x^* + y^*\|_1^n)^2 + \frac{1}{4}(\|x^* - y^*\|_1^n)^2\right\}$$

We consider now the Banach space X × X with the norm

$$(x,y) \to [(\|x\|_1^n)^2 + (\|y\|_1^n)^2]^{1/2}$$

and X is considered now as a subspace of X × X obtained by the imbedding

$$x \to \left(\frac{1}{2^{1/2}}x, \frac{1}{2^{1/2}}y\right) = E(x)$$

Since

$$\ker E^* = \{(y^*,-y^*) : y^* \in X^*\}$$

we have that $(X \times X)^*/\ker E^*$ is isometric with $(X^*,\| \|_1^{n+1})$. Thus for any $x^* \in X^*$,

$$\begin{aligned}\|x^*\|_1^{n+1} &= \inf\left\{\left\|\left(\frac{1}{2}2^{1/2}x^*,\frac{1}{2}2^{1/2}y^*\right) + (y^*,-y^*)\right\| : y^* \in X^*\right\}\\ &= \inf\left\{\left|\left(\left\|\frac{1}{2^{1/2}}x^* + \frac{1}{2^{1/2}}y^*\right\|_1^n\right)^2 + \left(\left\|\frac{1}{2^{1/2}}x^* - \frac{1}{2^{1/2}}y^*\right\|_1^n\right)^2\right|^{1/2} : y^* \in X^*\right\}\\ &= \inf\left\{\left|\frac{1}{2}(\|x^* + y^*\|_1^n)^2 + \frac{1}{2}(\|x^* - y^*\|_1^n)^2\right|^{1/2} : y^* \in X^*\right\}\end{aligned}$$

and this gives the relation for f_{n+1}^*. In a similar way we obtain the second relation.

THEOREM 2.9.43 Let X be a Banach space with two equivalent norms $\| \|_1$ and $\| \|_2$. Suppose that

1. $(X,\| \|_1)$ is uniformly convex.
2. $(X,\| \|_2)$ is uniformly smooth.

Then there exists an equivalent norm $\| \|_3$ on X such that $(X,\| \|_3)$ is uniformly convex and uniformly smooth.

Proof: Let f_0 and g_0 be the functions defined on X by

$$f_0(x) = \frac{1}{2}\|x\|_1^2 \qquad g_0(x) = \frac{1}{2}\|x\|_2^2$$

From the fact that $\| \ \|_1$ and $\| \ \|_2$ are equivalent norms we may suppose without loss of generality that for some $c > 0$,

$$g_0 \leq f_0 \leq (1 + c)g_0$$

From the hypothesis and Lemma 2.9.37 we get that f_0 is uniformly convex, and since X* is uniformly convex we get that g_0^* is uniformly convex. We now use the averaging procedure of Asplund and we obtain the functions h and h*. Then we know that h is uniformly convex and also h* is uniformly convex. Thus the corresponding norms for h and for h* are both uniformly convex and smooth. This proves the theorem.

An interesting proof for a (weaker) version of Asplund's result was proposed by K. John and V. Zizler (1979a). Interesting renormings were introduced by M. I. Kadets and V. Klee (1979a). In what follows we give some of them. For related results and more information we refer to V. D. Milman (1973), V. Zizler (1968a,b), K. John and V. Zizler (1979b) and M. M. Day (1973).

THEOREM 2.9.44 Let X be a Banach space and suppose that X* is separable. Then there exists an equivalent norm $\| \ \|_1$ on X with the following properties:
1. $(X^*, \| \ \|_1^*)$ is a strictly convex space ($\| \ \|_1^*$ is the dual norm of $\| \ \|_1$).
2. If $(f_n) \subset X^*$, $\|f_n\| \to \|f_0\|$, and $f_n \to f_0$ (weak*), them $\lim \|f_n - f\| = 0$.

Proof: Since X* is separable then X is again a separable space and thus we can consider a sequence (x_n) dense in $S(X) = \{x : x \in X, \|x\| = 1\}$. Since X* is separable we can choose a sequence (g_n) which is dense in S(X*). If we define

$$V_n = \left\{x : \|x\| \leq 1, \ |f(x)| \leq \frac{1}{n}\right\}$$

then the Minkowski functional of the V_n's defines a norm equivalent with the original norm of X. We denote the dual norm by w_n. This is clearly defined by

$$w_n(f) = \sup \{|f(x)| : x \in V_n\}$$

and satisfies the inequality

$$\frac{1}{n}\|f\| \leq w_n(f) \leq \|f\|$$

We now define new norms on X^*, w, and p by the relations

$$w(f) = \sum_{1}^{\infty} \frac{w_n(f)}{2^n}$$

$$p(f) = \left[\sum_{1}^{\infty} \frac{1}{2^{2n}} f(x_n)^2\right]^{1/2}$$

and using these the following norm on X^* may be introduced:

$$\|f\|_1 = w(f) + p(f)$$

This norm is obviously equivalent with the original norm of X^* and clearly X^* with this norm is strictly convex. Now, since w_n are weak* lower continuous, w has also this property. Thus, since $f \to f(x_n)$ are weak* lower semicontinuous, the norm p is weak* lower semicontinuous. But from the form of p we get that in fact it is weak* continuous on bounded subsets of X^*. Thus $\| \ \|_1$ is a lower semicontinuous equivalent norm on X^* and this determines a dual norm for X, say $\| \ \|_1$. Now since $(X^*, \| \ \|_1)$ is strictly convex, $(X, \| \ \|_1)$ is smooth.

It remains to show that if (f_n) is a sequence in X^* with the following properties:

$$\|f_n\|_1 \to \|f_0\|_1 \qquad f_n \rightharpoonup f_0 \text{ (weak*)}$$

then $\lim \|f_n - f_0\| = 0$. Now, since p is weak* continuous on bounded subsets of X^* we have that

$$\lim p(f_n) = p(f_0)$$

and since

$$\|f_n\|_1 \to \|f_0\|_1$$

we have that

$$\lim w(f_n) = w(f_0)$$

Suppose now that the assertion is not true. Thus there exists an $\varepsilon > 0$ and a subsequence [we can suppose without loss of generality that this is just (f_n)] such that

$$\|f_n - f_0\| \geq \varepsilon$$

Now, by the well-known diagonal procedure we can extract a subsequence [which we may suppose to be just (f_n)] such that

$$\lim_i w_n(f_i)$$

exists for each n; we denote it by a_n. Also, we may suppose that $\|f_0\| = 1$, and thus

$$w(f_0) = \lim_i w(f_i) = \lim_i \Sigma\, 2^{-n} w_n(f_i) = \Sigma_n \lim_i 2^{-n} w_n(f_i) = \Sigma_n 2^{-n} a_n$$

From the weak* lower semicontinuity of w_n we have that

$$w_n(f_0) = \underline{\lim}_i\, w_n(f_i) = a_n$$

and thus

$$w_n(f_0) = a_n = \lim_i w_n(f_i)$$

since

$$w_n(f_0) = \Sigma\, 2^{-n} a_n$$

Now the sequence (f_n) is dense in S(X*) and thus there exists (n_i) such that

$$\|f_0 - f_{n_i}\| < \frac{\varepsilon}{8}$$

Thus we have, for n_{i_0} fixed such that $1/n_{i_0} < \varepsilon/8$,

$$w_{n_{i_0}}(f_0 - f_{n_{i_0}}) \le f_0 - f_{n_{i_0}}$$

which implies that

$$w_{n_{i_0}}(f_0) \le w_{n_{i_0}}(f_{n_{i_0}}) + \|f_0 - f_{n_{i_0}}\| \le \frac{1}{n_{i_0}} + \frac{\varepsilon}{8} < \frac{\varepsilon}{4}$$

Since $\lim w_{n_{i_0}}(f_i) = w_{n_{i_0}}(f_0)$ there exists M such that for all $i \ge M$,

$$w_{n_{i_0}}(f_i) < \frac{\varepsilon}{4}$$

Now, if $f_{n_{i_0}}(x) = 0$ then $|f_0(x)| < \varepsilon/4$ and by Lemma 2.9.45 (which we prove later), for all $i \ge M$,

$$\|f_i - f_{n_{i_0}}\| < \frac{\varepsilon}{2} \qquad \text{or} \qquad \|f_i + f_{n_{i_0}}\| < \frac{\varepsilon}{2}$$

This implies that

$$\|f_i + f_{n_{i_0}}\| < \frac{\varepsilon}{2}$$

since, if the first inequality holds,

$$\|f_i - f_0\| \le \|f_i - f_{n_{i_0}}\| + \|f_{n_{i_0}} - f_0\| < \frac{\varepsilon}{2} + \frac{\varepsilon}{8}$$

which is a contradiction.

Now we prove that

$$\|f_i + f_{n_{i_0}}\| < \frac{\varepsilon}{2}$$

leads also to a contradiction. Let us consider $x \in X$ with $\|x\| \le 1$ such that

$$1 - \frac{\varepsilon}{2} < f_{n_{i_0}}(x)$$

Thus for each $i \in M$,

$$f_i(x) + f_{n_{i_0}}(x) \le \|f_i + f_{n_{i_0}}\| < \frac{\varepsilon}{2}$$

and

$$f_i(x) + (1 - \frac{\varepsilon}{2}) \le f_i(x) + f_{n_{i_0}}(x)$$

which imply that

$$f_i(x) = \varepsilon - 1$$

But $f_i \rightharpoonup f$ (weak*) and thus

$$\lim f_i(x) = f_0(x)$$

and

$$\frac{1}{2} < (1 - \frac{\varepsilon}{2}) \le f_{n_{i_0}}(x) - f_i(x) \le \|f_{n_{i_0}} - f_i\| < \frac{\varepsilon}{8}$$

Thus $\varepsilon > 4$ and this contradiction proves the theorem.

LEMMA 2.9.45 Let f,g be in $S(X^*)$ and suppose that for some $\varepsilon > 0$, $f(x) = 0$ and $\|x\| \le 1$ imply $|g(x)| < \varepsilon/2$. Then $\|f + g\| \le \varepsilon$ or $\|f - g\| \le \varepsilon$.

Proof: By the Hahn-Banach theorem, we find h in X^* such that $\|h\| \le \varepsilon/2$ and $h(x) = g(x)$ for all $x \in \ker f$. Then $g = h = \alpha f$ for some α. Since

$$|1 - |\alpha|| = |\|g\| - \|g - h\|| \le \|h\| \le \frac{\varepsilon}{2}$$

we have, for $\alpha > 0$,

$$\|f - g\| = \|(1 - \alpha)f - h\| \le |1 - \alpha| + \|h\| \le \varepsilon$$

and for $\alpha < 0$,

$$\|f + g\| = \|(1 + \alpha)f + h\| \le |1 + \alpha| + \|h\| \le \varepsilon$$

and the lemma is proved.

We give now a result for spaces which are isomorphic with locally uniformly convex spaces.

THEOREM 2.9.46 (M. I. Kadets, 1959) If X is a separable Banach space then there exists an equivalent norm $\|\ \|_1$ on X such that $(X,\|\ \|_1)$ is a locally uniformly convex space.

In fact we prove a slightly more general result. For this we need the following notion.

DEFINITION 2.9.47 A set Γ in X^* is called C-norming if for any $x \in X$,

$$\sup_{f\in\Gamma} \frac{f(x)}{\|f\|} \ge C\|x\|$$

Using the extension theorems it is easy to prove the following.

LEMMA 2.9.48 If X is a separable Banach space and F is a C-norming family of subspaces then there exists a countable subfamily $F_1 \subset F$ such that F_1 is also C-norming.

We recall that the characteristic $\eta(F)$ of a linear subset of X^* is defined by the relation

$$\eta(F) = \inf_{x\in X} \sup_{f\in F} \left\{\frac{|f(x)|}{\|f\|\|x\|} : x \ne 0,\ f \ne 0\right\}$$

[and thus the set is norming if $\eta(F) > 0$]. We say that the Banach space X is H_Γ for the set $\Gamma \subset X^*$ if whenever $\{x_n\} \in X$, $x \in X$, $\|x_n\| \to \|x\|$, and $x^*(x_n) \to x^*(x)$ for all $x^* \in \Gamma$, it follows that $\|x_n - x\| \to 0$. We write this as (X,H_Γ). (X is also said to have the Radon-Riesz property relative to Γ.)

THEOREM 2.9.49 (M. I. Kadets, 1957) Let X be a separable Banach space and Γ be a norming set in X^*. Then there exists on X an equivalent norm $\|\ \|_1$ such that $(X,\|\ \|_1)$ is locally uniformly convex and the following properties hold:

1. $\|x_n\|_1 \to \|x\|_1$.
2. If $f(x_n) \to f(x)$ for all $f \in \Gamma$, then $\lim \|x_n - x\| = 0$.

Proof: We consider a subfamily Γ_1 generated by a countable set of functionals, say (f_n), $\|f_m\| = 1$. Now the unit ball of X^* is compact in the $\sigma(X^*,X)$-topology (where the base of the topology consists of sets $\{x^* : x^*(x_1) < \varepsilon, \ldots, x^*(x_m) < \varepsilon, \varepsilon > 0, x_i \in X\}$) and metrizable by Aloglu's theorem [L. Aloglu (1940); see also N. Dunford and J. Schwartz (1959)]. Let d be that metric and set $K = \Gamma_1 \cap \{f : \|f\| \le 1\}$. The modulus of continuity is defined, as usual, by the relation

$$w(x,\delta) = \sup_{d(f,g)\le\delta} |f(x) - g(x)|$$

for all $x \in X$ and f,g in K. Clearly $x \to w(x,\delta)$ is a seminorm; define

$$w_0(x) = \sup\{|f(x)| : x \in K\} \qquad \ldots \qquad w_n(x) = w(x,1/m) \qquad \ldots$$

and since Γ_1 is norming, $w_0(x) \ge \eta\|x\|$ $[\eta(\Gamma_1) = \eta]$. Since for each f, $\lim f(x_n) = f(x)$ we have

$$w_m(x) = \underline{\lim}_{n\to\infty} w_m(x_n)$$

We now define the following norm:

$$x \to \|x\|_1 = \Sigma\, 2^{-n} w_n(x)$$

which is clearly equivalent with the original norm of X, since

$$\eta\|x\| \le \|x\|_1 \le 3\|x\|$$

Now we have that $(X,\|\ \|_1)$ is H_{Γ_1} (H_Γ). Since

$$\lim_{n\to\infty} w_m(x_n) = w_m(x)$$

for all $m = 0, 1, 2, 3, \ldots$ we obtain that $\{x_n(f)\}_{f \in K}$ is a bounded and equicontinuous set of functions. Indeed,

$$|f(x_n)| \leq w_0(x_n) \leq \|x_n\| \leq 1$$

and thus the boundedness is proved. For equicontinuity, let $\varepsilon > 0$, take m sufficiently large that

$$w_m(x) < \frac{\varepsilon}{2}$$

and choose $n_0(m,\varepsilon)$ such that $|w_m(x_n) - w_m(x)| < \varepsilon/2$ for all $n \geq n_0$. Then $w_m(x_n) < \varepsilon/2$ for all $n \geq n_0$, and by increasing m we extend this to all x_n. Thus

$$w\left(x_n, \frac{1}{m}\right) < \varepsilon \qquad n = 1, 2, 3, \ldots$$

This clearly proves the equicontinuity. Now, by the Arzela-Ascoli theorem the set $\{x_n(f)\}_{f \in K}$ is relatively compact. Thus some subsequence of (x_n) converges to x at all points of K. Since the convergence is weak* we have in fact that (x_n) converges uniformly on K to x. This implies that

$$\lim w_0(x_n - x) = 0$$

which gives that $\lim x_n = x$ in the norm of X.

Using this we can further introduce a locally uniformly convex norm on X. To this end we consider a new norm on X defined by the formula

$$x \to \|x\|_2 = [\|x\|_1^2 + \Sigma\ 2^{-n} f_n^2(x)]^{1/2}$$

which is obviously equivalent with the original norm on X. Now we have that $(X, \|\ \|_2)$ is H_{Γ_1}. Indeed, if (x_n) is a sequence in X such that

1. $\|x_n\|_2 \to \|x\|_2$
2. $f(x_n) \to f(x)$ for each $f \in \Gamma_1$

then we have

$$\underline{\lim}_n \|x_n\| \geq \|x\|_1$$

and

$$\underline{\lim}_n [\Sigma\ 2^{-m} f_m^2(x_n)]^{1/2} \geq [\Sigma\ 2^{-m} f_m^2(x)]^{1/2}$$

Comparing these, we get that

$$\lim \|x_n\|_1 = \|x\|_1$$

and since $(X, \| \|_1)$ is H_{Γ_1}, $\lim \|x_n - x\| = 0$.

Now we show that $(X, \| \|_2)$ is locally uniformly convex. Let (x_n) and x be elements in X such that

$$\|x_n\|_2 = \|x\|_2 = 1 \qquad \|x_n + x\|_2 \to 2$$

From these we have

$$\sum_{m=1}^{\infty} \Sigma\, 2^{-m} f_m^2(x_n - x) + \sum_{m=1}^{\infty} 2^{-m} f_m^2(x_n + x) = 2\left[\sum_{m=1}^{\infty} 2^{-m} f_m^2(x_n) + \sum_{m=1}^{\infty} 2^{-m} f_m^2(x)\right]$$

and thus

$$\lim_n \sum_{m=1}^{\infty} 2^{-m} f_m^2(x_n - x) = 0$$

Thus for each m, $\lim f_m(x_n - x) = 0$. Since $(X, \| \|_2)$ is H_Γ this implies that $x_n \to x$ in the norm of X.

For other results concerning the renormings of Banach spaces we refer to the works of M. I. Kadets (1967, 1968, 1969), M. I. Kadets-A. Pelczynski (1965), V. D. Milman (1970a,b), V. Zizler (1968a,b; 1969), K. John-V. Zizler (1979), M. A. Smith (1977a,b; 1978), I. Singer (1970), V. N. Nikolski (1948), V. Istrăţescu (1962), J. Lindenstrauss (1972), A. R. Lovaglia (1955), S. L. Troyanski (1971).

2.10 STRICT CONVEXITY AND APPROXIMATION THEORY

Let X be a Banach space and M be a closed linear subspace of X. For any $x \in X$ we define

$$d(x,M) = \inf \{\|x - m\| : m \in M\}$$

and an element $m_0 \in M$ is called the best approximation of x by means of elements of M if

$$d(x,M) = \|x - m_0\|$$

The set

$$\{m_0 : d(x,M) = \|x - m_0\|\} = P_M(x)$$

is closed, bounded, and convex (possibly empty). The set M is called a Chebyshev set if the mapping

$$x \to P_M(x)$$

is single valued. This mapping is called the metric projection on M. Of course, the mapping $x \to P_M(x)$ may be defined in a similar way for other classes of subsets of X. In this connection we mention that $x \to P_M(x)$ is sometimes called the *Chebyshev map*, or the *proximity map*, or the *best approximation operator*. We give now some properties of the metric projection for some classes of spaces connected with convexity and smoothness.

LEMMA 2.10.1 Let X be a Banach space and $(x_1,\ldots,x_n)$ be n elements, $n < \infty$. If

$$\sum_{i=1}^{n} |a_i|^2 \to \infty$$

then for each fixed $x \in X$,

$$f_x(a_1,\ldots,a_n) = \|x - a_1x_1 - \cdots - a_nx_n\|$$

tends to ∞ as $a_i \to \infty$.

Proof: Since

$$f(a_1,\ldots,a_n) \geq \|a_1x_1 + \cdots + a_nx_n\| - \|x\|$$

and the function

$$(a_1,\ldots,a_n) \to \|a_1x_1 + \cdots + a_nx_n\|$$

is continuous, the latter function has a minimum, say μ, on the set

$$\left\{(a_1,\ldots,a_n) : \sum_{i=1}^{n} |a_i|^2 \geq 1\right\}$$

Now, if $M > 0$ and

$$\sum_{i=1}^{n} |a_i|^2 > M^2 \frac{1}{\mu^2} \|x\|^2$$

then

$$f_x(a_1,\ldots,a_n) = \left\|\sum_{i=1}^{n} a_ix_i - x\right\| = \left(\sum_{i=1}^{n} |a_i|^2\right)^{1/2} \left\|\frac{\Sigma_1^n a_i}{\Sigma_{j=1}^n |a_j|^2} (x_i - x)\right\|$$

$$\geq \left(\sum_{i=1}^{n} |a_i|^2\right)^{1/2} \mu - \|x\| > M$$

and since M is arbitrary, the assertion of the lemma follows.

LEMMA 2.10.2 If M is a finite dimensional closed linear subspace of X, then $P_M(x)$ is nonempty for each $x \in X$.

Proof: Indeed, we can take $x_1, \ldots, x_n$ as a basis of M and we consider the function $f_x(a_1,\ldots,a_n)$ as above. Since the assertion of the lemma is obvious when x is in M we may suppose without loss of generality that x is not in M. Now we have the inequalities

$$|f_x(a_1,\ldots,a_n) - f_x(b_1,\ldots,b_n)| = \left|\left\|x - \sum_{i=1}^{n} a_i x_i\right\| - \left\|x - \sum_{i=1}^{n} b_i x_i\right\|\right|$$

$$\leq \sum_{i=1}^{n} |(a_i - b_i)|\|x_i\|$$

$$\leq \max_i |(a_i - b_i)| \sum_{i=1}^{n} \|x_i\|$$

and according to Lemma 2.10.1, $f_x(a_1,\ldots,a_n) \geq \|x\|$ outside of some ball $\Sigma_{i=1}^n |a_i|^2 = k$. But this ball is a compact set and f_x is a continuous function. Thus f attains its minimum $\tilde{\mu}$ at some point, say $(a_1^*,\ldots,a_n^*)$. Since $\tilde{\mu} \leq f_x(0,\ldots,0) = \|x\|$ we get that $\tilde{\mu}$ is the least value of f_x on the entire space M and this proves the lemma.

We give now a sufficient condition for the mapping $x \to P_M(x)$ to be single valued.

LEMMA 2.10.3 Suppose X is a strictly convex space. Then the metric projection P_M is single valued for each finite-dimensional subspace M of X.

Proof: Indeed, let us suppose that the assertion of the lemma is false. In this case, for some $x \in X$ there exist at least two elements m_1 and m_2 such that

$$d(x,M) = \|x - m_1\| = \|x - m_2\|$$

Now we have obviously that $(1/2)(m_1 + m_2)$ is in $P_M(x)$ and thus

$$\left\|x - \frac{1}{2}(m_1 + m_2)\right\| = \left\|\frac{1}{2}(x - m_1)\right\| + \left\|\frac{1}{2}(x - m_2)\right\|$$

and this, according to the strict convexity of X, implies that $m_1 = m_2$.

In his paper I. Singer (1960) considered in some detail the problem of the dimension of the set $P_M(x)$ (here, the dimension of a set is the dimension of the linear subspace which it generates), when M is a closed linear subspace of X. He introduced a class of spaces which gives a necessary and sufficient condition for $\dim P_M(x) = n$. We consider now this interesting class of spaces.

DEFINITION 2.10.4 (I. Singer, 1960) A Banach space X is called k-strictly convex iff for any k + 1 elements $x_0, \ldots, x_k$ of X the relation

$$\|x_0 + x_1 + \cdots + x_k\| = \|x_0\| + \|x_1\| + \cdots + \|x_k\|$$

implies that $x_0, \ldots, x_k$ are linearly dependent.

REMARK 2.10.5 (a) If k = 1 this definition gives the class of strictly convex spaces.

(b) If X is a k-strictly convex space, then for any $m \geq k$, X is m-strictly convex.

THEOREM 2.10.6 (I. Singer, 1960) For any Banach space X the following assertions are equivalent to k-strict convexity:

(1) For any k + 1 elements $x_0, \ldots, x_k$ of X, $\|x_i\| = 1$, $\|\Sigma_{i=0}^k x_i\| = \Sigma_{i=0}^k \|x_i\|$ implies that $x_0, \ldots, x_k$ are linearly dependent.

(2) For any k + 1 linearly independent elements $x_0, \ldots, x_k$ in X, all of norm 1,

$$\left\|\sum_{i=0}^{k} a_i x_i\right\| < 1 \qquad \text{for all } 0 < a_i < 1,\ \sum_{i=0}^{k} a_i = 1$$

(3) The set

$$S(X) = \{x : x \in X,\ \|x\| = 1\}$$

contains no convex subsets of dimension > k - 1.

(4) For any $x_0 \in X$ and $r > 0$ the set

$$S(x_0,r) = \{x : x \in X,\ \|x - x_0\| = r\}$$

contains no convex sets of dimension > k - 1.

Proof: The fact that the k-strict convexity implies (1) is obvious. Suppose now that (1) holds; then we prove (2). Indeed, if (2) does not hold then there exist $x_0, \ldots, x_k$ of norm 1 such that for some a_i in (0,1) whose sum is 1,

$$a_0 x_0 + \cdots + a_k x_k$$

has norm equal to 1. We may assume without loss of generality that

$$\frac{1}{a_0} = \max\left\{\frac{1}{a_i} : 1 \le i \le k\right\}$$

and thus

$$\sum_{i=0}^{k} \|x_i\| + \sum_{i=0}^{k}\left(\frac{1}{a_0} - \frac{1}{a_i}\right)\|a_i x_i\|$$

$$= \frac{1}{a_0}\left\|\sum_{i=0}^{k} a_i x_i\right\| \le \left\|\sum_{i=0}^{k} x_i\right\| + \left\|\sum_{i=0}^{k}\left(\frac{1}{a_0} - \frac{1}{a_i}\right)a_i x_i\right\|$$

$$\le \left\|\sum_{i=0}^{k} x_i\right\| + \sum_{i=0}^{k}\left(\frac{1}{a_0} - \frac{1}{a_i}\right)\|a_i x_i\|$$

whence

$$\left\|\sum_{i=0}^{k} x_i\right\| = \sum_{i=0}^{k} \|x_i\|$$

and this contradicts (1). Thus (1) → (2).

We show now that (2) → (3). Suppose that (2) holds and (3) does not. Thus $S(X)$ contains a convex set F of dimension $k - 1$. This gives that there exist $x_0, \ldots, x_k$ in F such that

$$x_1 = x_0 \qquad x_2 - x_0 \qquad \ldots \qquad x_k - x_0$$

are linearly independent. Then

$$\frac{1}{k+1}(x_0 + \cdots + x_k) \in F$$

and this contradicts (2).

Now suppose that (4) does not hold and (3) holds. In a similar way we obtain a contradiction with (3).

Now we prove that (4) implies the k-strict convexity of X. Indeed, if (4) holds and X is not k-strictly convex, then in X there exist points $x_0, \ldots, x_k$ such that

1. $\|x_0 + \cdots + x_k\| = \|x_0\| + \cdots + \|x_k\|$.
2. $x_0, \ldots, x_k$ are linearly independent.

Then we have

$$\sum_{i=0}^{k} \left\|\frac{x_i}{\|x_i\|}\right\| + \sum_{i=0}^{k}\left(\frac{1}{\|x_0\|} - \frac{1}{\|x_i\|}\right)\|x_i\|$$

$$= \frac{1}{\|x_0\|} \sum_{i=0}^{k} \|x_i\| = \left\| \sum_{i=0}^{k} \frac{x_i}{\|x_i\|} + \sum_{j=0}^{k} \left(\frac{1}{\|x_0\|} - \frac{1}{\|x_j\|} \right) x_j \right\|$$

$$\leq \left\| \sum_{i=0}^{k} \frac{x_i}{\|x_i\|} \right\| + \sum_{i=0}^{k} \left(\frac{1}{\|x_0\|} - \frac{1}{\|x_i\|} \right) \|x_i\|$$

whence

$$\left\| \sum_{i=0}^{k} \frac{x_i}{\|x_i\|} \right\| = k + 1$$

Now, each element $x_i/\|x_i\|$ is of norm 1 and the set $S(0,1)$ contains more than $k - 1$ linearly independent elements. This is a contradiction, and thus X is k-strictly convex.

Using the notion of k-strictly convex spaces we can prove the following result for $\dim P_M(x)$.

THEOREM 2.10.7 (I. Singer, 1960) A necessary and sufficient condition for the inequality

$$\dim P_M(x) \leq k$$

to hold for all $x \in X$ is that X be k-strictly convex.

Proof: The fact that the condition is sufficient follows from Theorem 2.10.6(4) and it thus remains to prove the necessity. Suppose that X is not k-strictly convex and the above inequality holds for all subspaces of X. In this case by Theorem 2.10.6(3) there exists a convex set F in S(X) which has dimension $> k - 1$. Let y and z be elements of F and $t < 0$ or $t > 1$. We have

$$\|ty + (1 - t)z\| \geq \left| |t|\|y\| - |(1 - t)|\|z\| \right| = \left| |t| - |(1 - t)| \right| = 1$$

and the convexity of F implies that $\|x\| \geq 1$ for all x in $\{ty + (1 - t)z : t \in \mathbb{R}\}$ and y,z in F. Let x_0 be arbitrary in F and set $G_0 = \text{span } F - x_0$. In this case we have

$$P_{G_0}(-x_0) = \{g_0 : g_0 \in G_0, \|-x_0 - g_0\| = \inf \{\|-x_0 - g\| : g \in G_0\}\}$$

$$= \{u_0 - x_0 \in \text{span } F - x_0 : \|-u_0\| = \inf \{\|-u\| : u \in \text{span } F\}\}$$

$$= \{u_0 - x_0 \in \text{span } F - x_0 : \|u_0\| = 1\} = \text{span } F \cap S(X) - x_0$$

which gives that $\dim P_{G_0}(-x_0) > k$ and the theorem is proved.

Suppose now that the Banach space X is in fact a Hilbert space H. In this case the metric projection has a very useful property stated in the following lemma.

LEMMA 2.10.8 Let H be a Hilbert space and M be a closed convex subset of H. Then $P_M(x)$ is nonempty, single valued, and satisfies the inequality

$$\|P_M(x) - P_M(y)\| \leq \|x - y\|$$

Proof: Since M is a closed convex subset of a Hilbert space, it is not difficult to see that $P_M(x)$ is nonempty and single valued. Since $P_M(x)$ and $P_M(y)$ are the best approximants of x and y, respectively, we have (where $\langle\ ,\ \rangle$ is the inner product)

$$\text{Re}\,\langle P_M(x) - x, P_M(y) - P_M(x)\rangle \geq 0$$

$$\text{Re}\,\langle y - P_M(y), P_M(y) - P_M(x)\rangle \geq 0$$

and adding we obtain that

$$\text{Re}\,\langle y - x, P_M(y) - P_M(x)\rangle + \text{Re}\,\langle P_M(x) - P_M(y), P_M(y) - P_M(x)\rangle \geq 0$$

or

$$\text{Re}\,\langle y - x, P_M(y) - P_M(x)\rangle \geq \|P_M(x) - P_M(y)\|^2$$

Now applying the Cauchy inequality to the left side, we have

$$\|x - y\|\|P_M(x) - P_M(y)\| \geq \|P_M(x) - P_M(y)\|^2$$

which obviously implies the assertion of the lemma.

NOTE 2.10.9 The property of the P_M expressed in the above inequality states that P_M is a nonexpansive mapping. For further details about nonexpansive mappings and their role in fixed point theory we refer to W. A. Kirk (1970) or V. I. Istrăţescu (1981b).

At this point it is worth mentioning that, in general, the mapping $x \to P_M(x)$ is nonlinear. It would be interesting to know other properties of the mapping P_M, such as continuity, differentiability, etc. We now give two results in this direction. First we need a notion introduced by Irving Glicksberg and Ky Fan (1958).

DEFINITION 2.10.10 A real Banach space is called an E-space if it is strictly convex and every weakly closed set in X is approximatively compact.

We recall that a set F in a metric space (X,d) is called approximatively compact if for any $x \in X$, every sequence $(x_n) \in F$ with the property that $\lim d(x_n,x) = d(x,F)$ has a cluster point in F.

THEOREM 2.10.11 Let M be a Chebyshev subspace of a Banach space. Then P_M is continuous when $\dim M < \infty$ or X is an E-space.

Proof: Let X be an E-space and $x_n \to x$. Since X is reflexive [see R. B. Holmes (1972a,b) for a proof of this result] then $P_M(x_n) \to P_M(x)$. But $(x - P_M(x_n))$ is a minimizing sequence, since

$$d(x,M) = \|x - P_M(x)\| \le \liminf \|x - P_M(x_n)\|$$
$$\le \limsup (\|x - x_n\| + \|x_n - P_M(x_n)\|) \le d(x,M)$$

But any space has the Radon-Riesz property and thus we have in fact the strong convergence.

For interesting necessary and sufficient conditions for continuity of P_M we refer to R. B. Holmes and B. Kripke (1968), E. Oshman (1970), R. B. Holmes (1972a,b, 1973).

In the geometry of Banach spaces there is a thesis formulated by the Soviet mathematician A. L. Garkavi: Every geometric question about Banach spaces has an equivalent expression as an approximation-theoretic question. The following result illustrates this.

THEOREM 2.10.12 (R. R. Phelps, 1960) A Banach space X is strictly convex and reflexive if and only if each closed subspace of X is Chebyshev [i.e., for each x, $P_M(x)$ is a single point].

Another illustration of this thesis is in F. Sullivan (1974). For further results we refer to F. Sullivan (1974), I. Singer (1974), R. R. Phelps (1960), A. L. Garkavi (1964), R. B. Holmes (1972a,b, 1973).

2.11 STRICT CONVEXITY AND FIXED POINT THEORY

Let us consider a Banach space X and let C be a closed, convex, and bounded subset of X. If $T : C \to C$ is a contraction mapping, i.e., there exists $k \in (0,1)$ such that for all x,y in C,

$$\|Tx - Ty\| \le k\|x - y\|$$

then, as is well known, there exists a unique point in C, say x_0, with the property that $Tx_0 = x_0$ (i.e., x_0 is a fixed point of T). Of course, an obvious generalization of the class of contraction mappings is the following one: The mapping $T : C \to C$ is called nonexpansive if for all x,y in C,

$$\|Tx - Ty\| \le \|x - y\|$$

A natural problem arises concerning the existence (and uniqueness) of fixed points for such mappings. The following example shows that, in general, fixed points do not exist.

EXAMPLE 2.11.1 (B. N. Sadovski, 1972) Let $X = c_0$, the space of all sequences converging to zero, where on c_0 the norm is the sup norm. Let C be the unit ball of c_0 and set $T : C \to C$,

$$Tx = (1, x_0, x_1, x_2, \ldots) \qquad x = (x_0, x_1, x_2, \ldots)$$

It is obvious that T is nonexpansive and has no fixed points in C.

Thus, without restrictions on the space or on the mappings, the existence of fixed points is not guaranteed. The first result in this direction was a theorem of F. E. Browder (1965c) which asserts that when the space X is uniformly convex then fixed points exist. This result was extended in several ways. For some of them see V. I. Istrăţescu (1981b). In what follows we present a result on fixed points for a class of mappings containing the nonexpansive mappings.

THEOREM 2.11.2 (K. Goebel, W. A. Kirk, and T. N. Shimi, 1973) Let C be a bounded, closed, and convex subset of a uniformly convex space, and let $T : C \to C$ be a continuous mapping satisfying the following property: for any x,y in C,

$$\|Tx - Ty\| \le a_1\|x - y\| + a_2\|x - Tx\| + a_3\|y - Ty\| + a_4\|x - Ty\| + a_5\|y - Tx\|$$

with $a_i \ge 0$ and $\Sigma_1^5 a_i = 1$. Then T has a fixed point in C. (The proof is given later in this section.)

This class of mappings obviously contains the class of nonexpansive mappings as well as the class of mappings [introduced by R. Kannan (1969, 1973)] $T : C \to C$:

$$\|Tx - Ty\| \le \frac{1}{2}(\|x - Tx\| + \|y - Ty\|)$$

The above class of mappings considered by Goebel, Kirk, and Shimi is contained in the following class of mappings.

DEFINITION 2.11.3 Let (X,d) be a complete metric space and M be a closed and bounded subset of X. Let $T : M \to M$ be a mapping. Then we say that T is nonexpanding on diameters if for any subset N of M, we have

$$d(TN) \leq d(N)$$

Consider now a set C as in Theorem 2.11.2. Then we can prove the following result.

THEOREM 2.11.4 Let $T : C \to C$ be continuous, $I - C$ closed, and T nonexpanding on diameters. Then T has a fixed point in C.

Proof: Let s be in $(0,1)$ and x_0 arbitrary in C. Consider the mapping T_s defined on C by the relation

$$T_s x = sTx + (1 - s)x_0$$

This has the property that for any bounded set M in C,

$$d(T_s M) \leq sd(M)$$

and thus T_s has a fixed point in C, say x_s. Since

$$sTxs + (1 - s)x_0 = x_s = sx_s + (1 - s)x_s$$

from the boundedness of C we get that

$$x_s - Tx_s \to 0 \qquad \text{for } s \to 1$$

Since $I - C$ is supposed closed we get that there exists a point $\bar{x}$ in C such that

$$T\bar{x} = \bar{x}$$

and the theorem is proved.

COROLLARY 2.11.5 (F. E. Browder, 1965c) If $T : C \to C$ is nonexpansive then there exist fixed points for T.

Proof: Indeed, the continuity of T is obvious and it is not difficult to see that $I - C$ is closed. Then the result follows from the above theorem.

COROLLARY 2.11.6 If T is as in Theorem 2.11.2 then the following relation holds:

$$\inf_{x \in C} \|x - Tx\| = 0$$

Proof: The assertion is contained in the proof of Theorem 2.11.4.

NOTE 2.11.7 The proof of this result in Goebel, Kirk, and Shimi (1973) uses the fact that C is a subset of a uniformly convex space, and thus seems to be valid only for such spaces. Our proof is valid for arbitrary Banach spaces.

Proof of Theorem 2.9.2: We sketch the proof. For details we refer to the paper of Goebel, Kirk, and Shimi. Let $\varepsilon > 0$ and consider the following sets:

$$C_\varepsilon = \{x : x \in C, \|x - Tx\| \leq \varepsilon\} \qquad D_\varepsilon = \{x : x \in C_\varepsilon, \|x\| \leq \bar{a} + \varepsilon\}$$

where

$$\bar{a} = \lim_{\varepsilon \to 0} a(C_\varepsilon) \qquad a(C_\varepsilon) = \inf \{\|x\| : x \in C_\varepsilon\}$$

From the continuity of T we get that these are closed sets and from Corollary 2.11.6 we obtain that these are nonempty sets. It is now obvious that the assertion of the theorem follows if we prove that $\cap C \neq \emptyset$. It is proved in Goebel, Kirk, and Shimi (1973) that

$$d(D_\varepsilon) \leq (\bar{a} + \varepsilon)\eta \frac{\bar{a} + \varepsilon - a(C_\varepsilon)}{\bar{a} + \varepsilon}$$

η the modulus of smoothness, which implies that $\lim_{\varepsilon \to 0} d(D_\varepsilon) = 0$. Then Cantor's theorem implies that $\cap D_\varepsilon \neq \emptyset$ and this gives that $\cap C_\varepsilon \neq \emptyset$.

CONJECTURE 2.11.8 If $T : C \to C$, C is as in Theorem 2.11.2 and T is a continuous nonexpanding diameters mapping, then T has a fixed point in C.

This seems to be a reasonable conjecture in virtue of the results in Theorems 2.11.2 and 2.11.4.

In an interesting note J. J. Moreau (1978) proved that for any nonexpansive mapping $T : C \to C$, where C is a closed, bounded subset of a Hilbert space, if $F(T) = \{x : Tx = x\}$ has nonempty interior then for any $x \in C$, $(T^n x)$ converges strongly to an element of F(T). We shall present an extension of this result to uniformly convex spaces [obtained by Bernard Beauzamy (1979)]. Our presentation refers to the so-called T-nonexpansive mappings first considered by Francesco Tricomi in 1916 for real functions.

DEFINITION 2.11.9 Let C be a closed, bounded, and convex subset of a Banach space X. A mapping $T : C \to C$ is called a T-nonexpansive mapping if the following assertions hold:

1. $F(T) = \{x : Tx = x\}$ is nonempty.
2. For each $p \in F(T)$ and $x \in C$, $\|Tx - p\| \leq \|x - p\|$.

We can now prove the following theorem, which contains Moreau's and Beauzamy's results.

THEOREM 2.11.10 If $T : C \to C$ is any continuous T-nonexpansive mapping with $F(T)$ having a nonempty interior, then for each $x \in C$, $(T^n x)$ converges strongly to an element of $F(T)$.

We note that since X is supposed uniformly convex,

$$\left\|\frac{x + y}{2}\right\|^2 \leq \frac{1}{2} \frac{1 - \delta(\|x - y\|)}{\max\{\|x\|, \|y\|\}(\|x\|^2 + \|y\|^2)} \tag{*}$$

where δ denotes the modulus of convexity of X.

Our proof of the theorem is adapted from Beauzamy's proof for the case of nonexpansive mappings. The proof will be given as a consequence of some lemmas.

For any $x \in C$ we define

$$m = \|p - x\|$$

where p is the center of $S_r(p,r)$, which is contained in $F(T)$. We consider now another number defined as follows:

$$\|p - a\| = \inf \|p - [tx + (1 - t)Tx]\|$$

i.e., a is the projection of the best approximation of p on the line through x and Tx.

LEMMA 2.11.11 Let z such that $\|z - p\| < 2m$. Then

$$\|p - z\|^2 \geq \|p - a\|^2 + \frac{1}{2}\|a - z\|^2 \delta\left(\frac{\|a - z\|}{2m}\right)$$

Proof: Since X is supposed uniformly convex, according to inequality (*) we have

$$\left\|p - \frac{1}{2}(z + a)\right\|^2 \leq \frac{1}{2}\left[1 - \delta\left(\frac{\|z - a\|}{2m}\right)\right](\|p - z\|^2 + \|p - a\|^2)$$

From the definition of a we get

$$\left\|- \frac{1}{2}(z + a)\right\|^2 \geq \|p - a\|^2$$

and

$$\frac{1}{2}(\|p - z\|^2 + \|p - a\|^2) \geq \left\|\frac{a - z}{2}\right\|^2$$

which imply the assertion of the lemma.

For $x \in C$ fixed we consider the function of a real variable,

$$t \to \|p - [tx + (1 - t)Tx]\|^2$$

which, as is easy to see, is a convex function. Let t_0 be the real number such that

$$a = t_0x + (1 - t_0)Tx$$

LEMMA 2.11.12 Let t_1 and t_2 be two points such that $t_i \leq t_0$ for $i = 1, 2$ or $t_i \geq t_0$ for $i = 1, 2$. Suppose that

$$\varphi(t) = \|p - tx - (1 - t)Tx\|$$

has the properties that $\varphi(t_1) \leq 4m^2$ and $\varphi(t_2) \leq 4m^2$, where m is defined above Theorem 2.11.11. Then

$$|\varphi(t_1) - \varphi(t_2)| \geq \frac{1}{2}|t_1 - t_2|^2\|x - Tx\|^2 \delta\left(\frac{|t_1 - t_2|\|x - Tx\|}{2m}\right)$$

Proof: Suppose that $t_0 \leq t_1 \leq t_2$. Since φ is a convex function we have

$$\varphi(t_2) - \varphi(t_1) \geq \varphi(t_0 + t_2 - t_1) - \varphi(t_0)$$

and

$$\varphi(t_0 + t_2 - t_1) \leq 4m^2$$

Using Lemma 2.11.11 we obtain

$$\varphi(t_2) - \varphi(t_1) \geq \varphi(t_0 + t_2 - t_1) - \varphi(t_0) \geq \frac{1}{2}|t_2 - t_1|^2\|x - Tx\|^2\delta(h)$$

where

$$h = \frac{|t_2 - t_1|\|x - Tx\|}{2m}$$

This proves the assertion of the lemma if $t_i \geq t_0$ and we can prove it similarly for $t_i \leq t_0$.

The following lemma is fundamental for the proof of the theorem.

LEMMA 2.11.13 Let $T : C \to C$ be a T-nonexpansive mapping. Then the following inequality holds:

$$\|p - x\|^2 - \|p - Tx\|^2 \geq \frac{r^2}{8m^2} \|x - Tx\|^2 \delta(h^-)$$

where

$$h^- = \frac{r}{4m^2} \|x - Tx\|$$

and m and r are defined above Theorem 2.11.11.

Proof: We have two cases to consider.

Case 1. $t_0 \leq 0$: In this case, by Lemma 2.11.12 for $t_1 = 0$ and $t_2 = 1$, we obtain

$$\varphi(1) - \varphi(0) \geq \frac{1}{2} \|x - Tx\|^2 \delta\left(\frac{\|x - Tx\|}{2m}\right)$$

Case 2. $t_0 \in (0,1)$: In this case let $a^- = (x - Tx)/2m$; clearly we have $\|a^-\| = 1$. This gives further,

$$\|p + ra^- - Tx\| \leq \|p + ra^- - x\|$$

and thus

$$\varphi - \frac{r}{2m} \geq \varphi\left(1 - \frac{r}{2m}\right)$$

Since the minimum of φ is attained at t_0 in (0,1), we get that

$$\varphi\left(- \frac{r}{2m}\right) \geq \varphi(0)$$

and

$$\varphi(1) \geq 1 - \frac{r}{2m}$$

Since

$$1 - \frac{r}{2m} \geq 0$$

we have that

$$t_0 \leq 1 - \frac{r}{2m}$$

Thus we can apply Lemma 2.11.12 for $t_1 = 1 - r/2m$, $t_2 = 1$, and hence

$$\varphi(1) - \varphi(0) \geq \varphi(1) - \varphi\left(1 - \frac{r}{2m}\right) \geq \frac{r^2}{8m^2} \|x - Tx\|^2 \delta\left(\frac{r}{4m^2} \|x - Tx\|\right)$$

and since the estimate in case 1 is optimal, the inequality of the lemma is proved.

Now we are in a position to prove Theorem 2.11.10. First we remark that for any x in C the sequence $(\|p - T^n x\|)$ is monotone and this implies that it is a Cauchy sequence. Lemma 2.11.13 implies that $(T^n x)$ is a Cauchy sequence. The continuity of T implies that $(T^n x)$ converges to a fixed point of T.

We show now by an example that nonexpansive mappings on strictly convex spaces or even on locally uniformly convex spaces do not necessarily have fixed points. To this end we consider the space c_0 with Day's norm,

$$x \to \|x\|^- = \sup \left[\sum_{i=1}^{\infty} \frac{1}{2^{2i}} x^2(\alpha_i) \right]^{1/2}$$

where the supremum is taken over all permutations α of the integers. We know that $(c_0, \| \|^-)$ is a locally uniformly convex space and thus strictly convex. As the set C we take the unit ball

$$C = \{x : \|x\|^- \le 1\}$$

and define the mapping $T : C \to C$ by the formula

$$Tx = (1, x_0, x_1, \ldots) \qquad x = (x_0, x_1, x_2, \ldots)$$

Obviously T is nonexpansive with respect to Day's norm and has no fixed points in C.

The Browder's fixed point theorem was extended to set-valued mappings. The following result was proved by T. C. Lim (1974) using asymptotic centers. For a proof we refer to Lim's paper.

THEOREM 2.11.14 Let X be a uniformly convex Banach space and C be a closed, convex, and bounded set in X. Let $T : C \to \mathcal{C}(C)$ be a mapping from C into the family of nonempty compact, convex subsets of C metrized with the Pompeiu-Hausdorff metric, satisfying the relation

$$d_{P-H}(Tx, Ty) \le \|x - y\|$$

where d_{P-H} means the Pompeiu-Hausdorff distance. Then T has a fixed point i.e., there exists a point x_0 in C such that x_0 is in Tx_0.

Suppose that C is a closed, convex, and bounded set in a Banach space and $T : C \to C$ is nonexpansive. Suppose that the set F(T) of fixed points of T is nonempty. An interesting problem is to determine the structure of this set. The following result was noted by F. E. Browder (1967).

THEOREM 2.11.15 Let T, C be as above with C a subset of a strictly convex space. Then F(T) is a convex set.

Proof: Let x_0 and x_1 be two different points of F(T). For $t \in (0,1)$ we consider the point $x_t = tx_0 + (1 - t)x_1$ and we show that $Tx_t = x_t$. Since we have

$$\|Tx_t - x_0\| = \|Tx_t - Tx_0\| \leq \|x_t - x_0\| = (1 - t)\|x_1 - x_0\|$$

$$\|Tx_t - x_1\| = \|Tx_t - Tx_1\| \leq \|x_t - x_1\| = t\|x_1 - x_0\|$$

we get that

$$\|x_1 - x_0\| = \|x_1 - Tx_t + Tx_t - x_0\| \leq \|Tx_t - x_0\| + \|Tx_t - x_1\| = x_1 - x_0$$

and by the strict convexity of X, Tx_t lies on the segment with endpoints x_0 and x_1. The above relations then give that $Tx_t = x_t$ and the theorem is proved.

NOTE 2.11.16 It may be of interest to know if the assertion of Theorem 2.11.15 remains true for other classes of mappings, for example, the class of T-nonexpansive mappings and the class of mappings considered in Theorem 2.11.2. It is obvious that the arguments in the proof of Theorem 2.11.15 are valid also for the case of T-nonexpansive mappings. In recent years the notion of normal structure has played an important role in the study of nonexpansive mappings.

DEFINITION 2.11.17 Let (X,d) be a complete metric space and C be a closed and bounded subset of X [i.e., $d(C) = \sup\{d(x,y) : x,y \in C\} < \infty$]. A point $x \in C$ is called diametral if

$$d(C) = \sup\{d(x,z) : z \in C\}$$

DEFINITION 2.11.18 (M. S. Brodskii and D. P. Milman, 1948) A bounded and closed set C in X is said to have a normal structure if whenever C_1 is any closed subset of C containing more than one point there exists a non-diametral point $x \in C_1$.

DEFINITION 2.11.19 We say that a Banach space has a normal structure if any bounded, convex, and closed subset of the space has a normal structure.

Concerning the sets with normal structure we can prove the following results.

THEOREM 2.11.20 Any compact convex set C in a Banach space X has a normal structure.

Proof: Suppose that C does not have a normal structure. Then there exists a closed and convex set $C_1 \subset C$, containing more than one point, with no diametral points. Thus for any $x_1 \in C_1$ there exists $x_2 \in C_1$ such that

$$d(C_1) = \|x_1 - x_2\|$$

since C_1 is compact. Since C_1 is convex, $(1/2)(x_1 + x_2)$ is in C_1, and thus, from the compactness, there exists x_3 in C_1 with the property that

$$d(C_1) = \|x_3 - \frac{1}{2}(x_1 + x_2)\|$$

and so on. We consider the sequence (x_n) constructed in this way; thus for all n the following relation holds:

$$d(C_1) = \|x_{n+1} - \frac{1}{n}(x_1 + \cdots + x_n)\|$$

Since

$$d(C_1) \le \left\|\frac{x_{n+1} - x_1}{n} + \frac{x_{n+1} - x_2}{n} + \cdots + \frac{x_{n+1} - x_n}{n}\right\| \le d(C_1)$$

then for all n,

$$d(C_1) = \|x_{n+1} - x_i\| \qquad i = 1, 2, 3, \ldots, n$$

This obviously implies that the sequence (x_n) has no convergent subsequences and thus contradicts the compactness property of C. The theorem is proved.

We show now that the class of UCED (uniformly convex in every direction) spaces have the normal structure property.

THEOREM 2.11.21 (V. Zizler, 1971) If X is a UCED Banach space then X has a normal structure.

Proof: It is sufficient to show that if C is a bounded convex subset in X and $d(C) > 0$, then there exists in C no diametral points. Since $d(C) > 0$ we can choose x and y in C, $x \neq y$. Then $(1/2)(x + y)$ is again in C. We show now that $(1/2)(x + y)$ is not diametral. Indeed, in the contrary case there exists a sequence (u_n) in C with the property that

$$\lim \|u_n - \frac{1}{2}(x + y)\| = d(C)$$

If we consider the sequences (U_n) and (V_n) defined as follows:

$$U_n = \frac{1}{2} \frac{u_n - x}{d(C)} \qquad V_n = \frac{1}{2} \frac{u_n - y}{d(C)}$$

then $\|U_n\| \le 1$, $\|V_n\| = 1$. Clearly $U_n - V_n = (1/2)(y - x)/d(C)$. This implies that X is not uniformly convex in every direction. The contradiction proves the theorem.

THEOREM 2.11.22 (V. Zizler, 1968a) If the Banach space X is WLUC then X has the normal structure property.

Proof: Essentially as above but instead of UCED we use the WLUC property.

REMARK 2.11.23 M. A. Smith (1978a) has remarked that the proof of Theorem 2.11.20 can be modified to obtain a slightly more general result, namely, that a Banach space has the normal structure property if it is uniformly convex in all but a countably infinite set of directions.

An interesting connection between the behavior of the modulus of convexity δ_X of a Banach space X and the property of having a normal structure was noted by K. Goebel. To formulate it we need the following definition.

DEFINITION 2.11.24 (K. Goebel, 1970) Let X be a Banach space with the modulus of convexity δ_X. The coefficient of convexity $\varepsilon_0(X)$ of X is defined by the formula

$$\varepsilon_0(X) = \sup \{t \in [0,2] : \delta_X(t) = 0\}$$

THEOREM 2.11.25 (K. Goebel, 1970) If the Banach space X has the property that $\varepsilon_0(X) < 1$ then X has a normal structure.

P. Gossez and E. Lami Dozo obtained, for the case of a Banach space with Schauder basis, a sufficient condition for the space to have a normal structure. W. L. Bynum (1974) obtained a result which extends this result and easily implies Goebel's theorem [however, using some deep results of R. C. James (1964b)]. For completeness we give Gossez and Lami Dozo's theorem.

THEOREM 2.11.26 (J. P. Gossez and E. Lami Dozo, 1969) Let X be a Banach space with the Schauder basis $(e_i)_1^\infty$. For each n we consider the operators $T_n : X \to X$, $S_n = I - T_n$ where

$$T_n x = x_1 e_1 + \cdots + x_n e_n$$

if $x = x_1 e_1 + \cdots + x_n e_n + x_{n+1} e_{n+1} + \cdots$. Suppose that there exists a strictly increasing sequence of integers (n_k) with the following property: for each $c > 0$ there exists $r > 0$ such that if $x \in X$ and k is an integer for which $\|T_{n_k}x\| = 1$ and $\|S_n x\| \leq c$, then $\|x\| \geq 1 + r$. Then each convex, weakly relatively compact subset K of X has a normal structure. W. L. Bynum's result is as follows.

THEOREM 2.11.27 (W. L. Bynum, 1974) Suppose that the Banach space X has the following property: there exists c in (0,1) and $r > 0$ such that for $a > 0$ and for each x in X, there is a seminorm ρ which is weakly continuous and a subadditive function σ such that the following relations hold:

1. For all $y \in X$, $\|y\| \leq \rho(y) + \sigma(y)$.
2. $|\rho(x) - \|x\|| < a$, $\sigma(x) < a$.
3. If $\rho(y) < 1 + a$, $\sigma(y) \geq c$, then $\|y\| \geq (1 + r)\rho(y)$.

Then every relatively compact convex subset of X has a normal structure.

To obtain the Gossez and Lami Dozo result from this theorem we remark that, if X has a Schauder basis, then we can set $\rho(x) = \|U_{n_k}x\|$, $\sigma(x) = \|V_{n_k}x\|$, where k is the integer from Theorem 2.11.26. Then the conditions in Bynum's result are obviously satisfied and the assertion follows.

Now for the proof of Bynum's result we need some facts related to normal structure, which are themselves of interest. First we give the following definition.

DEFINITION 2.11.28 A sequence (x_n) in a Banach space X is said to be diametral if

$$\lim d(x_{n+1}, \mathrm{conv}(x_1, \ldots, x_n)) = \mathrm{diam}((x_n)) = d((x_n))$$

where conv denotes convex hull. Then we have the following result.

THEOREM 2.11.29 (M. S. Brodski and D. P. Milman, 1948) A convex subset C of a Banach space X has a normal structure if and only if no diametral sequences exist in C.

Proof: Let C be in X, C convex and having diametral sequences. Let (x_n) be such a sequence and we consider the subset $C_1 = \mathrm{conv}\ (x_n)$. We show that C_1 does not have a normal structure. Indeed, for x arbitrary in C_1 there exists n_x such that x is in $\mathrm{conv}(x_1, \ldots, x_n)$ for all $n \geq n_x$. Then

$$\begin{aligned} d(C_1) &= \sup_{u,v \in C_1} (\|u - v\|) \geq \sup_{u \in C_1} \|u - x\| \geq \|x_{n+1} - x\| \\ &= d(x_{n+1}, \mathrm{conv}(x_1, \ldots, x_n)) \end{aligned}$$

which implies that

$$d(C_1) = \sup_{u \in C_1} \|u - x\|$$

and thus C_1 does not have a normal structure.

Conversely, if C_1 is a bounded convex subset of C with $d(C_1) > 0$ and all x in C_1 are diametral points, then we can construct a diametral sequence. Indeed, if $\varepsilon > 0$ with $\varepsilon < d(C_1)$ then we take x_1 arbitrary in C_1. Suppose now that we have chosen $x_1, \ldots, x_n$; then x_{n+1} will be a point in C_1 with the property that

$$\left\|x_{n+1} - \frac{1}{n}(x_1 + \cdots + x_n)\right\| \geq d(C_1) - \frac{\varepsilon}{n^2} \qquad (*)$$

The choice of x_{n+1} with this property is possible since all points of C_1 are diametral points. Thus we have the sequence (x_n); we show that it is a diametral sequence. Let x be in $\text{conv}(x_1,\ldots,x_n)$, and thus

$$x = \alpha_1 x_1 + \cdots + \alpha_n x_n \qquad \alpha_i \in [0,1],\ \Sigma\, \alpha_i = 1$$

Let $\alpha = \sup \alpha_i$; then we have

$$\frac{1}{n}(x_1 + \cdots + x_n) = \frac{1}{n\alpha} x + \sum_{i=1}^{n} \left(\frac{1}{n} - \frac{\alpha_i}{n\alpha}\right) x_i$$

and from (*) we obtain

$$d(C_1) - \frac{\varepsilon}{n^2} \leq \frac{1}{n\alpha} \|x - x_{n+1}\| + \sum_{i=1}^{n} \left(1 - \frac{1}{n\alpha}\right) d(C_1)$$

which implies that

$$d(C_1) \geq \|x - x_{n+1}\| \geq d(C_1) - \frac{\alpha\varepsilon}{n} \geq d(C_1) - \frac{\varepsilon}{n}$$

This obviously implies that

$$\lim d(x_{n+1}, \text{conv}(x_1,\ldots,x_n)) = d(C_1) = d((x_n))$$

and the theorem is proved.

PROPOSITION 2.11.30 Let (x_n) be a diametral sequence. Then every subsequence (x_{n_k}) of (x_n) is again diametral.

Proof: Indeed, we have the inequalities

$$d((x_n)) \geq d((x_{n_k})) \geq d(x_{n_{k+1}}, \text{conv}(x_{n_1},\ldots,x_{n_k})) = a_{n_{k+1}}$$

$$= d(x_{n_{k+1}}, \text{conv}(x_1,\ldots,x_{n_{k+1}-1})) = b_{n_{k+1}}$$

and for $k \to \infty$,

$$d((x_n)) \geq d((x_{n_k})) \geq \lim a_{n_{k+1}} \geq \lim b_{n_{k+1}} \geq d((x_n))$$

and the assertion follows.

THEOREM 2.11.31 (J. P. Gossez and E. Lami Dozo, 1969) Let C be a convex weakly compact subset in a Banach space X with the property that C has a diametral sequence. Then C contains a diametral sequence with the following properties:

1. $x_n \rightharpoonup 0$.
2. $\|x_n\| \leq 1$, $\lim \|x_n\| = 1$.
3. $d((x_n)) = 1$.

Proof: Using translations and multiplications by scalars we may suppose without loss of generality that (x_n) is a diametral sequence with the properties 1 and 2. Now since 1 is true, by a well-known theorem we can find (y_n) with y_n in $\text{conv}(x_1,\ldots,x_n)$ and $\lim \|y_n\| = 0$. This implies $0 \in \text{conv}(x_n)$ and thus $\|x_n\| \leq 1$. Since

$$d(x_{n+1}, \text{conv}(x_1,\ldots,x_n)) \leq \|x_{n+1} - y\| \leq \|x_{n+1}\| + \|y_n\|$$

and

$$\lim\inf \|x_{n+1}\| \geq \lim d(x_{n+1}, \text{conv}(x_1,\ldots,x_n)) = 1$$

we must have

$$\lim \|x_n\| = 1$$

The theorem is proved.

Proof of Theorem 2.11.27: Suppose that the assertion of the theorem is false. Then for some C_1 which is convex and relatively compact [and $d(C_1) > 0$] a normal structure is lacking. This implies by Theorem 2.11.30 that there exists a sequence (x_n) satisfying the following properties:

(a) (x_n) is a diametral sequence.

(b) $\|x_n\| \leq 1$, $\lim \|x_n\| = 1$.

Let $a > 0$ with $1 - 3a \geq c$; then by the property (b) of the sequence (x_n) there exists an integer k with $\|x_k\| > 1 - a$. From the properties of ρ and σ we get

$$1 - 2a < \rho(x_k) < 1 + a$$

and the weak continuity of ρ and the properties of (x_n) imply that for some m the following inequalities hold:

1. $\|x_m\| > 1 - a$
2. $\rho(x_m) < a$
3. $1 - 2a < \rho(x_m - x_k) < 1 + a$

Then, using the properties of σ, we have

$$\sigma(x_m) > 1 - 2a$$

Further,

$$\sigma(x_m - x_k) > 1 - 2a - \sigma(x_k) > c$$

and using 3 and condition 3 of Theorem 2.11.27 we obtain the following inequality:

$$\|x_m - x_k\| \geq (1 + r)(1 - 2a) \qquad (**)$$

But $d((x_n)) = 1$ and from (**) we see that for appropriate a, $d((x_n))$ is strictly greater than 1. This is a contradiction and the theorem is proved.

Now we remark that Goebel's result mentioned in Theorem 2.11.25 follows from this. Indeed, since $\varepsilon_0(X) < 1$, X is uniformly nonsquare, and by a theorem of James (1964b), X is reflexive. This obviously implies the assertion, using the result of Bynum.

W. L. Bynum (1972) presented an example of a Banach space X with $\varepsilon_0(X) = 1$ and X lacking a normal structure. Thus, in some sense, the result of Goebel is the best possible. Also, in the same paper there is an example of a uniformly nonsquare Banach space without normal structure. Some interesting connections between the behavior of the modulus of convexity and the fixed point property for nonexpansive mappings were discovered by J. Baillon and W. L. Bynum. To formulate them we need some interesting notions.

Let C be a bounded and closed convex set in a reflexive Banach space and let

$$r(x,C) = \sup\{\|x - y\| : y \in C\}$$

and

$$R(C) = \min\{r(x,C) : x \in C\}$$

Then to say that the Banach space X has a normal structure is equivalent to saying that for each C, $d(C) > R(C)$.

DEFINITION 2.11.32 (W. L. Bynum, 1980) The normal structure coefficient N(X) of X is defined as follows:

$$N(X) = \inf_C \frac{d(C)}{R(C)}$$

where the infimum is taken over all closed, bounded, convex subsets C of X with $d(C) > 0$.

DEFINITION 2.11.33 (W. L. Bynum, 1980) Let (x_n) be a bounded sequence in X. Then the asymptotic diameter of this sequence is the number defined as follows:

$$\lim_n \sup \{\|x_m - x_k\| : m \geq n, k \geq n\}$$

The bounded sequence coefficient BS(X) of X is the supremum of the set of all numbers $k > 0$ with the property that for each bounded sequence (x_n) with the asymptotic diameter a, then for some y in $\overline{\operatorname{conv}(x_n)}$ we have $\lim\sup \|x_n - y\| \leq a/k$. The weakly convergent sequence coefficient WCS(X) of X is defined like BS(X), but we consider only weakly convergent sequences.

DEFINITION 2.11.34 For any two Banach spaces X and Y which are isomorphic, the Banach-Mazur distance coefficient is $d(X,Y) = \inf_T \|T\|\|T^{-1}\|$ where the infimum is taken over all isomorphism mappings $T : X \to Y$.

An interesting result of W. L. Bynum relating the modulus of convexity and the number N(X) is as follows.

THEOREM 2.11.35 (W. L. Bynum, 1980) If X is a Banach space with the modulus of convexity $\delta_X(t)$ then the following inequality holds:

$$N(X) = [1 - \delta_X(1)]^{-1}$$

THEOREM 2.11.36 (W. L. Bynum, 1980) If X is uniformly convex and Y is a Banach space with the property that $d(X,Y) \leq WCS(X)$ [$d(X,Y)$ being the Mazur-Banach distance coefficient] then Y has the fixed point property for nonexpansive mappings with respect to closed, convex, and bounded subsets.

This result is related to the following theorem of J. Baillon.

THEOREM 2.11.37 Let $(X, \|\ \|)$ be a Hilbert space and suppose that on X there exists a new equivalent norm $\|\ \|_1$ with the property that

$$\|x\| \le \|x\|_1$$

Then $(X, \max\{\|\ \|, 2^{1/2}\|\ \|_1\})$ is a Banach space having the fixed point property for nonexpansive mappings with respect to bounded, closed, and convex subsets.

For other properties of N(X), BS(X), and WCS(X), and related results, we refer to Bynum (1980). An interesting approach to obtaining fixed point theorems for mappings which are lipschitzian was discovered by E. A. Lifschitz (1975) using the so-called coefficient $\chi(X)$ defined on any metric space (X,d) as follows:

$\chi(X) = \sup\{k > 0$: There exists $a > 1$ such that for all x,y in X and $r > 0$ with $d(x,y) \ge r$, there exists z in X such that

$$\{p : d(x,p) \le kr\} \cap \{q : d(q,y) \le ar\} \subseteq \{u : d(u,z) \le r\}\}$$

It is clear that $\chi(X) = 1$, and a result of E. A. Lifschitz (1975) asserts that if $T : X \to X$ is a lipschitzian mapping with the Lipschitz constant m then T has a fixed point in X if $m < \chi(X)$.

If X is a Banach space, we set $\chi_0(X) = \inf\{\chi(C)$: C a closed, bounded, and convex subset of X$\}$. Very interesting results relating $\chi_0(X)$ to $\varepsilon_0(X)$ and fixed point theory for lipschitzian mappings were obtained by B. Turett (1981) and D. J. Downing and B. Turett (to appear).

REMARK 2.11.38 It would be interesting to know if other classes of Banach spaces have a normal structure, for example, the spaces considered by Beck (1962, 1963). The following theorem of W. A. Kirk is one of the most useful existence theorems for fixed points of nonexpansive mappings.

THEOREM 2.11.39 (W. A. Kirk, 1965) Let X be a Banach space, C a weakly compact set in X, and $T : C \to C$ a nonexpansive mapping. If C has a normal structure then T has fixed points in C.

Proof: For each $x \in C$ we define

$r_x(C) = \sup\{\|x - y\| : y \in C\}$

$R(C) = \inf\{r_x(C) : x \in C\}$

$C_c = \{x : x \in C, R(C) = r_x(C)\}$

We remark that C_c is a nonempty closed convex set. Indeed, for each n we consider the set

$$C_n(x) = \left\{y : y \in C, \|x - y\| \le r(C) + \frac{1}{n}\right\}$$

which is a weakly closed convex set in X. The family of sets $\{C_n(x) : x \in C\}$

has the finite intersection property, as is easy to see. This implies that $C_n = \cap_{x \in C} C_n(x)$ is a weakly compact, convex, and nonempty subset of X. Now the sequence (C_n) is decreasing, and since $C_c = \cap_n C_n$, the assertion is proved.

Another important observation is that $d(C_c) < d(C)$. Indeed, since C has a normal structure we find $x \in C$ such that $r_x(C) < d(C)$. Let u,v be points in C_c. Then

$$\|u - v\| \leq r_u(C) = r(C)$$

which implies that

$$d(C_c) = \sup \{\|u - v\| : u,v \in C_c\} = r(C) = r_x(C) < d(C)$$

Now the proof of the theorem is as follows. We consider the family of all closed, convex, and nonempty subsets of C which are invariant under T. Clearly this family, say $\mathcal{F}$, is nonempty, and it is easy to see that with the order defined by inclusion, we can apply Zorn's lemma. Then we find a minimal element, say F, in $\mathcal{F}$. We show now that this is a singleton. Let us consider a point x in F. Since for all $y \in F$, $\|Tx - Ty\| \leq \|x - y\| = r(F)$ we get that T(F) is contained in the ball with center x and radius r(F). Since the image of a point in F or in the ball with center Tx and radius r(C) is in F or in the ball with center Tx and radius r(C), we obtain by the minimality of F that F is contained in the ball with center Tx and radius r(C). But $x \in F$ and thus $Tx \in F$, i.e., F_c (defined analogously to C_c) is invariant under T. By the first remark F_c is nonempty, closed, and convex, and thus it is in $\mathcal{F}$. If the set F is not a single point [i.e., $d(F) > 0$] then by the second observation, the diameter of F_c is strictly smaller than the diameter of F. Thus we have another element in $\mathcal{F}$ which is smaller than F. This is a contradiction and the theorem is proved.

COROLLARY 2.11.40 If X is a Banach space as in one of the Theorems 2.11.21 or 2.11.22 and C is a closed, convex, and bounded set in C, then any $T : C \to C$ which is nonexpansive has fixed points in C.

2.12 STRICT CONVEXITY IN PROBABILISTIC METRIC SPACES (PM-SPACES)

An interesting and important generalization of the notion of metric space was considered by Karl Menger in 1942, under the name *statistical metric spaces* (now called *probabilistic metric spaces*). For a detailed exposition of this theory up to 1975 we refer to V. I. Istrăţescu (1975a). Below, following A. N. Šerstnev (1963, 1964), we define probabilistic normed spaces and the notion of strict convexity in such spaces.

Let $\mathcal{D}$ be the set of all left continuous functions defined on $\mathbb{R}$ such that for each $f \in \mathcal{D}$,

1. $f(x) = 1$, $x \le 0$.
2. $f(\infty) = 0$ and f is a nonincreasing function.

On $\mathcal{D}$ we consider the natural order, i.e., $f \le g$ $(f,g \in \mathcal{D})$ iff $f(x) \le g(x)$ for all $x \in \mathbb{R}$, and $f < g$ if $f \le g$ and there exists an x_0 such that $f(x_0) < g(x_0)$. We denote by I the function in $\mathcal{D}$ with the property that $I(x) = 0$ for $x > 0$. Let E be a subset of $\mathcal{D}$ containing I.

DEFINITION 2.12.1 A triangle function μ on E is any associative and commutative composition law on E having the following properties:

1. I is a neutral element.
2. For each fixed $a \in E$, $\mu(a_1,a) \le \mu(a_2,a)$ whenever $a_1 \le a_2$.
3. For all a, b in E, $\mu(a,b) \le \mu_1(a,b)$, where

$$\mu_1(a,b)(x) = \inf_t \min \{a(tx) + b[(1 - t)x], 1\}$$

and the infimum is taken over all $t \in [0,1]$.

For examples, as well as for the connection of such a triangle function with the triangle functions used in the theory of Menger and Wald spaces, we refer to the above-quoted author's book. Now we are ready to define the notion of probabilistic normed space.

DEFINITION 2.12.2 A linear vector space L over K (the complex numbers or the real numbers) is called a probabilistic normed space if there exists a mapping $\| \; ; \; \|$ defined on L with values in $\mathcal{D}$ such that the following properties hold for any $x \in \mathbb{R}$:

1. $\|\varphi;x\| = I$ iff $\varphi = 0 \in L$.
2. $\|a\varphi;x\| = \|\varphi;\frac{x}{a}\|$.
3. $\|\varphi + \psi;x\| \le \mu(\|\varphi;x\|,\|\psi;x\|)$.

The mapping

$$\varphi \to \|\varphi; \;\|$$

is called a random norm on L or a probabilistic norm on L.

EXAMPLE 2.12.3 Let X be a real Banach space and define for each φ in X the random norm $\| \; ; \; \|$ by the formula

$$\|\varphi;x\| = \begin{cases} 1 & \text{if } x \le \|\varphi\| \\ 0 & \text{otherwise} \end{cases}$$

As a triangle function we can take the following:

$$(a,b)(x) = \inf_t \max \{a(tx),\ b[(1-t)x]\}$$

where the infimum is taken over all t in [0,1].

EXAMPLE 2.12.4 Let X be a linear space over K and suppose that on X there exists a countable family of norms $(\| \ \|_n)$ with the property that for all $n = 1, 2, 3, \ldots,$

$$\|\varphi\|_n \le \|\varphi\|_{n+1}$$

We consider now a random variable τ taking only positive integer values and define

$$\|\varphi;x\| = P(\|\varphi\|_\tau \ge x) \qquad P(\tau \ge N) > 0 \text{ for all } N$$

Then $(L,\|;\|)$ is a probabilistic normed space, where μ is defined as in Example 2.12.3.

If $(X,\|;\|)$ is a probabilistic normed space then we can define a topology on X by defining the neighborhoods of the zero element,

$$(U_{\varepsilon\delta})_{0<\varepsilon\le 1, 0<\delta<\infty}$$

by

$$U_{\varepsilon\delta} = \{\varphi \in L : \|\varphi;\delta\| < \varepsilon\}$$

In this way X is a topological vector space.

In his paper (1964) Šerstnev developed the theory of a best approximation in probabilistic metric spaces. Essentially the point of view is as follows. We consider a system of elements in the probabilistic normed space X, say $(\varphi_1,\ldots,\varphi_n,\ldots)$, and if φ is an element of X we are interested in the minimization of the random norm of

$$\varphi - c_1\varphi_1 - \cdots - c_n\varphi_n$$

with respect to the order in $\mathcal{D}$. The polynomials $c_1^*\varphi_1 + \cdots + c_n^*\varphi_n$ which attain the minimum are called the polynomials of best approximation.

THEOREM 2.12.5 Let X be a probabilistic normed space and $(\varphi_1,\ldots,\varphi_n)$ be a finite linearly independent system of elements of X. Then there exists a polynomial of best approximation for each element of X with respect to the system.

Proof: For each point $(c_1,\ldots,c_n)$ in K^n we define the polynomial

$$c_1\varphi_1 + \cdots + c_n\varphi_n$$

and we consider the following order in K^n:

$$(a_1,\ldots,a_n) \le (b_1,\ldots,b_n)$$

iff

$$\|\varphi - a_1\varphi_1 - \cdots - a_n\varphi_n;x\| \le \|\varphi - b_1\varphi_1 - \cdots - b_n\varphi_n;x\|$$

where φ is the given fixed element of X. We show that for K^n with this order we can apply Zorn's lemma.

Indeed, if $\{\|\varphi - c_1^s\varphi_1 - \cdots - c_n^s\varphi_n;x\|\}_{s\in T}$ is a chain then $\inf_{s\in T}\{\|\varphi - c_1^s\varphi_1 - \cdots - c_n^s\varphi_n;x\|\}$ is again of the form

$$\varphi - c_1\varphi_1 - \cdots - c_n\varphi_n$$

Let us consider the function on $\mathbb{R}$ defined by

$$x \to F(x) = \inf_{s\in T} \|\varphi - c_1^s\varphi_1 - \cdots - c_n^s\varphi_n;x\|$$

We show that there exists a sequence (x_i) in T such that

$$\lim_{s_p} \|\varphi - c_1^{s_n}\varphi_1 - \cdots - c_n^{s_n}\varphi_n;x\| = F(x) \tag{*}$$

at each point of continuity of F. For this we consider the set S = $[0,\infty) \cap \{x : F(x - 0) = F(x + 0)\}$; there exists a countable dense subset of S, say $(x_1,x_2,\ldots)$. For x_1 we find a sequence (s_1^m), $s_1^m \ge s_1^{m+1}$ for all m, such that

$$\lim_{m\to\infty} \|\varphi - c_1^{s_1^m}\varphi_1 - \cdots - c_n^{s_1^m}\varphi_n;x_1\| = F(x_1)$$

and similarly, for each x_k there exists (s_k^m). We define now $s_i = \min_{1\le j\le k} s_j^i$ which makes sense because we have a chain, and we show that (s_i) has the property (*).

Let $\varepsilon > 0$ and let x be any element in S. Then there exist x_p, $x_p \le x$, such that

$$0 = F(x_p) - F(x) < \frac{\varepsilon}{2}$$

and from the definition of the sequence (s_p^j) there exists $m_0 > p$ such that

$$0 = \|\varphi - \Sigma\, c_k^{s_p^m}\varphi_k ; x_p\| - F(x_p) < \frac{\varepsilon}{2} \qquad m > m_0$$

In this case, for $m > m_0$ we have

$$0 \le \|\varphi - \Sigma\, c_k^{s_m}\varphi_k ;\ x\| - F(x) < \|\varphi - \Sigma\, c_k^{s_m}\varphi_k ;\ x_p\| - F(x)$$

$$\le \|\varphi - \Sigma\, c_k^{s_p^m}\varphi_k ;\ x_k\| - F(x_p) + F(x_p) - F(x) \le \varepsilon$$

Thus the sequence (s_i) satisfies the assertion. Now since $\sup_{1\le m\le\infty} \max_{1\le k\le n} |c_k^{s_m}| < \infty$, the set $(\varphi - \Sigma\, c_k^{s_m}\varphi_k)_1^\infty$ is bounded. Indeed,

$$\|\varphi - \Sigma\, c_k^{s_m}\varphi_k ;\ x\| \le \|\varphi - \Sigma\, c_k^{s_1}\varphi_k ;\ x\|$$

for all m. We consider the subspace X_{n+1} generated by $\{\varphi, \varphi_1, \ldots, \varphi_n\}$. Clearly this is isomorphic with a finite dimensional space. This implies that there exists a sequence (m_p) such that $(c_k^{s_{m_p}})$ is convergent for $k = 1, 2, \ldots, n$. Thus we have

$$F(x) = \lim_{p\to\infty} \|\varphi - \Sigma\, c_k^{s_{m_p}}\varphi_k ;\ x\|$$

and from the definition of the topology on X, this means that

$$\lim_{p\to\infty} \|\varphi - \Sigma\, c_k^{s_{m_p}}\varphi_k ;\ x\| = \|\varphi - \Sigma\, c_k^o\varphi_k ;\ x\|$$

where

$$c_k^o = \lim c_k^{s_{m_p}}$$

The theorem is proved.

The following result gives some information about the set of best approximation polynomials (i.e., elements of the form $\Sigma\, a_k\varphi_k$).

THEOREM 2.12.6 For any φ in X the set of best approximation is a convex set.

Since the proof is simple it is omitted.

As in the case of Banach spaces it would be interesting to know when this set, i.e., the set of best approximation with respect to a subspace M_1 of X, is a single-valued function, and when it is continuous. Following Šerstnev, we call a system of linearly independent elements (we consider only systems with a finite number of elements) *Cebyshev* if the set of best approximation for each element in X is nonempty and is single-valued. In the following definition we consider a class of probabilistic normed spaces for which every finite system of linearly independent elements is Cebyshev.

DEFINITION 2.12.7 A probabilistic normed space is said to be strictly convex if whenever φ, ψ are elements of X such that

$$\|\varphi + \psi;\ x\| = \mu(\|\varphi;x\|,\|\psi;x\|)$$

then there exists $s > 0$ such that $\varphi = s\psi$.

THEOREM 2.12.8 In a strictly convex probabilistic normed space every finite system of linearly independent elements of X is Cebyshev.

Proof: Let φ be given and L be an arbitrary finite system of linearly independent elements of X. From the above theorem we know that the set of best approximation with respect to L for φ is nonempty and convex. Suppose that there exist two different elements of best approximation with respect to L. In this case, $(1/2)(p_1 + p_2)$, where p_1 and p_2 are the elements of best approximation, is again an element of best approximation. Thus we have

$$\|(\varphi - p_1) + (\varphi - p_2);\ x\| = \mu(\|\varphi - p_1;\ x\|,\|\varphi - p_2;\ x\|)$$

and from the strict convexity of X,

$$\varphi - p_1 = s(\varphi - p_2)$$

Since we may suppose without loss of generality that (L,φ) is also a linearly independent system we get that $s = 1$ and thus $p_1 = p_2$.

2.13 EXTREMAL STRUCTURE IN OPERATOR SPACES

Let us consider a compact Hausdorff space X and let C(X) be the Banach algebra of all continuous complex-valued functions on X with the sup norm:

$$f \to \|f\| = \sup \{|f(x)| : x \in X\}$$

Let us now consider the dual space of C(X) and let K be the set of all x* in [C(X)]* with the property that $x^*(1) = 1 = \|x^*\|$. Here 1 denotes the function in C(X) which has the value 1 at each point of X. A famous result of R. F. Arens and J. L. Kelley (1947) asserts that the set of extreme points of K is exactly the set of all nontrivial multiplicative functionals on C(X). We mention that in the paper of R. F. Arens and J. L. Kelley only the case of real scalars is considered. Later we give this result for C(X), following essentially the exposition in N. Dunford and J. T. Schwartz (1958). A purely algebraic proof of the same result for the case of certain commutative real algebras was given in the paper of J. Tate (1951).

Suppose now that we have another compact Hausdorff space Y and we consider the corresponding Banach algebra C(Y). Assume first that C(X) and C(Y) are real algebras and we consider the Banach space L(C(X),C(Y)) of all continuous linear operators defined on C(X) with values in C(Y). Let K(C(X),C(Y)) be the set of all T in L(C(X),C(Y)) with the property that T1 = 1. A. Ionescu Tulcea and C. Ionescu Tulcea proved that the extreme points of K(C(X),C(Y)) are exactly the operators T which are multiplicative, i.e., for all f,g in C(X), T(fg) = TfTg. This result has been extended to more general situations [R. R. Phelps (1963), R. M. Blumenthal, J. Lindenstrauss, and R. R. Phelps (1965), F. F. Bonsall, J. Lindenstrauss and R. R. Phelps (1966), J. A. Crenshaw (1969, 1974), G. Converse, I. Namioka and R. R. Phelps (1969), M. Sharir (1972, 1973, 1976). S. E. Mosiman and R. F. Wheeler (1972), S. Reich (1973), etc.].

We recall now some terminology from the theory of Banach algebras [for the theory of Banach algebras, see C. E. Rickart (1960)]. We say that an algebra over K ($\mathbb{R}$ or $\mathbb{C}$) has an involution (or that A is an involutive algebra, or simply involutive) if there exists a mapping

$$* : A \to A$$

with the following properties:

$$(x + y)^* = x^* + y^* \qquad (zx)^* = z^*x^* \qquad (xy) = y^*x^* \qquad (x^*)^* = x$$

for all $x,y \in A$ and $z \in K$. The algebra A is called self-adjoint if whenever $x \in A$ then $x^* \in A$. If A is an algebra of functions on a set S then an involution on A is defined by complex conjugation, i.e.,

$$f \to f^*(s)$$

where $f^*(x)$ denotes the complex conjugate of f(s).

If A and B are two algebras over K then any mapping

$$p : A \to B$$

is called multiplicative if $p(xy) = p(x)p(y)$ for all $x,y \in A$. Multiplicative mappings of algebras are called homomorphisms; if a homomorphism is injective and surjective then it is called an isomorphism.

Let A and B be two algebras of K-valued functions defined on the sets X and Y and let

$$K_0(A,B) = \{T : T \in L(A,B),\ T \geq 0,\ T1 \leq 1\}$$

and

$$K_1(A,B) = \{T : T \in L(A,B),\ T \geq 0,\ T1 = 1\}$$

(We recall that we say $T \geq 0$ if when $f \geq 0$ then $Tf \geq 0$.) In what follows we show, in essence, that under certain assumptions about A the set of extreme points of K_i $(i = 0, 1)$ are the only multiplicative operators.

If A and B are Banach function algebras on S [i.e., subalgebras of C(S)], S is a compact Hausdorff space, and A contains the constants and separates the points of S [this means that for any two different points s_0 and s_1 of S, there exists a function $f \in A$ such that $f(s_0) \neq f(s_1)$] we consider another important set,

$$K_2(A,B) = \{T : T \in L(A,B),\ T \geq 0,\ \|T\| = 1,\ T1 \leq 1\}$$

We give an example of a Banach algebra for which there exists extreme operators that are not multiplicative.

THEOREM 2.13.1 Let A and B be algebras of functions which contain the constants and such that every real function in A is bounded. Then every extreme point of $K_i(A,B)$, $i = 0, 1$, is multiplicative.

Proof: Since every extreme point of $K_1(A,B) = K_1$ [and a similar notation will be used for $K_0(A,B)$] is an extreme point of K_0 it suffices to consider the case of K_0. In what follows 1 denotes the function with value 1 at each point of the space where it is defined.

First we make a useful remark. If T is an extreme point of K_0 and $T \pm S \in K_0$ then $S = 0$. Indeed, this follows from the identity

$$T = \frac{1}{2}(T + S) + \frac{1}{2}(T - S)$$

We prove now that

$$Tf = T1Tf$$

for all $f \in A$. We define the following operator:

$$Sf = Tf - T1Tf$$

and we remark that $S \geq 0$ since $T1 \leq 1$. From this we get that $T + S \geq 0$. But

$$(T - S)f = T1Tf$$

and thus $T - S \geq 0$, and since

$$(T - S)1 = T1T1 = (T1)^2 \leq 1$$

we obtain that $T - S \in K_0$. From the above remark we get that $S = 0$ and the assertion $Tf = T1Tf$ is proved.

Now we prove the assertion of the theorem for arbitrary real functions. For this, according to the hypothesis, every real function being bounded and A containing the constants, we can suppose without loss of generality that $0 \leq g \leq 1$. In this case we have, for the operator S defined by the formula $Sf = T(fg) - TfTg$,

$$S1 = 0 \qquad (T + S)1 = T1 \leq 1$$

Now, if $f \geq 0$, then

$$(T + S)f = Tf + Sf = Tf + T(fg) - TfTg = Tf(1 - Tg) + TfTg \geq 0$$

and since $0 \leq g \leq 1$ we have that $Tg \leq T1 \leq 1$, and thus

$$(T - S)f = Tf - Sf = Tf - T(fg) + TfTg = Tf(1 - g) + TfTg \geq 0$$

Again, from the above remark we get that $S = 0$.

Now, for arbitrary g we set

$$Sg = T(fg) - TfTg$$

and we obtain easily that $Sg = 0$. Then $T \pm S \in K_0$, and from the remark, $S = 0$. Thus T is multiplicative and the theorem is proved.

THEOREM 2.13.2 Let A and B as above and suppose that K_0 or K_1 contains an extreme point. Then A is self-adjoint.

Proof: Let $A_R = \{\text{Re } f : f \in A\}$ and suppose that for some $g \in A$, g is not in $A_R + iA_R$. Then we take a subspace N of A such that:

$$A_R + iA_R \subset N$$
$$N \cap \{zg : z \in \mathbb{C}\} = \{0\}$$
$$A = N + \{zg : z \in \mathbb{C}\}$$

We define S in L(A,B) to be zero on N and S(zg) = z. In this case we have $T \pm S \in K_0$, and T being an extreme element of K_0, S = 0. This contradiction proves the theorem.

For the following theorem we suppose that A and B are sup norm algebras.

THEOREM 2.13.3 The following inclusion relation holds:

$$K_2(A,B) \subseteq K_1(A,B)$$

and the equality occurs iff A is self-adjoint.

Proof: In what follows we use the following more or less standard notation: for any $s \in X$, L_s denotes the functional defined by the relation

$$L_s(f) = f(s)$$

i.e., the evaluation at s (and, of course, for general algebras of functions we use the same notation for such a functional).

To prove the theorem we note that it suffices to prove only the inclusion $K_2(A) \subseteq K_1(A)$ $[K_2(A) = K_2(A,\mathbb{C})]$ since, if this holds, then for any $T \in K_2(A,B)$ and any $y \in Y$,

$$L_y \circ T \in K_2(A) \subseteq K_1(A)$$

and thus

$$(Tf)(y) = (L_y \circ T)(f) \geq 0$$

for $f \geq 0$. Since y is arbitrary in Y this implies that $T \geq 0$ and thus $T \in K_1(A,B)$. Let us consider $L \in K_2(A)$; then $|Lf| \leq \|f\|$ for all $f \in A$. We prove now that if $f \geq 0$ then $Tf \geq 0$. Indeed, let Lf = a + ib where a and b are real numbers. Consider a real number d such that db > 0. In this case we have

$$|L(f + (id)1)|^2 \leq a^2 + (b + d)^2$$

and

$$|L(f + (id)1)|^2 \leq (\|f\|^2 + d^2)$$

Now if d is sufficiently large we have

$$a^2 + b^2 + 2bd + d^2 > \|f\|^2 + d^2 = \|f + (id)1\|^2$$

which is a contradiction. Thus b = 0 and L(f) is real. Suppose now that

$L(f) < 0$; since $0 \le f \le \|f\|1$ we get

$$-\|f\|1 \le f - \|f\|1 \le 0$$

Thus

$$\|f - \|f\|1\| \le \|f\|$$

and

$$|L(f - \|f\|1)| = |L(f) - \|f\|| = \|f\| - L(f) > \|f\|$$

which is again a contradiction. Thus $L(f) \ge 0$.

Let $A = A_R + iA_R$. To prove that $K_1(A,B) \subseteq K_2(A,B)$, we show first that it suffices to prove only the special case of $K_1(A)$ and $K_2(A)$. Indeed, suppose that this is true, i.e., $K_2(A) \supseteq K_1(A)$, and let $T \in K_1(A,B)$. For any $y \in Y$, we consider L_y and we have clearly that $L_y \circ T \in K_1(A) \subseteq K_2(A)$. But y is arbitrary, so this implies that $T \in K_2(A,B)$, since

$$|(Tf)(y)| = |(L_y \circ T)(f)| \le \|f\|$$

Let L be in $K_1(A)$ and $f \in A_R$. Then $\|f\|1 \ge f \ge 0$ and thus

$$\|f\| - L(f) \ge 0$$

which gives that $L(f)$ is real. Now let $g(s) = L((sf + 1)^2) = s^2L(f^2) + 2sL(f) + 1$, and since $g(s) = 0$ for all $s \in \mathbb{R}$, the discriminant of this quadratic form is negative. Thus

$$[L(f)]^2 = L(f^2)$$

for each $f \in A_R$. (This is Tate's elegant argument.) Now let f be arbitrary in $A_R + iA_R$,

$$f = f_1 + if_2$$

with $f_i \in A_R$, $i = 1, 2$. Further we have

$$\|f\|^2 1 \ge |f|^2 = f_1^2 + f_2^2$$

and thus

$$\|f\|^2 = L(\|f\|^2 1) \ge L(f_1^2 + f_2^2) \ge [L(f_1)]^2 + [L(f_2)]^2$$
$$= |L(f_1) + iL(f_2)|^2 = |L(f_1 + if_2)|^2 = |L(f)|^2$$

which gives $\|L\| = 1$.

Now we show that if $A \setminus (A_R + iA_R)$ is nonempty then $K_1(A,B) \setminus K_2(A,B)$ is nonempty. Indeed, let N be such that:

$$A_R + iA_R \subset N \qquad N \cap \{zg : z \in \mathbb{C}\} = \{0\} \qquad A = N + \{zg : z \in \mathbb{C}\}$$

where $g \in A \setminus (A_R + iA_R)$. Define $T : A \to B$ by the relation

$$T(f + zg) = [f(x) + zd]1$$

where $x \in X$ is a fixed point and d is a positive number such that $\|g\| > d$. In this case it is clear that $T \in K_1(A,B)$, and since $d > \|g\|$, $T \notin K_2(A,B)$.

The following theorem essentially contains Tate's result (1951).

THEOREM 2.13.4 Any multiplicative element of $K_i(A)$, $i = 0, 1$, is an extreme point of $K_i(A)$ iff A is self-adjoint.

Proof: Let us suppose that L is a multiplicative element of K_i and suppose further that it is not an extreme point of K_i. Then there exist L_j, $j = 1, 2$, such that

$$L_j \in K_i \qquad L = \frac{1}{2}(L_1 + L_2)$$

Since $L_j \geq 0$ we get, using an argument like that above, that $L_j(f^2) \geq [L_j(f)]^2$ for all $f \in A_R$. Now, since

$$[L(f)]^2 = L(f^2) = \frac{1}{4} L_1(f) + \frac{1}{2} L_1(f)L_2(f) + \frac{1}{4} L_2(f)$$
$$= \frac{1}{2} L_1(f^2) + \frac{1}{2} L_2(f^2) \geq \frac{1}{2} [L_1(f)]^2 + \frac{1}{2} [L_2(f)]^2$$

we get that

$$[L_1(f) - L_2(f)]^2 \leq 0$$

Thus $L_1(f) = L_2(f)$ and L is an extreme point of K_1. The converse assertion follows immediately from Theorems 2.13.2 and 2.13.3.

The following result shows an interesting connection between extremeness and the multiplicative property of elements of $K_i(A,B)$ and $K_i(A)$.

THEOREM 2.13.5 Every element of $K_i(A,B)$, $i = 0, 1, 2$, is an extreme point of $K_i(A,B)$ iff every multiplicative element of $K_i(A)$, $i = 0, 1, 2$, respectively, is an extreme point of $K_i(A)$.

Proof: Let $T \in K_i(A,B)$, $i = 0, 1$, or 2, be multiplicative, and suppose that for some $S \in K_i(A,B)$, $T \pm S \in K_i(A,B)$. In this case the functional $L_y \circ T \pm L_y \circ S$ is in $K_i(A)$ and multiplicative. Then it is an extreme point and thus $L_y \circ S = 0$. Since $y \in Y$ is arbitrary this implies that $S = 0$. For the converse, suppose that $L \in K_i(A)$ is multiplicative. We set Tf equal to (Lf)1 which clearly is multiplicative and thus it is extreme. This implies that L is extreme in $K_i(A)$ and the theorem is proved.

COROLLARY 2.13.6 Suppose that A is self-adjoint and contains only bounded functions. Then $T \in K_i(A,B)$ is multiplicative iff it is extreme.

Proof: The assertion follows from Theorems 2.13.1, 2.13.3, 2.13.4, and 2.13.5.

We prove now the important result of R. F. Arens and J. L. Kelley (1947) on the characterization of the extreme points of the unit ball of the conjugate space of C(S), where S is a compact Hausdorff space and C(S) denotes the Banach space of all continuous complex-valued functions on S with the sup norm.

THEOREM 2.13.7 (Arens-Kelley, 1947) The extreme points of

$$B(C(S)^*) = \{x^* \in C(S)^* : \|x^*\| \le 1\}$$

are the functionals of the form

$$f \to zf(s)$$

where s is a fixed point in S and $|z| = 1$.

Proof: Let x^* be an extreme point of $B(C(S)^*)$, and suppose that for no $s \in S$ is $x^*(f)$ equal to $zf(s)$ for some $|z| = 1$. It is clear that $\|x^*\| = 1$, and from the Riesz-Kakutani theorem there exists a measure μ such that

$$x^*(f) = \int_S f(t)\ d\mu(t) \qquad 1 = |\mu|(S) = \|\mu\|$$

First we remark that we can suppose without loss of generality that μ is positive. Indeed, if μ is the measure representing the above extreme functional then we can consider the variation $|\mu|$ of μ. Then the functional

$$x_1^*(f) = \int_S f(t)\ d|\mu|(t)$$

is in $B(C(S)^*)$ and moreover it is extreme. Indeed, in the contrary case there exist two measures μ_1 and μ_2 such that

$$|\mu| = \frac{1}{2}(\mu_1 + \mu_2)$$

Clearly we have the relations

$$\|\mu_1\| = \|\mu_2\| = 1$$

and if $d\mu/d|\mu|$ is the Nicodym derivative then we have again

$$\mu = \frac{1}{2}(B\mu_1 + B\mu_2) \qquad \|B\mu_1\| = \|B\mu_2\| = 1 \qquad \frac{d\mu}{d|\mu|} = \beta$$

and thus x* is not extreme.

Now the proof of the Arens-Kelley theorem is as follows. From the compactness of S we get that there exist two disjoint open subsets V_1 and V_2 such that

$$0 < \mu(V_i) < 1$$

and we define the following functional:

$$F(f) = \mu(V_1) \int_{V_2} f(t)\, d\mu(t) - \mu(V_2) \int_{V_1} f(t)\, d\mu(t)$$

Then it is not difficult to see that the norm of this functional is equal to $\mu(V_1)\mu(V_2)$. We now define the functionals

$$F_1(f) = x^*(f) + F(f) \qquad F_2(f) = x^*(f) - F(f)$$

which are obviously continuous. Since

$$F_1(f) = \int_{S-V_1-V_2} f(t)\, d\mu(t) + [1 + \mu(V_1)] \int_{V_2} f(t)\, d\mu(t)$$
$$+ [1 - \mu(V_2)] \int_{V_1} f(t)\, d\mu(t)$$

we get that the norm of F_1 satisfies the inequality

$$\|F_1\| \le 1 - \mu(V_1) - \mu(V_2) + [1 + \mu(V_1)]\mu(V_2) + [1 - \mu(V_2)]\mu(V_1) = 1$$

Similarly we get that the norm of F_2 is not greater than 1. But

$$x^* = \frac{1}{2}(F_1 + F_2)$$

and this contradiction proves that there exists s such that x* is the evaluation functional at s. Since it is not difficult to see that any functional of the form

$$z \times (\text{evaluation at some point of } S)$$

is an extreme point of B, the Arens-Kelley theorem is proved.

THEOREM 2.13.8 (A. and C. Ionescu Tulcea) Let X and Y be two compact Hausdorff spaces, let $A = C(X)$, and B be a subalgebra of $C(Y)$ containing the constants. Then the following assertions hold:

1. $K_1(A,B) = K_2(A,B)$.
2. T is an extreme point in $K_1(A,B)$ iff it is multiplicative and is in $K_1(A,B)$.
3. $T \in L(A,B)$ is an extreme point of $K_2(A,B)$ iff there exists a continuous function $\phi : Y \to X$ such that

$$Tf = f \circ \phi$$

Proof: Since $A = C(X)$ is self-adjoint the equality $K_1(A,B) = K_2(A,B)$ follows. From Theorem 2.13.5 we get that (2) holds. We note also that assertion (3) implies assertion (2). We show now that the converse implication holds. Suppose now that T is multiplicative and let $y \in Y$ be arbitrary. Consider the functional

$$L_y \circ T$$

which is multiplicative and thus extreme. By the Arens-Kelley theorem it is the evaluation at some point of X, which we denote by $\phi(y)$. Thus we have a function

$$y \to \phi(y)$$

on Y with values in X. We remark that it is continuous because the topology on X is the same as the weak topology on X defined by $C(X)$ and because Tf is an element of $C(Y)$ for each $f \in C(X)$. From the form of ϕ we see that (2) holds. The theorem is proved.

REMARK 2.13.9 For interesting and important applications of Theorem 2.13.8 to lifting theory we refer to the monograph of A. and C. Ionescu Tulcea (1969).

The above result was extended by R. M. Blumenthal, J. Lindenstrauss, and R. R. Phelps (1965) to extreme operators of the unit ball of $L(C(X), C(Y))$ in the case when X is metrizable. Further related results are in A. Gendler (1976), F. F. Bonsall, J. Lindenstrauss and R. R. Phelps (1966), and M. Sharir (1972, 1973, 1976).

We present now an example given by J. Lindenstrauss, R. R. Phelps, and J. V. Ryff of an operator which is extreme and not multiplicative. This is an operator defined on the Banach algebra A of all analytic functions in $\{z : |z| < 1\}$ which are continuous on $\{z : |z| \leq 1\}$. This algebra may be the continuous functions on the unit circle with the Fourier coefficients of negative index zero. Another description of A is the closure of the complex polynomials in the variable $e^{i\theta}$ (of course, with the sup norm). As an excellent source concerning this algebra, we cite K. Hoffman (1962).

We denote the function with the value k at all points $e^{i\theta}$ by k. Let K be the set of all linear operators $T : A \to A$ such that

$$T1 = 1 \qquad \|T\| \le 1$$

It is clear that any homomorphism is in the above set K. We give now an example of an operator $T \in K$ which is extreme and not multiplicative. For each $f \in A_0 = \{g : g \in A, \int g(e^{i\theta})d\theta = 0\}$, we set

$$(Tf)(e^{i\theta}) = \frac{1}{2}\{[1 + \lambda(\theta)]f(e^{i\theta/2}) + [1 - \lambda(\theta)]f(-e^{i\theta/2})\}$$

with $\lambda(\theta) = \cos(\theta/2)$. If $p(z) = a_0 + a_1 z + \cdots + a_n z^n$ then

$$(Tp)(z) = \frac{1}{2}\Sigma\,(a_{2n-1} + 2a_{2n} + a_{2n+1})z^n$$

with $a_k = 0$ whenever $k \ne 0, 1, 2, \ldots, n$. From this we see that on the subspace of polynomials T is a contraction and $T1 = 1$. Thus T may be extended to an operator in K. Clearly T is not multiplicative. We show now that T is an extreme point in K. Indeed, suppose that there exists $U \in L(A,A)$ such that $T \pm U \in K$. Consider now the measure $\varepsilon(\theta)$ with unit mass at $e^{i\theta}$. We set

$$(T^* + U^*)(\varepsilon(\theta)) = \frac{1}{2}\,[1 + \lambda(\theta)]\varepsilon\left(\frac{\theta}{2}\right) + [1 - \lambda(\theta)]\varepsilon\left(\frac{\theta}{2} + \pi\right) + \mu$$

$$(T^* - U^*)(\varepsilon(\theta)) = \frac{1}{2}\,[1 + \lambda(\theta)]\varepsilon\left(\frac{\theta}{2}\right) + [1 - \lambda(\theta)]\varepsilon\left(\frac{\theta}{2} + \pi\right) - \mu + \nu$$

where μ is unknown and $\int f\,d\nu = 0$ for all $f \in A$ (i.e., ν annihilates A). We recall the famous F. and M. Riesz theorem which asserts that any Baire measure P with the property that

$$\int_{-\pi}^{\pi} e^{in\theta}\,dP(\theta) = 0 \qquad n = 1, 2, 3, \ldots$$

is absolutely continuous with respect to Lebesgue measure $\frac{1}{2}\,d\theta$. This clearly implies that ν is nonatomic. We write

$$\mu = \frac{1}{2}\left[\rho(\theta)\varepsilon\left(\frac{\theta}{2}\right) - \sigma(\theta)\varepsilon\left(\frac{\theta}{2} + \pi\right)\right] + r$$

where ρ and σ are to be determined. Since $\|T^* \pm U^*\| \le 1$ we obtain

$$\left\|\frac{1 + \lambda + \rho}{2}\,\varepsilon\left(\frac{\theta}{2}\right)\right\| + \left\|\frac{1 - (\lambda + \sigma)}{2}\,\varepsilon\left(\frac{\theta}{2} + \pi\right) + r\right\|$$

$$= \left|\frac{1 + \lambda + \rho}{2}\right| + \left|\frac{1 - (\lambda + \sigma)}{2}\right| + \|r\| \le 1$$

and we have the relation

$$\left|\frac{1 + (\lambda - \rho)}{2}\right| + \left|\frac{1 - (\lambda - \sigma)}{2}\right| + \|r + \nu\| \le 1$$

Then we have the relations

$$|1 + \lambda + \rho| + |1 - (\lambda + \sigma)| \le 2$$
$$|1 + \lambda - \rho| + |1 - (\lambda - \sigma)| \le 2$$

and then $r = 0$ and $\nu = 0$. This implies also that the remaining terms are real and of the same sign.

This implies that $T^* \pm U^*$ is of the form

$$(T^* \pm U^*)\varepsilon(\theta) = \frac{1}{2}\,[1 + \lambda(\theta) \pm \rho(\theta)]\varepsilon\left(\frac{\theta}{2}\right) + [1 - \lambda(\theta) \pm \rho(\theta)]\varepsilon\left(\frac{\theta}{2} + \pi\right)$$

and since $U = 0$ is admissible, taking $\rho = 0$ we find $T^*\varepsilon(\theta)$. Then for any other admissible U we have

$$U^*\varepsilon(\theta) = \frac{1}{2}\,\rho(\theta)\left[\varepsilon\left(\frac{\theta}{2}\right) - \varepsilon\left(\frac{\theta}{2} + \pi\right)\right]$$

Let f be the function $f(z) = z$; then

$$(Uf)(e^{i\theta}) = \rho(\theta)e^{i\theta/2}$$

which is an element of A. This implies that the Fourier coefficients of this function of negative index must be zero. This implies that Uf has the form

$$(Uf)(e^{i\theta}) = a_0 + a_1 e^{i\theta} = \rho(\theta)e^{i\theta/2}$$

Now if $\theta = 0$ and $0 = \pi$ we get $a_0 = a_1^*$; thus $\rho(\theta) = k \cos(\theta/2 + b)$, where $a_0 = (1/2)ke^{-ib}$. Since

$$\left|k \cos\left(\frac{\theta}{2} + b\right)\right| \le 1 - \cos\frac{\theta}{2}$$

we must have $b = \pi/2$, and thus $\rho(\theta) = k \sin(\theta/2)$. But

$$(T + U)f(e^{i\theta}) = \left(\cos\frac{\theta}{2} + k \sin\frac{\theta}{2}\right)e^{i\theta/2}$$

and since the norm of this function is not greater than 1, the derivative of

$$\cos\frac{\theta}{2} + k\sin\frac{\theta}{2}$$

is zero at $\theta = 0$, and this implies clearly that $k = 0$. But then $U = 0$, and thus T is extreme.

J. Lindenstrauss, R. R. Phelps, and J. Ryff gave an extension to the case of the algebra H^{∞}. As we mentioned above, F. F. Bonsall, J. Lindenstrauss, and R. R. Phelps extended some results about extreme operators to more general algebras of functions, namely, to algebras of real-valued functions defined on the sets X and Y or to algebras of complex-valued functions where these are assumed to be self-adjoint. The following two hypotheses were used to obtain more precise results:

(1) We say that the algebra A of real-valued functions on X satisfies the hypothesis H_* if whenever $f \in A$, $f \geq 0$, then $f(1 + f)^{-1} \in A$.

(2) We say that the algebra A of real-valued functions on X satisfies the hypothesis H_{**} if all f in A are bounded, i.e., for each f in A there exists a constant $M_f > 0$ such that $|f(x)| \leq M_f$.

For an algebra of functions A we denote by A_+ the set of all f in A which are nowhere negative. As above we consider the sets $K_0(A,B)$, $K_1(A,B)$, $K(A,B)$, where A, B are algebras of real-valued functions on X and Y, respectively. The following result is useful when the algebra A has the property that 1 is not in A (1 denotes, as usual, the function defined on X which has the value 1 at each point of X).

LEMMA 2.13.10 Let $T \in K_0(A,B)$ and g be a fixed element of A. We define the mapping $U_g : A \to B$ by the relation

$$U_g(f) = Tfg - TfTg$$

Then the following assertion holds: if $0 \leq g \leq 1$ then $T \pm U_g$ is in $K_0(A,B)$; if 1 is in A and $T \in K_1(A,B)$ then $T \pm U_g$ is in $K_1(A,B)$.

Proof: Since $1 - Tg \geq 0$ for any $f \geq 0$ we have

$$(T + U_g)(f) = Tf(1 - Tg) + Tfg \geq 0$$

and since $f - fg \geq 0$ we obtain that

$$(T - U_g)f = T(f - fg) + TfTg \geq 0$$

Now, if $0 \leq f \leq 1$ then $0 \leq fg \leq g$ and $Tf \leq 1$, which implies that

$$(T + U_g)f \leq Tf(1 - Tg) + Tg \leq 1 - Tg + Tg = 1$$

Also, $0 \le f - fg + g = f(1 - g) + g$, and thus

$$(T - U_g)f = Tf - Tfg + TfTg \le Tf - Tfg - Tg \le T(f - fg + g) = 1$$

The assertions of the lemma follows clearly from these and the remark that if T is in $K_1(A,B)$, $T1 = 1$, $1 \in A$, $U_g 1 = 0$.

LEMMA 2.13.11 Let A,B be as above and suppose further that either $K_0(A,B)$ or $K_1(A,B)$ has extreme points. Then $A = A_+ - A_+$.

Proof: Suppose that the assertion is false. Then we find f in A such that $f \notin A_+ - A_+$. We consider a linear subspace V containing $A_+ - A_+$ such that $A = V + \mathbb{R}f = V + \{sf : s \in \mathbb{R}\}$. We consider g in B, $g \neq 0$, and we define the mapping $U : A \to B$ by the relation

$$U(h + sf) = sg$$

Then clearly $U \neq 0$ and $U(A_+) = 0$. Then $tU \in K_0(A,B)$ and $tU \in K_1(A,B)$ for all t in $\mathbb{R}$. This implies that $K_0(A,B)$ and $K_1(A,B)$ have no extreme points. This is a contradiction.

We remark that if A has no bounded elements then $K_0(A,B) = K(A,B)$. Also, it is not difficult to see that a necessary and sufficient condition for the equality $K_0(A,B) = K(A,B)$ is that no bounded elements exist in A. Now using the above lemmas, in the presence of some additional hypotheses about A, we can characterize the extreme points of $K_0(A,B)$.

THEOREM 2.13.12 Suppose that A satisfies either the hypothesis H_* or H_{**} and let $T \in K_0(A,B)$ be an extreme element of $K_0(A,B)$. Then T is multiplicative.

Proof: Indeed, if g is in A and A satisfies the hypothesis H_{**} then we may suppose that $0 \le g \le 1$ and we define the operator $U_g : A \to B$ by the formula

$$U_g(f) = Tfg - TfTg \tag{*}$$

which, according to the above lemmas, is the zero operator. This implies obviously that $Tfg = TfTg$ whenever $g \ge 0$. But by Lemma 2.13.11, $A = A_+ - A_+$ and the assertion follows.

Now, if A satisfies the hypothesis H_*, we take h in A_+ and then $h/(1 + h)$ is in A and obviously $0 \le h \le 1$. Consider the operator U_g, $g = h/(1 + h)$. We obtain $Tfg = TfTg$ for all g. But $h = g + gh$, and thus

$$T(fh) = T(f(g + gh)) = TgT(f + fh) = TfTg + TfhTg$$
$$Th = Tg + TgTh$$

which imply that

$$\begin{aligned} TfTh + TfhTh + (Tf + Tfh)Th &= (Tf + Tfh)(Tg + TgTh) \\ &= TfTg + TfhTg + TfThTg + TfhTgTh \\ &= Tfh + TfhTh \end{aligned}$$

Comparing the first and the last term in the above sequence of equalities we get that $TfTh = Tfh$ for all f in A and h in A_+. But $A = A_+ - A_+$ and thus $Tfh = TfTh$ for all f,h in A.

COROLLARY 2.13.13 Let A, B be algebras of real-valued functions on X and Y, respectively, with the following properties:

1. $1 \in A$, $1 \in B$.
2. A satisfies either the hypothesis H_* or H_{**}.

Then every extreme point of $K_1(A,B)$ is a multiplicative operator.

Proof: In this case we have that $K_1(A,B) \subset K_0(A,B)$ and all the extreme points of $K_1(A,B)$ are extreme points of $K_0(A,B)$.

EXAMPLE 2.13.14 Let $A = \{p : p \text{ is a polynomial with real coefficients}\}$. Then $K_1(A,A)$ contains an extreme operator which is not multiplicative.

Proof: Define the operator T by the formula

$$(Tp)(x) = \frac{x}{x^2 + 1} p(x^2 + 1) + \frac{x^2}{x^2 + 1} p(x^2 + 1)(x + 1)$$

We show that T has the following properties:

1. T is an extreme point of $K_1(A,A)$.
2. T is not multiplicative.

Indeed, the second assertion follows immediately if we consider the polynomials $p(x) = x$, $q(x) = x$. Then

$$(Tp)(x) = (Tq)(x) = \frac{x^2 + 1}{x^2 + 1} + \frac{x^2}{x^2 + 1} (x^2 + 1)(x + 1) = 1 + x^2(x + 1)$$

and

$$\begin{aligned} T^*(pq)(x) = T(x^2) &= \frac{(x^2 + 1)^2}{x^2 + 1} + \frac{x^2}{x^2 + 1} (x^2 + 1)^2(x + 1)^2 \\ &= x^2 + 1 + x^2(x^2 + 1)(x + 1)^2 \end{aligned}$$

and thus

$$T(pq)(x) \neq Tp(x)Tq(x)$$

Now we show that T is an extreme point $K_1(A,A)$. If $S : A \to A$ has the property that $T \pm S$ is in $K_1(A,A)$ we show that $S = 0$. For each fixed x we define

$$p \to x_x^*(p) = (Tp)(x)$$

If we denote

$$x_1 = x^2 + 1 \qquad x_2 = (x^2 + 1)(x + 1)$$

then clearly

$$x_x^*(p) = ap(x_1) + (1 - a)p(x_2)$$

for $a \geq 0$. If we set

$$y_{S,x}^*(p) = (Sp)(x)$$

we see that $T \pm S \in K_1(A,A)$ implies that

$$|y_{S,x}^*(p)| = x_x^*(p)$$

Thus $y_{S,x}^*$ is a combination of the functionals

$$p \to p(x_1)$$

$$p \to p(x_2)$$

because $p(x_1) = p(x_2) = 0$ implies that $y_{S,x}^*(p) = 0$ since $|y_{S,x}^*(p^2)| \leq x_x^*(p^2) = 0$. Then $T \pm S \in K_1(A,A)$ implies that $y_{S,x}^*(1) = 0$, and thus for all $n > 1$,

$$y_{S,x}^*((np + 1)^2) = 1 \geq |y_{S,x}^*(n^2p^2 + 2np + 1)| = |2ny_{S,x}^*(p)|$$

Thus for all $n > 1$,

$$|y_{S,x}^*(p)| \leq \frac{1}{n}$$

and this obviously implies that

$$y_{S,x}^*(p) = 0$$

Then for each x in $\mathbb{R}$ there exists a(x) and b(x) such that

$$y^*_{S,x}(p) = a(x)p(x^2 + 1) + b(x)p((x^2 + 1)(x + 1))$$

and since $y^*_{S,x}(1) = 0$ we obtain

$$a(x) = -b(x)$$

Thus for all p in A,

$$(Sp)(x) = a(x)p(x^2 + 1) - a(x)p((x^2 + 1)(x + 1))$$

which implies that $(Sp)(0) = 0$.

If $x \neq 0$ then we can find positive polynomials p and q such that the following properties hold:

$$p(x^2 + 1) = 0 \qquad p((x^2 + 1)(x + 1)) \neq 0$$

$$q(x^2 + 1) \neq 0 \qquad q((x^2 + 1)(x + 1)) = 0$$

Applying $T \pm S$ to these polynomials we get the inequalities

$$a(x) \pm \frac{x^2}{x^2 + 1} \geq 0$$

$$a(x) \pm \frac{1}{x^2 + 1} \geq 0$$

Thus

$$(x^2 + 1)a(x) \leq 1$$

$$(x^2 + 1)a(x) \leq x^2$$

which obviously implies that $Q(x) = x(x^2 + 1)a(x)$ is a polynomial satisfying the inequalities

$$|Q(x)| \leq |x| \tag{*}$$

$$|Q(x)| = x^3 \tag{**}$$

From these inequalities we obtain that $Q = 0$. Indeed, from (**) we obtain that Q is a polynomial with no terms in x^4, x^5, ..., etc., and thus

$$Q = q_0 + q_1 x + q_2 x^2 + q_3 x^3$$

and from (*) we obtain that $q_2 = q_3 = 0$. Then (**) gives $q_1 = 0$. Thus Q is constant and this implies that $a(x) = 0$. Thus S is the zero operator and T is an extreme point.

This example shows that something like the hypotheses H_* and H_{**} must be imposed on A to obtain information about the extreme points in $K_1(A,B)$. The above results show a strong connection between extreme points of certain convex sets and multiplicative operators. The results below make this thesis more plausible. First we give the following.

LEMMA 2.13.15 Let T be a positive operator on A. Then the following inequalities hold:
$(Tfg)^2 \leq Tf^2Tg^2$.
For all $f \geq 0$, $(Tf^2)^2 \leq TfTf^3$.

Proof: To prove the first inequality we consider the function $f - tg$ for t in $\mathbb{R}$, and thus the positivity of T implies that

$$0 \leq T(f - tg)^2 = Tf^2 - 2tTfg + t^2Tg^2$$

and the fact that this is positive for all t obviously implies our first inequality. For the second inequality we consider the function $f(f - t)^2$ for t in $\mathbb{R}$ and thus again the positivity of T implies that

$$0 \leq T(f^3 - 2tf^2 + t^2f) = Tf^3 - 2tTf^2 + t^2Tf$$

which, as in the case of (*), implies that (**) holds.

THEOREM 2.13.16 Let $m : A \to \mathbb{R}$ be a nontrivial multiplicative linear functional on A and suppose that $A = A_+ - A_+$. Then m lies on an extreme ray of $K(A,\mathbb{R})$.

Proof: First we remark that if

$$m = m_1 + m_2$$

with $m_i \geq 0$ then $m_1(f) = m(f)$ for all f in A_+. For this we need the following inequality, which is an immediate consequence of the Hölder inequality: if t,s are numbers in [0,1] then

$$(ts)^{1/2} + [(1 - t)(1 - s)]^{1/2} \leq 1$$

and equality holds if and only if $t = s$.

First we remark that for positive f, $m(f) = 0$ implies that $m_1(f) = 0$, and the assertion follows if we show that $m_1(f) = m_1(g)$ for all f,g in A_+ such that $m(f) = m(g) = 1$. We consider f with the property that $m(f) = 1$, and thus

$$m_1(f) + m_2(f) = 1$$

Set

$$t = m_1(f) \qquad s = m_1(f^3)$$

and thus

$$m(f^2) = 1 = m_1(f^2) + m_2(f^2) = (ts)^{1/2} + [(1 - t)(1 - s)]^{1/2} = 1$$

which implies that t = s, i.e.,

$$m_1(f) = m_1(f^3)$$

But

$$m_1(f^2) = [m_1(f)m_1(f^3)]^{1/2} = (ts)^{1/2} = t = m_1(f)$$

and similarly, we obtain that

$$m_1(g) = m_1(g^2) \qquad m_2(f) = m_2(f^2) \qquad m_2(g) = m_2(g^2)$$

Since $(f - g)^2 = 0$ and $m(f - g) = 0$ we get that

$$m_1(f - g)^2 m_1(g^2) = 0 = m_1((fg - g^2)^2)$$

Thus

$$m_1(fg) = m_1(g^2)$$
$$m_1(fg) = m_1(f^2)$$

which imply that

$$m_1(g) = m_1(g^2) = m_1(f^2) = m_1(f)$$

and similarly,

$$m_2(g) = m_2(g^2) = m_2(f^2) = m_2(f)$$

Thus $m(f) = m(g)$ and the assertion is proved as well as the theorem.

THEOREM 2.13.17 Let m be a multiplicative functional on A and suppose that $A = A_+ - A_+$. If m is in $K_0(A,\mathbb{R})$, and for some bounded element of A, m is not zero, then m is an extreme point of $K(A,\mathbb{R})$ when A satisfies H_*; in the contrary case m is on an extreme ray of $K_0(A,\mathbb{R})$.

Proof: From the above theorem we know that m lies on an extreme ray of $K_0(A,\mathbb{R})$. If m is zero on all bounded elements of A then tm is in $K_0(A,\mathbb{R})$ for all t in $\mathbb{R}^+$. Thus m is on an extreme ray of $K_0(A,\mathbb{R})$. If now, m is nonzero at some bounded element of A, say f, then we show that m is an extreme point of $K_0(A,\mathbb{R})$. Indeed, if this is not the case we find m_1 and m_2 in $K_0(A,\mathbb{R})$ such that

$$m = \frac{1}{2}(m_1 + m_2)$$

By the above result we have that $m_i = k_i m$, $k_i \geq 0$.

Suppose now that for some $\varepsilon > 0$ there exists a function g in A with the following properties:

$$0 \leq g \leq 1 \qquad m(g) > 1 - \varepsilon$$

Then

$$1 = m_i(g) = k_i m(g) = k_i(1 - \varepsilon)$$

and thus k_i are in [0,1]. But $2m = m_1 + m_2$, so $k_1 + k_2 = 1$, and thus $k_1 = k_2 = 1$ and m is an extreme point.

To complete the proof it remains to show that such a function g exists. Since f is a bounded element we may suppose that $f \geq 0$ (taking eventually f^2) and $m(f) > 0$. From the hypothesis H_* we conclude that $f/(1 + f)$ is in A and taking eventually the function nf we may assume without loss of generality that f satisfies the inequality

$$m\left(\frac{f}{1 + f}\right) \leq 1 - \frac{\varepsilon}{2} \qquad \varepsilon > 0$$

Since f is bounded we can find constants k and K such that

$$0 \leq f \leq k < K$$

We consider now the function

$$s_n = \sum_{i=0}^{n} \frac{f}{1 + K}\left(\frac{K - f}{1 + K}\right)^i$$

and using the multiplicativity of m we obtain that the following relations hold:

$$m(f(K - f)^i) = m(f)[K - m(f)]^i$$

If we set

$$t = (1 + K)^{-1}[K - m(f)]$$

then t is in (0,1) and

$$m(x_n) = \frac{m(f)}{1 + K} \sum_{i=0}^{n} t^i = \frac{m(f)}{1 + K} \frac{1 - t^{n+1}}{1 - t} = \frac{m(f)}{1 + m(f)} (1 - t^{n+1})$$

But t being in $(0,1)$, $\lim t^n = 0$, and thus for some N_ε, for all $n \geq N_\varepsilon$, $m(s_n) > 1 - \varepsilon$. Then for some $n \geq N_\varepsilon$ we can take s_n as an example of a function g.

THEOREM 2.13.18 Let T be in $K_1(A,B)$ and suppose that T is multiplicative. Then T is an extreme point of $K_1(A,B)$.

Proof: This follows from the fact that for all y in Y,

$$y \to (Tf)(y)$$

is in $K_1(A,\mathbb{R})$ and multiplicative. From this we obtain easily that T is an extreme point of $K_1(A,B)$.

THEOREM 2.13.19 Let T be in $K_0(A,B)$ and suppose that the following assertions hold:

1. T is multiplicative.
2. $A = A_+ - A_+$.
3. A satisfies either H_* or H_{**}.

Then T is an extreme point of $K_1(A,B)$.

Proof: For each y in Y we define the functional

$$y \to (Tf)(y) = L_y(f)$$

which is clearly multiplicative. If $\{L_y : y \in Y\}$ is a subset of the set of extreme points of $K_0(A,\mathbb{R})$ then we see easily that T is in fact an extreme point. If A satisfies the hypothesis H_* then we can show that L_y is not identically zero on the set of bounded elements of A. This implies that L_y is extreme in $K_0(A,\mathbb{R})$, and if H_* holds we can prove similarly that L_y is extreme. Then T is an extreme point and the theorem is proved.

REMARK 2.13.20 There exists an algebra A and an operator T in $K(A,A)$ such that the following assertions hold:

1. T lies on an extreme ray of $K(A,A)$.
2. T is not of the form kS, where k is a constant and S is a multiplicative operator.

Proof: See F. F. Bonsall, J. Lindenstrauss, and R. R. Phelps (1966), p. 176.

For some related results we refer to J. A. Crenshaw (1969, 1972), and S. E. Mosiman and R. F. Wheeler (1972), as well as to the literature quoted there.

Let X, Y be two compact Hausdorff spaces and let C(X), C(Y) denote the algebras of continuous functions defined on X and Y, respectively. R. M. Blumenthal, J. Lindenstrauss, and R. R. Phelps (1965) considered, in connection with the extreme operators, the following class of operators.

DEFINITION 2.13.21 An operator T in L(C(X),C(Y)) is called *nice* if

$$T^* : Y \to [C(X)]^*$$

defined by the formula

$$T_y^*(f) = (Tf)(y)$$

has the property that T_y^* is an extreme point of the unit ball of $[C(X)]^*$, for each y in Y.

It is easy to see that the following proposition holds.

PROPOSITION 2.13.22 If T is nice then T is extreme.

The problem posed by R. M. Blumenthal, J. Lindenstrauss, and R. R. Phelps related to Proposition 2.13.22 is as follows:

PROBLEM 2.13.23 If T is an extreme operator then is T nice?

There are several results on this problem; we list now those that are known (more precisely, known to this author): The answer is yes if:

1. X is a metric space, and C(X), C(Y) are real algebras [R. M. Blumenthal, J. Lindenstrauss, and R. R. Phelps (1965)].

2. C(X), C(Y) are real algebras, Y is a metric space, and X is an Eberlein-compact space [D. Amir and J. Lindenstrauss (1968)].

3. C(X), C(Y) are real algebras, Y is separable, and T is weakly compact [H. H. Corson and J. Lindenstrauss (1966)].

4. T is compact, and C(X), C(Y) are either real or complex algebras [R. M. Blumenthal, J. Lindenstrauss, and R. R. Phelps (1965); P. D. Morris and R. R. Phelps (1970)].

5. X is dispersed (i.e., contains no connected set with more than one point; also called hereditarily disconnected), and C(X), C(Y) are either real or complex algebras [M. Sharir (1972)].

6. Y is extremally disconnected (i.e., a compact Hausdorff space such that the closure of every open set is open), and C(X), C(Y) are either real or complex algebras [M. Sharir (1972)].

7. C(X), C(Y) are real algebras, and $\|T_y^*\| = 1$ for all $y \in Y$ [R. M. Blumenthal, J. Lindenstrauss, and R. R. Phelps (1965)].

8. C(X), C(Y) are complex algebras, and $\|T_y^*\| = 1$ for all $y \in Y$ [A. Gendler (1976)].

9. X is metric and Y is basically disconnected (i.e., a compact metric space such that the closure of every open F_σ set is open) [A. Gendler (1976)].

Also, M. Sharir (1977) showed that if C(X), C(Y) are complex algebras and X is a nondispersed compact Hausdorff space then there exist a compact Hausdorff space Y and an operator T in L(C(X),C(Y)) which is extreme but not nice. Further, in the case of real algebras, M. Sharir constructed a very interesting nonmetrizable Eberlein-compact space such that the same assertion holds as for the complex algebras. Thus we see that Problem 2.13.23 has, in general, a negative answer for both the real and the complex case. For the construction of these counterexamples we refer to Sharir's papers.

In connection with the above problem and results on the extreme points of the unit ball of L(C(X),C(Y)), it would be interesting to consider another pair of function spaces, namely, functions satisfying a Lipschitz condition. Let us suppose now that X,Y are compact metric spaces and α is in (0,1). We denote by $C^\alpha(X)$ the set of all functions in C(X) with the property that

$$m_\alpha(f) = \sup \frac{|f(s) - f(t)|}{d^\alpha(t,s)} < \infty$$

Then $C^\alpha(X)$ is a Banach algebra (with identity) with respect to the norm

$$\|f\|_\alpha = \|f\|_X + m_\alpha(f)$$

where

$$f \to \|f\|_X = \sup \{|f(s)| : s \in X\}$$

We can consider a very interesting subalgebra in C(X), namely, the subalgebra denoted by $C^{\alpha,0}(X)$ and consisting of all f in $C^\alpha(X)$ with the property that

$$\lim_{t \to s} \frac{|f(t) - f(s)|}{d^\alpha(t,s)} = 0$$

Then we can consider the unit balls of the Banach spaces $L(C^\alpha(X),C^\alpha(Y))$ and $L(C^{\alpha,0}(X),C^{\alpha,0}(Y))$, and the problem is to determine the extreme points of these unit balls as well as the nice operators.

We conjecture that for these spaces all extreme operators are nice. (It may be of interest to note that the embeddings

$$C^{\alpha}(X) \to C(X)$$

and

$$C^{\alpha,0}(X) \to C(X)$$

are compact, as follows immediately from the Arzela-Ascoli theorem.) From the theory of Banach algebras we know that the Banach algebra C(X) for X a compact Hausdorff space is the standard model for all commutative Banach algebras satisfying the following conditions:

1. A has an identity.
2. A is a B*-algebra.

Since the fact that A is isometric and isomorphic to C(X) for some X plays a crucial role, it is of interest to consider Banach algebras with an isometric involution and to try to extend some results for C(X) to this more general setting. [See S. Watanabe (1971) and M. S. Espelie (1973).] In what follows we give some of these results.

We consider A be an involutive Banach algebra and suppose that the involution is isometric, i.e., for each x in A, $\|x\| = \|x^*\|$. We suppose further that A is commutative. We say that an algebra A as above is a *Banach *-algebra*. If A* denotes the dual of A then we consider the following sets:

$$\hat{P}_A = \{f : f \in A^*,\ f(xx^*) \geq 0 \text{ for all } x \text{ in } A\}$$

$$P_A = \{f : f \in \hat{P}_A,\ \|f\| \leq 1\}$$

$$M'_A = \{f : f \in P_A,\ f(xy) = f(x)f(y) \text{ for all } x,y \text{ in } A\}$$

$$M_A = \{f : f \in M'_A,\ f \text{ nontrivial}\}$$

If A is an involutive algebra then any element x in A with $x^* = x$ is called hermitian. A hermitian element is called positive if the spectrum $\sigma(x)$ lies in $\mathbb{R}^+$. Using this we may define a partial order in the hermitian part $H(A) = \{x : x \in A, x^* = x\}$ of A by setting $h \geq k$ if and only if $h - k$ is positive. It is not difficult to see that if A = L(H), the Banach algebra of all bounded linear operators on a Hilbert space H, then the partial order defined above coincides with the usual order relation defined on the set of hermitian operators. If B is another involutive Banach algebra then any homomorphism $\varphi : A \to B$ is called a *-homomorphism if for all a in A, $\varphi(a^*) = [\varphi(a)]^*$. If A is an involutive Banach algebra then we say that the algebra is hermitian if the spectrum of every hermitian element in H(A) is real. Sometimes we say that A has a hermitian involution.

Let A, B be two commutative Banach algebras with isometric involution. An operator $T : A \to B$ is said to be positive if for every a in A there exists a finite family of elements $b_1, \ldots, b_n$ in B such that

$$Ta = b_1 b_1^* + \cdots + b_n b_n^*$$

Let us consider further the set $\hat{P}_{(A,B)}$, the cone of all positive operators $T : A \to B$, and let $P_{(A,B)} = \{P : P \in \hat{P}_{(A,B)}, \|P\| \le 1\}$. The set of all multiplicative operators defined on A with values in B is denoted by $M_{(A,B)} = \{T : T \in P_{(A,B)}, T(ab) = TaTb$ for all a,b in A$\}$. The main question about these sets is as follows:

QUESTION 2.13.24 For what pairs of commutative Banach algebras with isometric involution is it true that $M_{(A,B)}$ is exactly the set of extreme points of the set $P_{(A,B)}$?

In what follows some answers to this question are given for classes of Banach algebras with isometric involution. We suppose that the following property holds: for every T in $P_{(A,B)}$,

$$\|T\| = \|Te\|$$

where e is the identity of A. Let A be a commutative Banach algebra. An element a in A is said to be quasi-nilpotent if $\sigma(a) = (0)$. The set of all quasi-nilpotent elements of A is called the radical of A. We say that A is semisimple if the radical of A reduces to the element 0.

THEOREM 2.13.25 (M. S. Espelie, 1973) Let A and B be commutative *-algebras with the following properties:

1. A has an identity element e.
2. B is semisimple and hermitian.

Then for T in $P_{(A,B)}$, $T \in M_{(A,B)}$ implies that T is an extreme point of $P_{(A,B)}$.

Proof: Suppose that T in $P_{(A,B)}$ is multiplicative and not an extreme point of $P_{(A,B)}$. Thus there exists S in L(A,B) such that

$$T + S \in P_{(A,B)} \qquad T - S \in P_{(A,B)}$$

If h is any multiplicative functional on B then hT is an extreme point of P_A and since $hT + hS \ge 0$, $\|hT - hS\| \le 1$, which implies that for all a in A, $hS(a) = 0$. Since B is supposed to be semisimple this implies further that $Sa = 0$ since h was arbitrary. But then $S = 0$ and thus T is an extreme point.

The following is a partial converse of the result in Theorem 2.13.25.

THEOREM 2.13.26 (M. S. Espelie, 1973) Let A, B be two commutative Banach *-algebras and suppose that the following conditions are satisfied:

1. A has an identity e.
2. $T \in P_{(A,B)}$ implies $\|T\| = \|Te\|$.

Then

$$TeTab = TaTb$$

For the proof of this result we need two preliminary lemmas.

LEMMA 2.13.27 Let A be a commutative Banach algebra with isometric involution * and let b be a hermitian element of A with $\|b\| < 1$. Then for any a in A the element aa* - aba* is of the form xx* for some x in A.

Proof: If A has an identity then the element e - b is hermitian and positive. We can define the square root $(e - b)^{1/2} = u$. Then u is again a hermitian element and clearly

$$e - b = u^2$$

This implies that

$$aa^* - aba^* = a(1 - b)a^* = au^2a^*$$

and if we set

$$x = au$$

then this element satisfies the required relation.

If A does not have an identity, then we consider the Banach algebra A_1 where we adjoin an identity, and A is a maximal ideal in A. Then we can consider the element u as above and this is clearly hermitian (we note that A_1 has a natural involution induced by the involution of A). Since for any a in A and y in A_1, ay is in A, we obtain that aa* - aba* is of the above form and the lemma is proved.

LEMMA 2.13.28 Suppose that A, B are commutative Banach *-algebras and $T \in P_{(A,B)}$. Then the operator S : A → B defined by the relation

$$Sa = Tb_1T(ab_2) - Tb_2T(ab_1)$$

has the property that T + S and T - S are positive operators when $b_i \in A$, i = 1, 2, and $b_i = c_ic_i^*$, i = 1, 2, with c_i elements in A.

Proof: First we remark that for any a in A we have, for the operator T + S (and similarly for the operator T - S),

$$\begin{aligned}(T + S)(aa^*) &= T(aa^*) + Tb_1T(B_2aa^*) - Tb_2T(b_1aa^*)\\ &\geq T(aa^*) - Tb_2T(b_1aa^*)\\ &= T(aa^* - b_1aa^*) + T(b_1aa^*) - Tb_2T(b_1aa^*)\\ &\geq T(b_1aa^*) - Tb_2T(b_1aa^*) \geq 0\end{aligned}$$

For we may suppose without loss of generality that $\|b_1\| < 1$ and Lemma 2.13.27 asserts that $aa^* - b_1aa^*$ is positive; and further, we may suppose without loss of generality that $\|Tb_2\| < 1$ and then Lemma 2.13.27 assures that $T(b_1aa^*) - Tb_2T(b_1aa^*)$ is positive. Thus the lemma is proved, since for the operator T - S we may argue in a similar way.

Proof of Theorem 2.13.26: If T is an extreme point of $P_{(A,B)}$, b is an element of A of the form cc*, and $\|b\| < 1$, then we define the operator S by the formula

$$Sa = \frac{1}{2}\,[TbTa - TeT(ab)]$$

This operator is like the operator considered in Lemma 2.13.28 and thus $T + S \geq 0$ and $T - S \geq 0$. But from the hypothesis on operators in $P_{(A,B)}$ we have

$$\begin{aligned}\|T + S\| &= \|Te + Se\| \leq \|Te\| \leq 1\\ \|T - S\| &= \|Te - Se\| = \|Te\| \leq 1\end{aligned}$$

and from the fact that T is an extreme point we must have S = 0. But this means that the required equality of the theorem holds, since every element b of A is of the form

$$a_1 + a_2 + a_3 + a_4$$

where a_i or $-a_i$ is of the form xx* for some x in A. The assertion of the theorem follows for all a and b.

The following result applies to the case when the second Banach algebra is a B*-algebra.

THEOREM 2.13.29 (M. S. Espelie, 1973) Let A be a commutative Banach *-algebra and let B be a B*-algebra. We suppose further that A has an identity element and B is commutative. Then every extreme point of $P_{(A,B)}$ is multiplicative.

Proof: From the above theorem we have that for any extreme point T of $P_{(A,B)}$,

$$TeTab = TaTb$$

and clearly the assertion of the theorem follows if we prove that for any a in A $Ta = TeTa$ when T is extreme. As above, we consider the following operator on A:

$$Sa = \frac{1}{2}(TeTa - Ta)$$

which clearly is well defined, and as in Lemma 2.13.28 we can prove that the following relations hold:

$$T + S \geq 0 \qquad T - S \geq 0$$

We show now that

$$\|T + S\| \leq 1 \qquad \|T - S\| \leq 1$$

Since B is assumed to be a B*-algebra, from $T + S \geq 0$ and $T - S \geq 0$ we get that

$$\|T + S\| = \|Te + Se\| \qquad \|T - S\| = \|Te - Se\|$$

Since

$$\|T + S\| = \|Te + Se\| \leq \frac{1}{2}\|Te\| + \frac{1}{2}\|(Te)^2\|$$

we have

$$\|T + S\| \leq \frac{1}{2}\|Te\| + \frac{1}{2}\|Te\|^2 \leq 1$$

For $T - S$ first we note that $0 \leq Te - Se = (3/2)Te - (1/2)(Te)^2$, and since B is a B*-algebra, the image of Te under the Gelfand transform is positive and the values are in (0,1). Thus, since

$$\frac{3}{2}t - \frac{1}{2}t^2 \leq 1$$

holds for all t in (0,1), from the isometry realized by the Gelfand transform we obtain that $\|T - S\| \leq 1$. This implies, since T is an extreme point, that $S = 0$. But $S = 0$ is exactly the multiplicative property of T.

Suppose now that A is a Banach algebra which has no identity element but has a so-called approximate identity. This notion is given below.

DEFINITION 2.13.30 A net $(e_\alpha)_{\alpha \in I}$ is called an approximate identity for A if the following conditions are satisfied:

1. $\|e_\alpha\| \leq 1$, $\|e_\alpha\| > 0$ for all α in I.
2. For each x in A, $\lim_\alpha \|e_\alpha x - x\| = 0$.

For examples of Banach algebras having an approximate identity we refer to books about Banach algebras. Under the assumption that A is a commutative Banach *-algebra and satisfies with the Banach *-algebra B the condition that any $T \in P_{(A,B)}$ has the property that

$$\|T\| = \lim \|Te_\alpha\|$$

M. S. Espelie (1973) proved that every extreme point T of $P_{(A,B)}$ satisfies the identity: for any a,b in A,

$$TaTb = \lim Te_\alpha Tab$$

and if B is a B*-algebra then every extreme point of $P_{(A,B)}$ is multiplicative. For examples of algebras which indicate the need for hypotheses like those considered above, we refer to M. S. Espelie (1973). It would be interesting to know if similar results hold when we consider locally multiplicatively convex algebras.

In closing this account about extremality in algebras we note that G. Converse, I. Namioka, and R. R. Phelps (1969) and S. Reich (1973) considered the problem of characterizing the extreme operators which are in the unit ball of invariant operators under a left amenable topological semigroup [for the terminology concerning semigroups we refer to F. Greenleaf (1970)].

2.14 STRONG EXTREMALITY STRUCTURE IN OPERATOR ALGEBRAS

In what follows we consider a subclass of the class of extreme operators, namely, the class of strongly extreme operators.

DEFINITION 2.14.1 (V. I. Istrăţescu, 1983b) Let X and Y be two Banach spaces and T be an operator in the unit ball of the Banach space L(X,Y). We say that T is a strongly extreme operator if for each $a > 0$ the number $d_S(T,a)$ is strictly positive, where $d_S(T,a)$ is exactly $d_{(X,Y)}(T,a)$ with

$$d_{(X,Y)}(T,a) = \inf (\max \{\|T \pm aR\| : \|R\| = 1\}) - 1$$

DEFINITION 2.14.2 (V. I. Istrăţescu, 1983b) The operator T in the unit ball of L(X,Y) is called strongly nice if every strongly extreme point of the unit ball of Y* is transformed by T* into a strongly extreme point of the unit ball of X*.

Then we have the following.

THEOREM 2.14.3 T in L(X,Y) is strongly extreme if it is strongly nice.

Proof: Indeed, if T is strongly nice (and we suppose that the family of strongly extreme points of the unit ball of Y* is nonempty) and the contrary assertion is true then for some $a > 0$, $d_{(X,Y)}(T,a) = 0$. Then there exists a sequence of operators (R_n), $\|R_n\| = 1$, such that

$$\lim_{n\to\infty} (\max \|T \pm aR_n\|) = 1$$

Then we obtain that also

$$\lim_{n\to\infty} (\min \|T \pm aR_n\|) = 1$$

This implies that

$$\lim \|T + aR_n\| = 1 = \lim \|T - aR_n\| = 1$$

But this implies that some strongly extreme points of the unit ball of Y* are not transformed by T* into extreme points of the unit ball of X*, since obviously we have the relations:

$$\lim (T^* + aR_n^*) = 1 = \lim (T^* - aR_n^*)$$

Similar to the problem about the connection between nice operators and extreme operators, we have the problem concerning strongly extreme operators and strongly nice operators: Is every strongly extreme operator strongly nice? We use an example given by R. M. Blumenthal, J. Lindenstrauss, and R. R. Phelps to show that not every strongly extreme operator is strongly nice. Of course, this shows that not every extreme operator is a nice operator. The main observation used here is that the operator considered by R. M. Blumenthal, J. Lindenstrauss, and R. R. Phelps (1965) is an element of a compact set. Then it is easy to see that in the case of a compact convex set the extreme points and the strongly extreme points of a convex set form the same family. Now the example is as follows.

We consider g be a monotonic increasing function on [0,1] such that g is discontinuous at each rational point and continuous elsewhere. Let M be the subset in $\mathbb{R}^4$ defined as the union of the following sets:

$$M_1 = \{\varepsilon(\cos t, \sin t, \cos g(t + 0), \sin g(t + 0))\}$$
$$M_2 = \{\varepsilon(\cos t, \sin t, -\cos (t - 0), -\sin g(t + 0))\}$$
$$M_3 = \{\varepsilon(\cos t, \sin t, \cos g(t - 0), \sin g(t - 0))\}$$
$$M_4 = \{\varepsilon(\cos t, \sin t, -\cos g(t - 0), -\sin g(t - 0))\}$$

where $\varepsilon = \pm 1$. Obviously M is symmetric with respect to the origin of $\mathbb{R}^4$

and it is a compact set. We denote the convex hull of M by M'. This is again a compact set and has a nonempty interior (the compactness of M' may be proved using the result of Mazur, or directly, since the set is in $\mathbb{R}^4$).

We consider X to be the four-dimensional real space which has as the unit ball the polar M° of M'. We show now that certain points in $\mathbb{R}^4$ are in M'. First we remark that for t irrational (cos t, sin t, 0, 0) are in M°. For this we remark that if (u,v,w,z) is in M, $u \cos t + v \sin t \leq 1$ and the equality occurs if and only if u = cos t, v = sin t, and (w,z) is equal to ±(cos g(t), sin g(t)). In this case, from (cos t, sin t, 0, 0) ± (u,v,w,z) ∈ M', it follows that u = v = 0, since the inner product of these with (cos t, sin t, 0, 0) is 1, and the point (cos t, sin t, w, z) is in the subset of M' consisting of all points whose inner product with (cos t, sin t, 0, 0) is 1. We have that w = a cos g(t), z = a sin g(t) for some $|a| \leq 1$.

Now we define the operator T on X with values in $C_{[0,1]}$ by the formula

$$F_T(t) = (\cos t, \sin t, 0, 0)$$

where we recall that if T ∈ L(C(X),C(Y)) then $F_T(k)$ is the functional on C(X) defined by

$$F_T(k) : f \to (Tf)(k)$$

For each t we have that $F_T(t) \pm (0, 0, \cos g(t), \sin g(t)) \in M$, which implies that $F_T(t)$ is in M' and $F_T(t)$ is not an extreme point of M'. Indeed, if there exists h : [0,1] → x* which is continuous and $F_T(t) \pm h(t) \in M'$ for t in [0,1], then for t irrational, h(t) = (0, 0, a cos g(t), sin g(t)) with $|a(t)| \leq 1$. We consider now a rational point r in [0,1], and h being continuous, for each sequence $t_n \to r$ we have that $h(t_n) \to h(r)$. We can suppose without loss of generality that t_n is an irrational point for each n, and from the fact that g is discontinuous at r we must have $a(t_n) \to 0$. Then h(r) = 0 and thus h is zero at each rational point. From the continuity it follows that h is zero at all points of [0,1] and thus identically zero on [0,1]. Thus T is an operator which is extreme and nonnice. Also, from the compactness we see that it is strongly extreme and not strongly nice.

3
Complex Strict Convexity

The present chapter presents some problems related to the maximum principle for holomorphic vector-valued functions defined on domains of the complex plane. The fundamental role of holomorphic functions in the theory of linear operators is now well known. It is worth mentioning that the use of holomorphic functions in the theory of linear operators goes back to H. Poincaré, Fr. Riesz, and G. Giorgi. Important contributions to the theory of holomorphic (analytic) functions with values in Banach spaces were given by N. Wiener, L. Fantappié, N. Dunford, I. M. Gelfand, A. E. Taylor, and many others. For a detailed account of this theory we refer to the books of E. Hille and R. S. Phillips (1957), and N. Dunford and J. Schwartz (1958).

The main theme, in some sense, of this chapter is to determine when a maximum modulus principle holds for holomorphic functions, and connected with this, to develop a theory of a class of Banach spaces associated strongly with the validity of such maximum principle. At the same time, we mention some classes of Banach spaces or operators on Banach spaces which are, in some sense, the analogue of the extreme operators or strongly extreme operators.

3.1 HOLOMORPHIC FUNCTIONS WITH VALUES IN BANACH SPACES

Let Λ be an open nonempty subset of the plane.

DEFINITION 3.1.1 The function $f : \Lambda \to \mathbb{C}$ ($\mathbb{C}$ the set of all complex numbers) is said to have a derivative at $z_0 \in C$ if

$$\lim \frac{f(z) - f(z_0)}{z - z_0}$$

exists for $z \to z_0$. The limit (if it exists) is denoted by $f'(z_0)$.

DEFINITION 3.1.2 We say that the function $f : \Lambda \to C$ is holomorphic (or analytic) in Λ if $f'(z)$ exists for all z in Λ.

DEFINITION 3.1.3 An entire function is a function which is holomorphic in the whole plane.

THEOREM 3.1.4 (Liouville's Theorem) If an entire function is bounded then it is constant.

We recall that by a region in the complex plane we mean any (nonempty) connected and open subset. Then the following holds.

THEOREM 3.1.5 (Maximum Modulus Theorem) Let $f : \Lambda \to C$, where Λ is a region and f is holomorphic in Λ, and let z_0 be a point in Λ. Then either f is constant in Λ or for each neighborhood U of z_0 there exists a point z_1 in U with the property that

$$|f(z_0)| < |f(z_1)|$$

[This means, of course, that $z \to |f(z)|$ does not possess a local maximum at any point of Λ.]

Let γ be a closed path in Λ (i.e., a piecewise continuously differentiable curve for which the initial point coincides with the endpoint) and let $f : \Lambda \to C$ be a holomorphic function in Λ. Then we have

THEOREM 3.1.6 (Cauchy's Formula for Holomorphic Complex-Valued Functions) If γ is convex and z is a point not in the range of the path then

$$f(z) = \frac{1}{2}\pi i \operatorname{ind}_\gamma(z) \int_\gamma \frac{f(\xi)}{\xi - z}\, d\xi$$

where $\operatorname{ind}_\gamma(z)$ is defined as

$$\operatorname{ind}_\gamma(z) = \frac{1}{2}\pi i \int_\gamma \frac{d\xi}{\xi - z}$$

Now we recall some definitions concerning holomorphy for functions defined on open subsets of the plane and with values in a Banach space.

DEFINITION 3.1.7 Let $f : \Lambda \to X$, where X is a Banach space. Let z_0 be a point of Λ. We say that f has a derivative at z_0 if

$$\lim \frac{f(z) - f(z_0)}{z - z_0}$$

exists for $z \to z_0$. If the limit exists then it is denoted by $f'(z_0)$. We say that f is holomorphic in Λ if f has derivative at each point of Λ.

REMARK 3.1.8 We have in this case the so-called strong holomorphy of functions on Λ with values in X.

Another notion of holomorphy may be defined using the elements of the dual space of X^* of X.

DEFINITION 3.1.9 Let $f : \Lambda \to X$, where X is a Banach space. We say that f is weakly holomorphic at z_0 if for all x^* in X^* the function defined on Λ with values in $\mathbb{C}$,

$$z \to x^*(f(z))$$

is holomorphic at z_0. We say that f is weakly holomorphic in Λ if it is weakly holomorphic at each point of Λ.

Of course, it is obvious that every function f which is holomorphic in the sense of Definition 3.1.7 is weakly holomorphic.

It is a very interesting and important result that the above notions of holomorphy coincide. This remarkable fact was proved by N. Dunford and for operator-valued functions by E. Hille [see E. Hille and R. S. Phillips (1957)]. We remark that most of the theorems from the theory of analytic functions can be extended in a suitable way to holomorphic functions defined as in Definition 3.1.7. In particular, a Cauchy formula holds for functions in this class.

In order to formulate an important property of the function

$$z \to f(z)$$

where f is a holomorphic function (in the sense of Definition 3.1.7) defined on an open set Λ with values in a Banach space X, we recall the notion of subharmonic function.

DEFINITION 3.1.10 A real-valued function h defined on an open subset Λ of the complex plane and with values in $[-\infty,\infty]$ is said to be subharmonic in Λ if the following assertions hold:

1. $-\infty \le h(z) < \infty$ for all $z \in \Lambda$.
2. h is upper semicontinuous in Λ.
3. For any z_0 in Λ and any r such that $\{z : |z - z_0| \le r\} \subset \Lambda$,

$$h(z_0) \le \frac{1}{2}\pi \int_{-\pi}^{\pi} h(z_0 + re^{is})ds$$

4. For all z_0 and r the integral in (3) is not $-\infty$.

It is not difficult to see that if h is subharmonic in Λ and φ is a monotonically increasing function on $\mathbb{R}$ (the set of all real numbers) then $\varphi_0 h$ is again a subharmonic function in Λ.

It is well known that a subharmonic function defined on a connected open set Λ has no maximum except for the case when this function is constant. Suppose now that $f : \Lambda \to X$ is a holomorphic function defined on Λ with values in the Banach space X. Then from the Cauchy formula we get that for any z_0 and any r with the property that $\{z : |z - z_0| \leq r\} \subset \Lambda$, the following inequality is true:

$$\|f(z_0)\| \leq \frac{1}{2}\pi \int_{-\pi}^{\pi} \|f(z_0 + re^{is})\| \, ds$$

Then obviously we have the following very important property of holomorphic functions defined on open subsets of the complex plane and with values in Banach spaces.

THEOREM 3.1.11 If $f : \Lambda \to X$ is a holomorphic function on an open connected subset of the complex plane and with values in a Banach space then the function

$$z \to \|f(z)\|$$

is a subharmonic function in Λ.

Proof: Indeed, for any z_0 and r such that $\{z : |z - z_0| \leq r\} \subset \Lambda$ we have, from the Cauchy formula, that

$$\|f(z_0)\| \leq \frac{1}{2}\pi \int_{-\pi}^{\pi} \|f(z_0 + re^{is})\| \, ds$$

and the other properties of subharmonic functions are obvious.

COROLLARY 3.1.12 If $f : \Lambda \to X$ is a holomorphic function defined on a connected open subset Λ of the complex plane then f has no maximum in Λ unless $\|f\|$ is constant.

Proof: Obvious from Theorem 3.1.11.

From Theorem 3.1.11 and Corollary 3.1.12 we obtain the following result, known as the *maximum principle*, for holomorphic functions defined on open connected subsets of the complex plane and with values in Banach spaces.

THEOREM 3.1.13 (Maximum Principle) Let $f : \Lambda \cup \partial\Lambda \to X$, where Λ is an open connected subset of the complex plane and $\partial\Lambda$ denotes the boundary of Λ. Suppose that f is continuous on $\Lambda \cup \partial\Lambda$ and holomorphic in Λ. If $M = \sup \{\|f(z)\| : z \in \partial\Lambda\}$ then either $\|f(z)\| = M$ for all z in Λ or $\|f(z)\| < M$ for all $z \in \Lambda$.

Of course, we may ask, as in the case of complex-valued holomorphic functions, about the behavior of f(z) for z in Λ. The following example shows that $z \to \|f(z)\|$ may have a minimum in Λ other than zero.

EXAMPLE 3.1.14 Let us consider the Banach space m of all bounded sequences of complex numbers, $z = (z_1, z_2, \ldots, z_n, \ldots)$, with the following norm:

$$z \to \|z\| = \sup_i |z_i|$$

We consider the elements $m_1 = (1,0,0,\ldots)$, $m_2 = (0,1,0,0,\ldots)$ and $\Lambda = \{z : |z| < 1\}$. On $\Lambda \cup \partial\Lambda$ we define the function f by the formula

$$f(z) = m_1 + zm_2$$

It is clear that f is continuous on $\Lambda \cup \partial\Lambda$ and holomorphic in Λ. Also, for each z in Λ we have

$$\|f(z)\| = 1$$

and obviously $z \to f(z)$ is not constant in Λ.

DEFINITION 3.1.15 Let X be a Banach space. We say that the strong maximum principle holds for X if whenever $f : \Lambda \cup \partial\Lambda \to X$ is a holomorphic function defined on an open connected subset Λ of the complex plane and continuous on $\Lambda \cup \partial\Lambda$, and $z \to \|f(z)\|$ is constant, then necessarily $z \to f(z)$ is constant.

We give now an example of a Banach space for which the strong maximum principle holds. Let X be a uniformly convex space and Λ be an arbitrary open connected subset of the complex plane. Suppose further that $f : \Lambda \cup \partial\Lambda \to X$ is continuous and holomorphic in Λ. If $z \to \|f(z)\|$, z in Λ, is a constant function, then f is a constant function. Indeed, we consider for an arbitrary but fixed point z_0 of Λ a functional in X^* with the property that $x^*(f(z_0)) = f(z_0)$, $\|x^*\| = \|f(z_0)\|$. We define now the following function on Λ:

$$z \to \frac{x^*(f(z))}{\|f(z_0)\|}$$

Then clearly this is a complex-valued holomorphic function on Λ and for each z we have the following inequality:

$$|x^*(f(z))| \leq \|f(z)\|\|f(z_0)\| = \|f(z_0)\|^2$$

If $z = z_0$ we have

$$|x^*(f(z_0))| \leq \|f(z_0)\|^2$$

and thus by the maximum principle for complex-valued functions, $x^*(f(z_0)) = x^*(f(z))$. Since X is strictly convex this implies that for each z there exists $c(z)$ with the property that

$$f(z) = c(z)f(z_0)$$

and then the complex-valued function

$$z \to c(z)$$

is holomorphic in Λ. Of course, from the property of f we must have that $|c(z)| = 1$, and thus $z \to c(z)$ is a constant function, which is clearly equal to 1. Then, finally, we have that f is constant.

The example given above (3.1.14) and the preceding example concerning a large class of Banach spaces for which the strong maximum principle holds suggest the following problem:

PROBLEM 3.1.16 Does there exist a class of Banach spaces for which the strong maximum principle holds? If the answer is yes, how can we characterize this class?

From the above example we conclude that such a class exists and we show, in what follows, that this class may be characterized using an extension (in some sense) of the notion of extreme point of a convex set. To do this we remark that if C is a convex set in a Banach space then a point x is an extreme point for C if and only if for y with the property that $x + y \in C$ and $x - y \in C$, then necessarily $y = 0$.

Then we can define the *complex extreme points* as follows.

DEFINITION 3.1.17 Let C be a convex set in the Banach space X. The point u of C is said to be a complex extreme point for C if $u + zv \in C$ whenever $z \in \mathbb{C}$, $|z| \leq 1$, and v in X, then $v = 0$.

It is easy to see that the following assertion holds.

e next result we refer to C. A. Akeman and B. Russo (1970)
1).

3.2.6 If A is a von Neumann algebra with the predual A_d [see
(1968)] then this is a complex strictly convex Banach space.

e now some characterizations of complex strictly convex spaces.

2.7 (I. I. Istrăţescu and V. I. Istrăţescu, 1979) The Banach
complex strictly convex if and only if for each pair of elements
X, if for all $s \in \mathbb{R}$,

$$^{is}v\| \leq \|u\| \qquad [v,u] = 0$$

eans a semi-inner product on X in the sense of Lumer (1961).]
It is clear that we may suppose in the above theorem that
pose first that X is a complex strictly convex space and u, v
f elements on X satisfying the conditions: for all $s \in \mathbb{R}$

$$^{is}v\| \leq \|u\| \qquad [v,u] = 0$$

e, for $|z| \leq 1$ we have that (if $z = re^{is}$),

$$1 = [u,u] \leq [u + zv, u] \leq \|u + zv\|\|u\|$$

$$\|re^{is}v + re^{is}(e^{-is}u) + (1 - r)e^{is}(e^{-is}u)\|$$

$$r\|v + e^{is}u\| + 1 - r \leq 1$$

s that $u + zv = 1$ for all z with $|z| \leq 1$. Then $v = 0$.
sely, suppose that X satisfies the condition and the assertion
rem that X is complex strictly convex is false. Then for some
$_0\| = 1$, there exists y_0 in X such that for all z in {z :
$x_0 + zy_0\| \leq 1$. We show that this implies that $[y_0,x_0] = 0$.
have the inequalities

$$zy_0, x_0]| \leq |1 + z[y_0,x_0]| \leq 1$$

$x_0]$ is not zero, z being arbitrary in $\{z : |z| \leq 1\}$, this obvi-
a contradiction. Thus $[y_0,x_0] = 0$ and then $y_0 = 0$. This com-
proof of the theorem.

llowing result may be proved in a similar way.

PROPOSITION 3.1.18 If u is an extreme point of the convex set C in X then it is a complex extreme point of C.

Using the complex extreme points we define below, following I. I. Istrăţescu and V. I. Istrăţescu (1979) (the paper was actually written in 1974) the class of complex strictly convex spaces and their relation to the strong maximum principle. Further, we define the class of points which we call *strongly complex extreme points*, and some properties of the associated class of Banach spaces are given. Also, we consider complex extremality in the space of bounded linear operators defined on a Banach space X with values in the Banach space Y.

3.2 COMPLEX STRICTLY CONVEX SPACES

In what follows we consider a class of Banach spaces which is suggested by the class of strictly convex spaces.

DEFINITION 3.2.1 (I. I. Istrăţescu and V. I. Istrăţescu, 1979) Let X be a Banach space. We say that X is a complex strictly convex space if any point x, $\|x\| = 1$, is a complex extreme point of the closed unit ball of X, i.e., x is a complex extreme point of the set $\{x : x \in X, \|x\| \leq 1\}$.

From the definition we see that the following proposition is true.

PROPOSITION 3.2.2 If X is a complex strictly convex space then any closed linear subspace X_1 of X is a complex strictly convex space.

Let us consider a measure space $(\Lambda,\mathcal{B},P)$ where for simplicity we suppose that P is a probability measure [i.e., $P(\Lambda) = 1$]. Then, as usual, we consider the Banach spaces $L^p(\Lambda,\mathcal{B},P) = L^p$ for $1 \leq p \leq \infty$. The following result furnishes a large class of Banach spaces which are complex strictly convex spaces.

PROPOSITION 3.2.3 (E. Throp and R. Witley, 1967) For $1 \leq p < \infty$ every Banach space L^p is a complex strictly convex space.

Proof: Since for $1 < p < \infty$ the L^p spaces are uniformly convex the assertion follows from Proposition 3.1.18. It remains to consider only the case $p = 1$. For this we need the following result.

LEMMA 3.2.4 Let f,g be in L^1. Then $\|f + g\| = \|f\| + \|g\|$ if and only if $f = hg$ [a.e. (P)] and h is a positive function on $S(f) \cap S(g)$. (Here and in what follows, for any f in L^1, S(f) is the subset

$$\{s : s \in \Lambda, f(x) \neq 0\}$$

Proof: Since $\|f + g\| = \|f\| + \|g\|$, in integral form this means that

$$\int_S |f(s) + g(s)|\ dP(s) = \int_S |f(s)|\ dP(s) + \int_S |g(s)|\ dP(s)$$

$$f(s) + g(s) = f(s) + g(s) \qquad \text{a.e. on } S(f) \cap S(g)$$

and thus

$$f(s) = h(s)g(s) \qquad \text{a.e. on } S(f) \cap S(g)$$

Further, we get that

$$|[1 + h(s)]g(s)| = [1 + |h(s)|]|g(s)| \qquad \text{a.e. on } S(f) \cap S(g)$$

This implies obviously that h is positive on $S(f) \cap S(g)$. It is obvious that for a pair of functions f, g satisfying the relation $f = hg$ (a.e.) on $S(f) \cap S(g)$ with h positive on $S(f) \cap S(g)$, the relation $\|f + g\| = \|f\| + \|g\|$ holds (of course we suppose that the functions are in L^1). This completes the proof of the lemma.

Proof of Proposition 3.2.3: Let us consider f in L^1 and $\|f\| = 1$. Suppose on the contrary that f is not a complex extreme point of $\{z : \|z\| \le 1\}$, and thus for some g in L^1 and all complex numbers z, $|z| \le 1$, we have the relation

$$\|f + zg\| \le 1$$

Then we remark that in fact we have the equality, i.e., for all $|z| \le 1$,

$$\|f + zg\| = 1$$

We show now that

$$S(g) \subseteq S(f) \cup N$$

where $P(N) = 0$.

Indeed, we have the relations

$$\begin{aligned} 2 = 2\|f\| &= 2 \int_{S(f)} |f(s)|\ dP(s) = \int_{S(f)} |f + g|\ dP + \int_{S(f)} |f - g|\ dP \\ &= \int_{S(f)} (|f + g| + |f - g|)\ dP + 2 \int_{S\setminus S(f)} |g(s)|\ dP(s) \\ &= 2 + 2 \int_{S\setminus S(f)} g(s)\ dP(s) \end{aligned}$$

which implies that

$$\int_{S\setminus S(f)} |g(s)|\ dP(s) = 0$$

and this obviously implies our assertion about S(f) and S(g).

Since for each z, $|z| \le 1$,

$$f = \frac{1}{2}(f + zg) + \frac{1}{2}(f - zg)$$

we can apply Lemma 3.2.4 to find a func

properties:

1. h_z is positive on $S(f + zg) \cap$
2. $f + zg = hz(1 - zg)$ a.e. in S(

We consider now the function sign f (de

$$\text{sign } f(s) = \begin{cases} \dfrac{f^*(s)}{f(s)} & \text{if } f(s) \ne \\ 0 & \text{in the con} \end{cases}$$

where z* is the complex conjugate of z)

have that

$$z \text{ sign } f = \frac{(h_z - 1)f}{h_z + 1}$$

a.e. on $S(f + zg) \cap S(f - zg)$. Now if

$$s \in S(f + zg)$$

whenever s is in S(f). Then for such a

that

$$2f^*(s)g(s) = \frac{[h_z(s) - 1]\ f(s)^2}{h_z(s) + 1}$$

Since the right side is real, we must h

Consider now the Banach algebra A

$\{z : |z| \le 1\}$ which have all the Fourie

2, 3, ... It is known that we may co

tinuous functions on the closed unit di

open unit disk. Let H^1 be the closure

on $\{z : |z| \le 1\}$ for which the integral

$$\frac{1}{2}\pi \int_{-\pi}^{\pi} |f(e^{is})|\ ds < \infty$$

Then we have, using the above result an

lowing proposition holds.

PROPOSITION 3.2.5 The Hardy space H^1 i

For th

(Theorem 3.

PROPOSITION

J. Dixmier

We giv

THEOREM 3.

space X is

u and v of

$\|u +$ e

then v = 0.

(Here [,] m

Proof:

u = 1. Sup

is a pair o

$\|u +$ e

In this cas

$\|u\|^2$ =

=

≤

This implie

Conver

of the theo

x_0 in X, $\|x$

$|z| \le 1\}$, $\|$

Indeed, we

$|[x_0 +$

and if $[y_0,$

ously gives

pletes the

The fo

THEOREM 3.2.8 (I. I. Istrăţescu and V. I. Istrăţescu, 1979) The Banach space X is complex strictly convex if and only if the following property holds: if u,v in X are such that

$$u + zv = 1 \qquad z \in \mathbb{C}, \ |z| \le 1$$
$$\|u\| = 1 \qquad [v,u] = 0$$

then $v = 0$.

The next theorem gives a sufficient condition for a Banach space to be a complex strictly convex space. It is suggested by a similar condition given by J. R. Holub (1975) for the case of strictly convex spaces.

DEFINITION 3.2.9 For any pair of elements (x_1,x_2) in X we set

$$<<x_1,x_2>> = \{y : y \in X, \ y = a_1x_1 + a_2x_2, \ \text{Im } a_1 = \text{Im } a_2\}$$
$$C_{<<x_1,x_2>>} = \{y : y \in X, \ y = b_1x_1 + b_2x_2, \ (\text{Re } b_1)(\text{Re } b_2) \ge 0\}$$

where for any complex number $z = x + iy$, $x = \text{Re } z$, $y = \text{Im } z$.

Then we have:

THEOREM 3.2.10 (V. I. Istrăţescu, 1979) For the Banach space X to be complex strictly convex it is sufficient that for every pair of normalized elements (x_1,x_2) of X ($\|x_1\| = 1 = \|x_2\|$) the points of $<<x_1,x_2>>$ that are equidistant from x_1 and x_2 are in $C_{<<x_1,x_2>>}$.

Proof: Suppose on the contrary that X is not a complex strictly convex space. Then there exists x with $\|x\| = 1$ and a nonzero element y in X with the property that $\|x + zy\| = 1$ for all z, $|z| \le 1$. We consider two complex numbers z_1 and z_2 with the following properties:

$$|2z_2 - z_1| < 1 \qquad |z_i| < 1 \qquad i = 1, 2$$

It is easy to see that such a pair exists. We define elements of X by the formulas

$$x_1 = x + z_1y \qquad x_2 = x + z_2y$$

Clearly $\|x_1\| = 1 = \|x_2\|$ and the point $u = (z_1 - z_2)y$ is equidistant from x_1 and x_2. Indeed, we have

$$x_1 - u = x + z_1y - (z_1 - z_2)y = x + z_2y$$

and

$$x_2 - u = x + z_2y - (z_1 - z_2)y = x + (2z_2 - z_1)y$$

and the assertion follows since these are in $\{h : h \leq 1\} \subset X$. But

$$x_1 - x_2 = (z_1 - z_2)y$$

is not in $C_{<<x_1,x_2>>}$. This contradiction proves the theorem.

3.3 COMPLEX STRICTLY CONVEX SPACES AND THE STRONG MAXIMUM PRINCIPLE

As we remarked in Sec. 3.1, for a holomorphic function defined on an open connected subset of the complex plane with values in a Banach space a maximum principle like that valid for complex-valued holomorphic functions does not hold. Thus it is of interest to characterize those Banach spaces for which the strong maximum principle holds. This result is given in the following theorem.

THEOREM 3.3.1 (E. Throp and R. Witley, 1967) A necessary and sufficient condition that the strong maximum principle holds for a Banach space is that X be a complex strictly convex space.

Proof: The "only if" part of the proof is easy and is as follows. Suppose that X is not a complex strictly convex space. Then there exists x, $\|x\| = 1$, and y nonzero in X such that for all z with $|z| \leq 1$, $\|x + zy\| = 1$. The function

$$f(z) = x_0 + yz$$

has the property that for all z with $|z| \leq 1$, $\|f(z)\| = 1$. Obviously f is analytic in the complex plane and not identically zero. Also, $\|f(0)\| = \|x_0\| = \|f(z)\|$ for all z with $|z| \leq 1$. The assertion is thus proved.

For the "if" part of the proof of the theorem we need several results which also are of interest in themselves.

LEMMA 3.3.2 (J. Buckholtz, 1966a,b; G. Szegö, 1924) Let $P_n : \mathbb{C} \to \mathbb{C}$ be a polynomial defined by

$$P_n(z) = \sum_{j=0}^{n} \frac{(nz)^j}{j!} = e^{nz} - \sum_{j=n+1}^{\infty} \frac{(nz)^j}{j!}$$

Then if $|z| \leq 1$ and $|ze^{1-z}| \leq 1$, the polynomial P_n has the property that $P_n(z) \neq 0$.

Proof: For any $z \in \mathbb{C}$ the following relation holds:

$$1 - e^{-nz}P_n(z) = e^{-nz} \sum_{j=n+1}^{\infty} \frac{(nz)^j}{j!} = e^{-n}(ze^{1-z})^n \sum_{j=n+1}^{\infty} \frac{n^j z^{j-n}}{j!}$$

Thus if $|z| \le 1$ and $|ze^{1-z}| \le 1$ we have

$$|1 - e^{-nz}P_n(z)| \le e^{-n} \sum_{j=n+1}^{\infty} \frac{n^j}{j!} = 1 - e^{-n}P_n(1) < 1$$

since obviously $P_n(1) > 0$. This inequality implies that $P_n(z) \neq 0$ for all z such that $|z| \le 1$ and $|ze^{1-z}| \le 1$.

From this lemma we obtain the following result.

COROLLARY 3.3.3 There exists $c \in (0,1)$ such that for all $|z| \le c$, $P_n(z) \neq 0$ for all n.

Proof: This follows from the fact that the origin belongs to the sets $|z| < 1$ and $|ze^{1-z}| < 1$, both open sets.

For other interesting properties of the constant c we refer to J. Buckholtz (1966).

LEMMA 3.3.4 Let c be the constant of Corollary 3.3.3. Then for any integer n there exists complex numbers $z_1, \ldots, z_n$ having the following properties:

1. $|z_j| < 1/c$, $j = 1, 2, \ldots, n$.
2. $\Sigma\, z_j = n$.
3. $\Sigma\, z_j^p = 0$, $p = 2, 3, \ldots, n$.

Proof: Let us consider the polynomial

$$Q_n(z) = z^n - nz^{n-1} + \frac{n^2}{2!} z^{n-2} + \cdots + \frac{(-1)^n}{n!} n^n$$

which has n roots, say $z_1, \ldots, z_n$. This implies clearly that $z_1, \ldots, z_n$ satisfy the relations (2) and (3). Now we note that Q_n and P_n satisfy the relation

$$Q_n(z) = z^n P_n\left(\frac{-1}{z}\right)$$

and then (1) follows from Corollary 3.3.3.

Let $D_r = \{z : |z| < r\}$, $D = D_1$.

LEMMA 3.3.5 If $f : D \to X$ is holomorphic and $f'(0) \neq 0$ then for c as in Corollary 3.3.3, the image of D_c under the mapping

$$z \to f(0) + f'(0)z$$

belongs to the closure of the convex hull of $f(D)$.

Proof: We consider the holomorphic mapping

$$g(z) = f(z) - f(0)$$

and note that $g'(0) = f'(0)$. Clearly the convex hull of $f(D)$ is the translate of the convex hull of $g(D)$ by $f(0)$. Since $g(0) = 0$ and g is holomorphic, it has the form

$$g(z) = a_1 z + a_2 z^2 + \cdots + a_n z^n + \cdots$$

Consider now the complex numbers $z_1, \ldots, z_n$ given by Lemma 3.3.4. For any $|z| \leq c$ we have

$$\frac{1}{n} \sum_{j=1}^{\infty} g(z_j z) = a_1 z + h_n(z)$$

with

$$nh_n(z) = a_{n+1}\left(\sum_{j=1}^{n} z_j^{n+1}\right) z^{n+1} + a_{n+2}\left(\sum_{j=1}^{n} z_j^{n+2}\right) z^{n+2} + \cdots$$

Then

$$|h_n(z)| \leq \sum_{p=n+1}^{\infty} |a_p| |cz|^p$$

and thus for any z in D_c,

$$\lim h_n(z) = 0$$

Thus

$$\lim \frac{1}{n} \sum_{j=1}^{n} g(z_j z) = a_1 z$$

for all $z \in D_c$. But

$$\frac{1}{n} \sum_{j=1}^{n} g(z_j z)$$

is in the convex hull of $g(D)$ and thus we get that $a_1 z$ is in the convex hull of $g(D)$. The lemma is proved.

LEMMA 3.3.6 Let $f : D \to X$ be a nonconstant holomorphic mapping. Then there exists an element $b \neq 0$ such that the image of D_c by the holomorphic mapping

$$f_1(z) = f(0) + bz$$

belongs to the closed convex hull of $f(D)$.

Proof: Let g be the function defined in the proof of Lemma 3.3.5, and

$$g(z) = \sum_{j=n}^{\infty} a_j z^j \qquad a_n \neq 0$$

Let $e_1, \ldots, e_n$ be the nth roots of unity. For $z \in D$ we have

$$\begin{aligned}\frac{1}{n}\sum_{j=1}^{n} g(e_j z^{1/n}) &= a_n z + a_{n+1}(e_1 + \cdots + e_n) z^{(n+1)/n} + \cdots \\ &\quad + a_{2n} z^2 + \cdots + a_{3n} z^3 + \cdots \\ &= a_n z + a_{2n} z^2 + a_{3n} z^3 + \cdots\end{aligned}$$

and the function

$$h(z) = \frac{1}{n}\sum_{j=1}^{n} g(e_j z^{1/n})$$

is then holomorphic in D. Clearly $h'(0) \neq 0$, so h satisfies the conditions of Lemma 3.3.5. Thus the image of D_c under the holomorphic mapping

$$z \to a_n z$$

belongs to the closed convex hull of $h(D)$. But $h(D)$ is in the convex hull of $g(D)$, and since $g(z) = f(z) - f(0)$, the assertion of the lemma follows.

Now the "if" part of the proof of Theorem 3.3.1 follows from the next lemma.

LEMMA 3.3.7 Let C be a closed and convex subset of X. Suppose that $f : D \to X$ is holomorphic. If $f(D) \subset C$ and if $f(0)$ is a complex extreme point of C then f is constant.

Proof: Suppose that the assertion of the lemma is false. Then for some $b \in X$, $b \neq 0$, and c as in Corollary 3.3.3,

$$f(0) + bD_c$$

is in C and thus

$$\Delta(f(0), cb) \subset C$$

where

$$\Delta(x,y) = \{z \in X : z = x + \xi y, \ |\xi| \leq 1\}$$

This implies that f(0) is not a complex extreme point. This contradiction proves the lemma as well as Theorem 3.3.1.

REMARK 3.3.8 For the case of Hilbert spaces, a simple proof of Theorem 3.3.1 was found by Shmuel Agmon [see E. Vesentini (1970)]. The proof is as follows: If <,> denotes the scalar product on H, and $f : D \to H$ is holomorphic with the property that f(z) is constant on D, we pick a point z_0 in D and consider the complex-valued function

$$g(z) = \langle f(z), f(z_0)\rangle$$

which is obviously holomorphic. Now from the Cauchy inequality we get

$$|g(z)| = |\langle f(z), f(z_0)\rangle| \leq \|f(z_0)\|^2$$

and since

$$g(z_0) = \langle f(z_0), f(z_0)\rangle = \|f(z_0)\|^2$$

the function g(z) is constant; in fact, $g(z) = \|f(z_0)\|^2$. This implies that the elements f(z) and $f(z_0)$ are linearly dependent. Thus for each $z \in D$ there exists c(z) such that

$$f(z) = c(z)f(z_0)$$

and this implies that $z \to c(z)$ is holomorphic. But $|c(z)| = 1$ and thus $c(z) = 1$. Thus $f(z) = f(z_0)$ and the proof is complete.

The above proof extends easily to the case of uniformly convex spaces. Indeed let [,] denote the semi-inner product on the uniformly convex space X. Pick a point z_0 in D and set

$$g(z) = [f(z), f(z_0)]$$

which is obviously holomorphic. As above, we get that

$$f(z) = c(z)f(z_0)$$

and we conclude that $c(z) = 1$.

3.4 PRODUCTS AND QUOTIENTS OF COMPLEX STRICTLY CONVEX SPACES

We now give some properties of the product and quotient spaces of complex strictly convex spaces.

THEOREM 3.4.1 Let X, Y be two complex Banach spaces and consider the product X × Y. Then if the norm on X × Y is defined by the relation

$$\|(x,y)\| = (\|x\|^2 + \|y\|^2)^{1/2}$$

or

$$\|(x,y)\|_p = (\|x\|^p + \|y\|^p)^{1/p}$$

X × Y is a complex strictly convex space.

Proof: We consider only the first case. The other is similar and thus it is omitted. First we remark that a semi-inner product may be defined on X × Y using the semi-inner products on X and Y, respectively:

$$[(x,y),(x^*,y^*)] = [x,x^*] + [y,y^*]$$

Now let $u,v \in X \times Y$ be such that

1. $\|u + zv\| = 1$, $\|z\| \le 1$.
2. $\|u\| = 1$.
3. $[v,u] = 0$.

If $u = (x_0,y_0)$, $v = (x,y)$, then

$$\begin{aligned} 1 &\le (\|x_0\|^2 + \|y_0\|^2)^{1/2} = [y_0 + zx, x_0] + [y_0 + zy, y_0] \\ &\le \|x_0 + zx\|\|x_0\| + \|y_0 + zy\|\|y_0\| \\ &\le (\|x_0 + zx\|^2 + \|y_0 + zy\|^2)^{1/2}(\|x_0\|^2 + \|y_0\|^2)^{1/2} = 1 \end{aligned}$$

and thus there exists a constant k such that

$$\|x_0 + zx\| = k\|x_0\| \qquad \|y_0 + zy\| = k\|y_0\|$$

Obviously, from (1), k = 1 and then the complex strict convexity of X and Y imply that x = 0 and y = 0, i.e., v = 0, and the theorem is proved.

More generally we mention the following extension of this result, which can be proved similarly.

THEOREM 3.4.2 Let $(X_n)_1^\infty$ be complex strictly convex spaces and let $\ell^2((X_n))$ be the space of all sequences $x = (x_n)$, $x_n \in X_n$, with the property that

$$\|x\|^2 = \Sigma \|x_n\|^2 < \infty$$

Then $\ell^2((X_n))$ is a complex strictly convex space.

More generally we can prove the following result: if $(X_n)_1^\infty$ is a sequence of Banach spaces and $\ell^p((X_n))$ denotes the Banach space of all sequences $x = (x_n)$, x_n in X_n, with $\|x\| = (\Sigma_{n=1}^\infty \|x_n\|^p)^{1/p} < \infty$, then $(X_n)_1^\infty$ is a complex strictly convex space if and only if for each n, X_n is a complex strictly convex space.

Also, from the results of C. A. McCarthy (1967) together with the polar decomposition of compact operators (R. Schatten, 1960) we can prove that all the Schatten-von Neumann ideals C_p, $1 \leq p < \infty$, are complex strictly convex spaces.

We give now a result concerning the quotient spaces. Let X be a complex Banach space and let L be a closed subspace of X. The quotient space X/L is the Banach space of all translates of L in X, with the norm

$$\|x + L\| = \inf \{\|x + \ell\| : \ell \in L\}$$

The problem for complex strictly convex spaces and quotient spaces can be formulated as follows: To what extent is the complex strict convexity of X transmitted to its quotient spaces?

In what follows, using some results of Klee (1959) (or adapted from his work, as given in Chap. 2) we prove the following result.

THEOREM 3.4.3 If X is a complex strictly convex space and L is a reflexive closed subspace of X then X/L is a complex strictly convex space.

We need two results. The first one is Lemma 2.2.4, which is valid also for complex spaces. The second is as follows.

LEMMA 3.4.4 Let X, Y be two complex Banach spaces and suppose that C is a complex strictly convex set (i.e., every point of ∂C is a complex extreme point with respect to C). If $T : X \to Y$ is a bounded bijective operator then $C_1 = TC$ is a complex strictly convex set.

Proof: Let $x_1 \in \partial C_1$ and suppose that for all z with $|z| \leq 1$ and some $y \in Y$, $x_1 + zy \in C_1$. Then there exists x^* in ∂C such that $Tx^* = x_1$, and also, we find y^* such that $Ty^* = y$. Clearly $T(x^* + zy^*) = x_1 + zy$. Then $x^* + zy^*$ is in C and thus $y^* = 0$. This gives that $y = 0$ and the assertion is proved.

Proof of Theorem 3.4.3: If U and V denote the open unit balls of X and X/L, and T is the canonical map of X onto X/L, we have that

$T(\text{cl } U) = \text{cl } V$

and Lemma 3.3.4 gives the assertion of the theorem.

3.5 COMPLEX UNIFORMLY CONVEX SPACES AND COMPLEX SMOOTH CONVEX SPACES

In what follows we define some quantities for complex Banach spaces and we study them for the complex strictly convex spaces.

DEFINITION 3.5.1 (V. I. Istrăţescu, 1983b) For any Banach space X the complex modulus of convexity, denoted by δ_X^c, is the function defined in [0,2] with values in [0,1] by the following formula:

$$\delta_X^c(t) = \inf \min \{1 - \frac{1}{2}\|x + y\|, 1 - \frac{1}{2}\|x + iy\| : \|x\| = \|y\| = 1, \|x - y\| \geq t, \|x - iy\| \geq t\}$$

The Banach space X is said to be complex uniformly convex if for any $t > 0$, $\delta_X^c(t) > 0$. For the Banach space X the complex modulus of smoothness is the function $\rho_X^c(\tau)$, $\tau \geq 0$, defined by the formula

$$\rho_X^c(\tau) = \sup \max \left\{\left\|\frac{x + y}{2}\right\| + \left\|\frac{x - y}{2}\right\| - 1, \left\|\frac{x + iy}{2}\right\| + \left\|\frac{x - iy}{2}\right\| - 1 : \|x\| = 1, \|y\| = \tau\right\}$$

The Banach space is said to be complex uniformly smooth if $\rho_X^c(\tau)/\tau \to 0$ as $\tau \to 0$.

We have the following result.

THEOREM 3.5.2 The Banach space X is complex uniformly convex if and only if the following property holds: if $(x_n),(y_n)$ are sequences of unit vectors in X with

$$\lim_{n\to\infty} \max \{\|x_n + y_n\|, \|x_n + iy_n\|\} = 2 \qquad (*)$$

then

$$\lim_{n\to\infty} \min \{\|x_n - y_n\|, \|x_n - iy_n\|\} = 0 \qquad (**)$$

Since the proof can be modeled on the proof of the corresponding characterization of a uniformly convex space we omit it.

From this result we have:

COROLLARY 3.5.3 If X is a complex uniformly convex space then it is a complex strictly convex space.

Proof: Suppose that the assertion is not true. Thus for some x with $\|x\| = 1$, some $y \neq 0$ with $\|y\| = 1$, and all z with $|z| \leq 1$, $\|x + zy\| = 1$. Thus we have the relations

$$\|x + y\| = 1 \qquad \|x + iy\| = 1 \qquad \|x - y\| = 1 \qquad \|x - iy\| = 1$$

If we set $u_n = x + (1 - 1/n)y$, $v_n = x - (1 - 1/n)y$, then $x_n = u_n/\|u_n\|$, $y_n = v_n/\|v_n\|$ satisfies the condition (*) of Theorem 3.5.2. Then (**) holds for the sequences (u_n), (v_n), i.e.,

$$\lim_{n\to\infty} \min \{\|u_n - v_n\|, \|u_n - iv_n\|\} = 0$$

It is easy to see that this is impossible. Thus X is a complex strictly convex space.

Concerning the possible relations between the complex modulus of convexity and the complex modulus of smoothness we can prove, using the method of J. Lindenstrauss (1963), the following result.

THEOREM 3.5.4 Let X be a Banach space. Then

$$\rho^c_{X^*}(\tau) \geq \sup_{0\leq s\leq 2} \left[\frac{ts}{2} - \delta^c_X(s)\right]$$

It may be of interest to know if other characterizations from the class of uniformly convex spaces can be extended to the complex case, especially the characterizations using the duality mappings and Mielnik spaces. For some results in this direction we refer to V. I. Istrăţescu (1983b).

3.6 STRONGLY COMPLEX EXTREME POINTS IN BANACH SPACES AND RELATED CLASSES OF BANACH SPACES

In what follows we consider the analogue of the function introduced by V. D. Milman for the study of some geometrical problems of Banach space theory, namely, the analogue of the function $d_X(x,a)$, $a > 0$.

DEFINITION 3.6.1 (V. I. Istrăţescu, 1983b) Let X be a Banach space and C be a convex subset in X. For x in X,

$$\|x - C\| = \inf \{\|x - y\| : y \in C\}$$

A point s of C is a complex strongly extreme point of S if and only if for every $a > 0$,

$$d_S^c(s,a) = \inf \max \{\|s + ay - S\|, \|(s - ay) - S\|, \|(s + iay) - S\|, \|(s - iay) - S\|\} > 0$$

If $S = \{x : \|x\| \leq 1\}$ then we write simply

$$d_S^c(s,a) = d_X^c(s,a) = \inf (\max \{\|s \pm ay\|, \|s \pm iay\| : \|y\| = 1\} - 1)$$

DEFINITION 3.6.2 The function $a \to d_X^c(s,a)$ is called the modulus of complex extremeness at the point s. The Banach space X is said to be complex strongly strictly convex if for all s with $\|s\| = 1$ and all $a > 0$, $d_X^c(s,a) > 0$.

We give now an example of a point which is a complex strongly convex point. For this we need a complex version (in some sense) of the notion of denting point of a set.

DEFINITION 3.6.3 A point x of a convex set C is said to be a complex denting point if and only if for every $\varepsilon > 0$, $x \notin \text{conv cl } (C \setminus \{y: y \in C, \min (\|x - y\|, \|x - iy\|) < \varepsilon\})$.

Then we have the following result.

PROPOSITION 3.6.4 If x is a complex denting point of the convex set C then x is a complex strongly extreme point of C.

Proof: Suppose that the assertion of the proposition is false. Thus we find a sequence (y_n), $\|y_n\| = 1$, such that if a_n is defined by the formula

$$a_n = \max \{\|x \pm ay_n\|, \|x \pm iay_n\|\}$$

then

$$\lim a_n = 0$$

Further, from the definition of the function $d_S^c(s,a)$ it follows that we can find sequences (z_n^j), (w_n^j), $j = 1, 2$, of elements of C with the following properties:

$$x - ay_n - z_n^1 \to 0 \qquad x + ay_n - z_n^2 \to 0 \qquad x - iay_n - iw_n^1 \to 0$$

$$x + iay_n - w_n^2 \to 0$$

We note that if u_i, v_i, $i = 1, 2$, are elements of C then for any t in [0,1] we have the inequality

$$\|t(u_1 - v_1) + (1 - t)(u_2 - v_2)\| \leq \max \{\|u_1 - v_1\|, \|u_2 - v_2\|\}$$

If we set

$$u_1 = x + ay_n \qquad v_1 = z_n^1 \qquad u_2 = x - ay_n \qquad v_2 = z_n^2$$

and again

$$u_1 = x + iay_n \qquad v_1 = w_n^1 \qquad u_2 = x - iay_n \qquad v_2 = w_n^2$$

we get that

$$x - \frac{1}{2}(z_n^1 + z_n^2) \to 0 \qquad x - \frac{1}{2}(w_n^1 + w_n^2) \to 0$$

Then

$$\|x - z_n^1\| \to a \qquad \|x - z_n^2\| \to a \qquad \|x - iw_n^1\| \to a \qquad \|x - iw_n^2\| \to a$$

This implies that x is not a complex denting point of C.

For the class of Banach spaces which are complex strongly strictly convex spaces, we can prove the following result.

THEOREM 3.6.5 Let X be a Banach space with the property that the norm and weak* convergence of sequences agree on the boundary of the unit ball. Then every complex extreme point of the unit ball of X is a complex strongly extreme point.

For the proof we need the following lemmas.

LEMMA 3.6.6 Let X be a Banach space and suppose that x is a point with $\|x\| = 1$ and

$$\lim_{n\to\infty} \max \{\|x \pm ay_n\|, \|x \pm iay_n\|\} = 1 \qquad \|y_n\| = 1$$

for some fixed $a > 0$. Then

$$\lim_{n\to\infty} \min \{\|x \pm ay_n\|, \|x \pm iay_n\|\} = 1 \qquad \|y_n\| = 1$$

Proof: Taking a subsequence and eventually renumbering (if necessary) we may assume that

$$\min \{\|x \pm ay_n\|, \|x \pm iay_n\|\} = \|x - ay_n\| \text{ or } \|x - iay_n\|$$

Since

$$1 = \|x\| \leq \frac{\|x + ay_n\|}{2} + \frac{\|x - ay_n\|}{2} \leq \frac{1}{2} + \frac{\|x + ay_n\|}{2}$$

in the first case, and

$$1 = \|x\| \le \frac{\|x + iay_n\|}{2} + \frac{\|x - iay_n\|}{2} \le \frac{1}{2} + \frac{\|x + iay_n\|}{2}$$

in the second case. Further, if the assertion of the lemma is false, then for some $\varepsilon > 0$ we may suppose that (for all n)

$$\|x - ay_n\| > 1 - \varepsilon$$

or

$$\|x - iay_n\| > 1 - \varepsilon$$

in the second case. This implies in the first case that

$$1 \le \frac{1}{2} + \frac{\varepsilon}{4} - \frac{\varepsilon}{2} = 1 - \frac{\varepsilon}{4}$$

and a similar inequality holds in the second case. This is clearly a contradiction and the lemma follows.

LEMMA 3.6.7 Let x, (y_n) be as in Lemma 3.6.6. Then there exist sequences of unit vectors (u_n^i), i = 1, 2, 3, 4, such that for some $k > 0$,

$$\|u_n^i - u_n^j\| \ge k \qquad \|u_n^i - x\| \ge \frac{k}{2} \qquad x = \lim \frac{1}{4}(u_n^1 + u_n^2 + u_n^3 + u_n^4)$$

Proof: We set

$$u_n^1 = \frac{x - ay_n}{\|x - ay_n\|} \qquad u_n^2 = \frac{x + ay_n}{\|x + ay_n\|} \qquad u_n^3 = \frac{x - iay_n}{\|x - iay_n\|}$$

$$u_n^4 = \frac{x + iay_n}{\|x + iay_n\|}$$

for each n and these sequences satisfy the requirements of the lemma.

Proof of Theorem 3.6.5: Let x in X with $\|x\| = 1$ and $d_X^c(x,a) = 0$ for some $a > 0$. Then we find sequences (u_n^i) of unit vectors such that the conditions of Lemma 3.6.7 are satisfied. Now using the property of the space about the weak* convergence we get that x is not a complex extreme point. This gives the assertion of the theorem.

We close this section with some remarks. Let X be a Banach space and set

$$a \to d_X^c(a) = \inf \{d_X^c(x,a) : \|x\| = 1\}$$

CONJECTURE 3.6.8 The Banach space X is a complex uniformly convex space if and only if $d_X^c(a) > 0$ for all $a > 0$.

As suggested by the characterization of uniformly nonsquare Banach spaces of R. C. James (1964b), we consider the following class of Banach spaces.

DEFINITION 3.6.9 The Banach space X is said to be complex uniformly nonsquare if $d_X^c(1) > 0$.

3.7 THE RENORMING PROBLEM FOR COMPLEX STRICTLY CONVEX SPACES

As in the case of strictly convex spaces we can consider the following problem.

PROBLEM 3.7.1 Let X be a Banach space. What are necessary and sufficient conditions for X to have an equivalent norm, say $\| \|_1$, such that $(X, \| \|_1)$ is a complex strictly convex space?

The answer to this problem is given in the following result, which is the analogue of Klee's theorem for strict convexity.

THEOREM 3.7.2 The Banach space X has an equivalent norm $\| \|_1$ with the property that $(X, \| \|_1)$ is a complex strictly convex space if and only if there exists a complex strictly convex space Y and a one-to-one continuous linear mapping T defined on X with values in Y.

Proof: It is obvious that the condition is necessary since we can take in this case X = Y equipped with the new equivalent norm. Now we show that the condition is sufficient. First we remark that the functions

$$z \to \|x + zy\| \qquad z \to \|T(x + zy)\|$$

are obviously subharmonic functions. Then we can define the norm $\| \|_1$ on X by the formula

$$x \to \|x\|_1 = \left(\frac{1}{2}\|x\|^2 + \frac{1}{2}\|Tx\|^2\right)^{1/2}$$

Then, as is easy to see from the subharmonicity of the above functions, $(X, \| \|_1)$ is a complex strictly convex space. (Of course, it is obvious that the above norm is equivalent to the original norm of X.)

For other results concerning complex strictly convex spaces and interpolation spaces we refer to V. I. Istrăţescu (1979, 1978); these papers also give some results on complex extremality in operator spaces.

3.8 COMPLEX STRICTLY CONVEX SPACES AND OPERATOR THEORY

In what follows we apply a result from the theory of complex strictly convex spaces to a problem in operator theory, namely, the problem of generalizing Fuglede's well-known theorem. For the results about operators that we need we refer to the author's book (1981) and for the results concerning the ideals of completely continuous operators we refer to the excellent book of R. Schatten (1960); for an exposition of ideal theory see also the author (1983c).

We recall some notions and results. Let X be a complete inner product space and L(X) be the set of all bounded linear operators on X; K(X) denotes all the compact elements of L(X). It is known that L(X) is a B*-algebra with the involution

$$* : T \to T^*$$

where T* is the adjoint of T (i.e., the complete inner product adjoint of T).

John von Neumann conjectured that if for a normal operator N in L(X) and an element T in L(X) the equality

$$TN = NT$$

holds, then

$$TN^* = N^*T$$

This conjecture was proved by B. Fuglede and the result has attracted the attention of many people who have presented some extensions. These are (to the best knowledge of the author) concerned with the case where normality is replaced by subnormality or hyponormality for the operator N and, on the other hand, where certain classes of compact operators are considered instead of the operator T. Recently, R. L. Moore and T. T. Trent (1981) have proved a sort of extension of Fuglede's theorem using essentially the complex strictly convex spaces. In what follows we present this extension.

First we give the following lemma.

LEMMA 3.8.1 (V. I. Istrăţescu, 1983) Let X be a complete inner product space and I an ideal of compact operators with the ideal norm p. We suppose that the following properties hold:

1. (I,p) is a complex strictly convex space.
2. K, T are in L(X).
3. K is in I and T is hyponormal.
4. TK = KT*.

Then

$$T^*K = KT.$$

Proof: The idea of the proof is as in the Moore-Trent paper. Since T is hyponormal for all complex numbers a and b the operator aT + b is again hyponormal. Thus for all x in X,

$$\|(aT + b)^*x\| \le \|(aT + b)x\|$$

which implies, by the factorization theorem, that there exists a contraction operator $C_{a,b}$ with the property that

$$(aT + b)^* = C_{a,b}(aT + b) \qquad (+)$$

Now we take b = 1 and a such that $|a| < 1/\|T\|$. We define the following holomorphic function for $|a| < 1/\|T\|$:

$$f(a) = (aT + 1)K(aT + 1)^{-1}$$

Using (+) we have further that

$$f(a) = C_{\bar{a},1}(\bar{a}T + 1)K(aT + 1)^{-1} = C_{\bar{a},1}KC_{a,1}$$

where we used property 4. This clearly implies that $p(f) \le p(K)$, and since $p(f(0)) = p(K)$, from the fact that (I,p) is a complex strictly convex space, we conclude that f(a) is constant. This obviously implies the assertion of the lemma.

COROLLARY 3.8.2 (R. L. Moore and T. T. Trent, 1981) Let X and L(X) be as in Lemma 3.8.1 and suppose that (I,p) is one of the Schatten-von Neumann ideals C_p ($1 \le p < \infty$). Let T, K be as in Lemma 3.8.1 with $(I,p) = C_p$. Then TK = KT.

Proof: The spaces C_p for $1 \le p < \infty$ are complex strictly convex spaces and thus the assertion follows.

REMARK 3.8.3 It is worth noting that complex strictly convex ideals of completely continuous operators may be constructed, using, for example, the Lorentz sequence spaces $\ell_{r,s}$ and d(a,p). For details concerning uniform convexity, strict convexity, etc., of Lorentz spaces we refer to Z. Altshuler (1975) and P. G. Cassaza and Bor-Luh Lin (1977), and for ideals constructed with Lorentz sequence spaces S, to the papers by S. Kwapien (1968), B. S. Mitiagin (1968), and S. V. Kisliakov (1977).

NOTE 3.8.4 (Added in Proof, July 1983) Professor J. R. Partington (Cambridge University) has informed the author of the existence of strictly convex ideal norms on the ideal of compact operators.

Appendix

Classes of Operators and the Geometric Structure of Banach Spaces*

It is well known that for bounded linear operators defined on a Banach space with values in another Banach space, a central role in the study of the structure of operators is played by the structure of their range. In what follows we are interested in classes of operators and relations with the geometric structure of the underlying Banach spaces. In particular, we consider the following problem: To what extent can we define notions such as uniformly convex operator, uniformly smooth operator, Radon-Riesz operator, etc., and further, if such classes are definable is there duality between certain ones? We mention that the problem was first considered by B. Beauzamy in his Ph.D. thesis under L. Schwartz (1976), where he defines the class of uniformly convex operators (and the related class of uniformly convexifiable operators).

In what follows we introduce the class of smooth operators as well as the modulus of smoothness, and we show that the well-known duality theorem of Lindenstrauss extends to the case of operators. We mention that the notion of modulus of convexity of an operator is implicit in the work of Beauzamy. Further, we consider other classes of operators suggested by properties of the identity mapping in some Banach spaces. Thus we consider the classes of Radon-Riesz operators, Kadec-Klee operators, nearly uniformly convex operators, and Grothendieck operators. As a by-product we consider two classes of spaces, namely, the class of k-uniformly convex spaces and k-uniformly smooth spaces, for which we prove a duality result.

---

*Some of the results in this Appendix were presented at the seminar of Prof. Dr. G. Neubauer on the geometry of Banach spaces in summer 1983, University of Konstanz. The author takes the opportunity to thank Professor Neubauer for his interest and discussions.

A.1 UNIFORMLY CONVEX OPERATORS AND UNIFORMLY SMOOTH OPERATORS

Let X,Y be two Banach spaces and L(X,Y) be the set of all bounded linear operators defined on X with values in Y, with the usual norm.

DEFINITION A.1.1 (B. Beauzamy, 1976) The operator T in L(X,Y) is said to be uniformly convex if given $\varepsilon > 0$ there exists $\delta(\varepsilon) > 0$ such that whenever x,y are in X with $\|x\| = 1$, $\|y\| = 1$, and $\|Tx - Ty\| \geq \varepsilon$ then

$$\|x + y\| \leq 2[1 - \delta(\varepsilon)]$$

The modulus of convexity of an operator T in L(X,Y) is defined as follows.

DEFINITION A.1.2 For T in L(X,Y) the modulus of convexity is the function defined on [0,2] with values in [0,1] such that

$$\delta_T(\varepsilon) = \inf\left\{1 - \frac{1}{2}\|x + y\| : \|x\| = 1, \|y\| = 1, \|Tx - Ty\| \geq \varepsilon\right\}$$

Since the function δ_T is nondecreasing we see that T is uniformly convex iff for each $\varepsilon > 0$, $\delta_T(\varepsilon) > 0$.

The class of smooth operators is defined as follows.

DEFINITION A.1.3 The operator T in L(X,Y) is said to be smooth if for each $\varepsilon > 0$ there exists $\delta(\varepsilon) > 0$ such that if $\|x\| = 1$ and $\|Ty\| < \delta$ then

$$\|x + Ty\| + \|x - Ty\| < 2 + \varepsilon\|Ty\|$$

DEFINITION A.1.4 The modulus of smoothness of an operator T in L(X,Y) is the function on $[0,\infty)$ defined by

$$\rho_T(\tau) = \sup\{\|y + \tau Tx\| + \|y - \tau Tx\| : \|x\| = 1, \|y\| = 1, x \in X, y \in Y\}$$

The basic connection between the class of uniformly convex operators and uniformly smooth operators is reflected in the following result valid for arbitrary operators.

THEOREM A.1.5 For any operator T in L(X,Y) the following relation holds:

$$\rho_{T^*}(\tau) = \sup_{0\leq\varepsilon\leq 2}\left[\frac{\varepsilon\tau}{2} - \delta_T(\varepsilon)\right]$$

Proof: Let x,y be elements in X with the property that $\|x\| = \|y\| = 1$ and $\|Tx - Ty\| \geq \varepsilon$. Then we take in X* and Y*, respectively, elements x* and y* (of norm 1) such that

$$x^*(x + y) = \|x + y\| \qquad y^*(Tx - Ty) = \|Tx - Ty\|$$

Then we have obviously,

$$2\rho_{T^*}(\tau) \leq \|x^* + \tau T^*y^*\| + \|x^* - \tau T^*y^*\| - 2$$
$$= \langle x^*, x + y\rangle + \tau\langle y^*, Tx - Ty\rangle - 2 \leq \|x + y\| + \tau\|Tx - Ty\| - 2$$

Thus

$$2 - \|x + y\| \geq \tau\varepsilon - 2\rho_{T^*}(\tau)$$

or

$$2\delta_T(\varepsilon) \geq \tau\varepsilon - 2\rho_{T^*}(\tau)$$

Then clearly we have the inequality

$$\delta_T(\varepsilon) + \rho_T(\tau) \geq \frac{\varepsilon\tau}{2}$$

which implies that

$$\rho_{T^*}(\tau) \geq \sup \left[\frac{\varepsilon\tau}{2} - \delta_T(\varepsilon)\right]$$

We now prove the converse inequality. Let us consider x^* in X^* and y^* in Y^*, $\|x^*\| = \|y^*\| = 1$, and let $r > 0$ be arbitrary. Then we find x,y in X such that

$$(x^* + \tau T^*y^*)(x) \geq \|x^* + \tau T^*y^*\| - r \qquad \|x\| = 1$$
$$(x^* - \tau T^*y^*)(y) \geq \|x^* - \tau T^*y^*\| - r \qquad \|y\| = 1$$

and then

$$\|x^* + \tau T^*y^*\| + \|x^* - \tau T^*y^*\| \leq (x^* + \tau T^*y^*)(x) + (x^* - \tau T^*y^*)(y) + 2r$$
$$= x^*(x + y) + \tau y^*(Tx - Ty) + 2r$$
$$\leq \|x + y\| + \|\tau Tx - Ty\| + 2r$$
$$\leq 2 + 2 \sup \left[\frac{\varepsilon\tau}{2} - \delta_T(\varepsilon)\right] + 2r$$

or

$$\|x^* + \tau T^*y^*\| + \|x^* - \tau T^*y^*\| - 2 \leq 2 \sup \left[\frac{\varepsilon\tau}{2} - \delta_T(\varepsilon)\right] + 2r$$

But r is arbitrary, and thus

$$\rho_T(\tau) = \sup \left[\frac{\varepsilon\tau}{2} - \delta_T(\varepsilon)\right]$$

and the duality relation is proved.

It is not difficult to see that an operator is uniformly smooth iff

$$\frac{\rho_T(\tau)}{\tau} \to 0$$

for $\tau \to 0$. Also, from the duality relation given above it follows that for T to be uniformly convex it is necessary and sufficient that T* be uniformly smooth.

Now using the moduli defined above and a notion of Lindenstrauss (who has considered it for Banach spaces), we say that an operator $T \in L(X,Y)$ is more convex than the operator $S \in L(X,Y)$ if

$$\delta_T(\varepsilon) \geq \delta_S(\varepsilon) \qquad \varepsilon > 0$$

and we say that T is smoother than S if for every $\tau > 0$,

$$\rho_T(\tau) \leq \rho_S(\tau)$$

Interesting and important results in the theory of uniformly convex spaces are those of M. Kadec and J. Lindenstrauss concerning unconditionally convergent series and the divergence of series. We mention Kadec's theorem: If $\Sigma\, x_n$ is an unconditionally convergent series in the uniformly convex space X then the series

$$\Sigma\, \delta_X(\|x_n\|)$$

converges. Here $\delta_X(\ )$ is the modulus of convexity of the Banach space X. We can prove the following extension of this result.

THEOREM A.1.6 Let $T \in L(X,Y)$ be a uniformly convex operator and $\Sigma\, x_n$ an unconditionally convergent series. Then the series

$$\Sigma\, \delta_T(\|x_n\|)$$

converges.

We conjecture that an analogue of Lindenstrauss' result can be formulated and proved for smooth operators.

We close this section by introducing certain numbers associated with bounded linear operators. For any T in L(X,Y) (X,Y are Banach spaces) we set

$$\alpha_T(X,Y) = \inf \{p : \Sigma\, T\alpha_n x_n \text{ converges for every choice of } \alpha_n = \pm 1$$
$$\text{implies } \Sigma\, \|x_n\|^p < \infty\}$$

$\beta_T(X,Y) = \sup \{p : \Sigma\, \alpha_n Tx_n$ diverges for every choice of $\alpha_n = \pm 1$ implies $\Sigma \|x_n\|^p = \infty\}$

For $X = Y, T = I$, these numbers were introduced by Lindenstrauss (1963), who has obtained some inequalities for them. It may be of interest to know the value of such numbers for certain classes of operators, for example, absolutely p-summing operators.

A.2 UNIFORMLY KADEC-KLEE OPERATORS AND NEARLY UNIFORMLY CONVEX OPERATORS

The notion of uniformly Kadec-Klee Banach space and nearly uniformly convex spaces was defined by R. Huff (1980) as follows:

DEFINITION A.2.1 A Banach space X is said to be uniformly Kadec-Klee if the following property holds: For each $\varepsilon > 0$ there exists $\delta(\varepsilon) > 0$ such that whenever (x_n) is a sequence in the closed unit ball of X with the properties

$$\|x_n - x_m\| \geq \varepsilon \qquad n \neq m$$

$$x_n \rightharpoonup x \qquad (\rightharpoonup \text{ means weak convergence})$$

then

$$\|x\| \leq 1 - \delta(\varepsilon)$$

DEFINITION A.2.2 The Banach space X is said to be nearly uniformly convex if for each $\varepsilon > 0$ there exists $\delta(\varepsilon) > 0$ such that whenever (x_n) is a sequence in the closed unit ball of x and

$$\|x_n - x_m\| \geq \varepsilon \qquad n \neq m$$

then there exists an integer N and $a_1, \ldots, a_n$ in (0,1) such that

$$\|a_1x_1 + \cdots + a_nx_n\| \leq 1 - \delta(\varepsilon)$$

It is known that every nearly uniformly convex space is uniformly Kadec-Klee. It may be of interest to mention another fact related to nearly uniformly convex spaces, namely, that there exists a nearly uniformly convex space that does not have the Banach-Saks property but whose dual has the Banach-Saks property.

Following the results of Kadec and Klee on the connection between weak and strong convergence, we define the following class of operators.

DEFINITION A.2.3 The operator $T \in L(X,Y)$ is called a uniformly Kadec-Klee operator if for each $\varepsilon > 0$ there exists $\delta(\varepsilon) > 0$ such that whenever (x_n) is a sequence in the closed unit ball of X and

$$\|Tx_n - Tx_m\| \geq \varepsilon \qquad n \neq m$$

$$x_n \rightharpoonup x$$

then

$$\|x\| \leq 1 - \delta(\varepsilon)$$

If X is a uniformly Kadec-Klee Banach space then clearly the identity and every isometric operator is in the above class. We can prove the following useful property of these operators.

THEOREM A.2.4 Let $T \in L(X,Y)$ be a uniformly Kadec-Klee operator and suppose that (x_n) is a sequence of elements in the closed unit ball of X such that

$$x_n \rightharpoonup x \qquad \|x_n\| \to \|x\|$$

Then

$$\lim Tx_n = Tx$$

Proof: Suppose that the assertion of the theorem is false. Then, as we may suppose without loss of generality that $x \neq 0$ and $Tx \neq 0$, we may assume that $\|x\| = 1$. Since (Tx_n) does not converge to Tx we find an $\varepsilon > 0$ such that

$$\|Tx_n - Tx_m\| \geq \varepsilon \tag{*}$$

holds or this holds for some subsequence. Clearly we may suppose (changing the notation eventually) that (*) holds. But this clearly implies that the norm of x is strictly less than 1. This contradiction proves the theorem.

A subclass of this class of operators may be defined as follows.

DEFINITION A.2.5 The operator $T \in L(X,Y)$ is called nearly uniformly convex if for each $\varepsilon > 0$ there exists $\delta(\varepsilon)$ with the property that whenever (x_n) is a sequence in X such that

PROPOSITION 3.1.18 If u is an extreme point of the convex set C in X then it is a complex extreme point of C.

Using the complex extreme points we define below, following I. I. Istrăţescu and V. I. Istrăţescu (1979) (the paper was actually written in 1974) the class of complex strictly convex spaces and their relation to the strong maximum principle. Further, we define the class of points which we call *strongly complex extreme points*, and some properties of the associated class of Banach spaces are given. Also, we consider complex extremality in the space of bounded linear operators defined on a Banach space X with values in the Banach space Y.

3.2 COMPLEX STRICTLY CONVEX SPACES

In what follows we consider a class of Banach spaces which is suggested by the class of strictly convex spaces.

DEFINITION 3.2.1 (I. I. Istrăţescu and V. I. Istrăţescu, 1979) Let X be a Banach space. We say that X is a complex strictly convex space if any point x, $\|x\| = 1$, is a complex extreme point of the closed unit ball of X, i.e., x is a complex extreme point of the set $\{x : x \in X, \|x\| \leq 1\}$.

From the definition we see that the following proposition is true.

PROPOSITION 3.2.2 If X is a complex strictly convex space then any closed linear subspace X_1 of X is a complex strictly convex space.

Let us consider a measure space $(\Lambda,\mathcal{B},P)$ where for simplicity we suppose that P is a probability measure [i.e., $P(\Lambda) = 1$]. Then, as usual, we consider the Banach spaces $L^p(\Lambda,\mathcal{B},P) = L^p$ for $1 \leq p \leq \infty$. The following result furnishes a large class of Banach spaces which are complex strictly convex spaces.

PROPOSITION 3.2.3 (E. Thr545 and R. Witley, 1967) For $1 \leq p < \infty$ every Banach space L^p is a complex strictly convex space.

Proof: Since for $1 < p < \infty$ the L^p spaces are uniformly convex the assertion follows from Proposition 3.1.18. It remains to consider only the case $p = 1$. For this we need the following result.

LEMMA 3.2.4 Let f,g be in L^1. Then $\|f + g\| = \|f\| + \|g\|$ if and only if $f = hg$ [a.e. (P)] and h is a positive function on $S(f) \cap S(g)$. (Here and in what follows, for any f in L^1, S(f) is the subset

$$\{s : s \in \Lambda, f(x) \neq 0\}$$

Proof: Since $\|f + g\| = \|f\| + \|g\|$, in integral form this means that

$$\int_S |f(s) + g(s)| \, dP(s) = \int_S |f(s)| \, dP(s) + \int_S |g(s)| \, dP(s)$$

$$f(s) + g(s) = f(s) + g(s) \qquad \text{a.e. on } S(f) \cap S(g)$$

and thus

$$f(s) = h(s)g(s) \qquad \text{a.e. on } S(f) \cap S(g)$$

Further, we get that

$$|[1 + h(s)]g(s)| = [1 + |h(s)|]|g(s)| \qquad \text{a.e. on } S(f) \cap S(g)$$

This implies obviously that h is positive on $S(f) \cap S(g)$. It is obvious that for a pair of functions f, g satisfying the relation $f = hg$ (a.e.) on $S(f) \cap S(g)$ with h positive on $S(f) \cap S(g)$, the relation $\|f + g\| = \|f\| + \|g\|$ holds (of course we suppose that the functions are in L^1). This completes the proof of the lemma.

Proof of Proposition 3.2.3: Let us consider f in L^1 and $\|f\| = 1$. Suppose on the contrary that f is not a complex extreme point of $\{z : \|z\| \leq 1\}$, and thus for some g in L^1 and all complex numbers z, $|z| \leq 1$, we have the relation

$$\|f + zg\| \leq 1$$

Then we remark that in fact we have the equality, i.e., for all $|z| \leq 1$,

$$\|f + zg\| = 1$$

We show now that

$$S(g) \subseteq S(f) \cup N$$

where $P(N) = 0$.

Indeed, we have the relations

$$\begin{aligned} 2 = 2\|f\| = 2 \int_{S(f)} |f(s)| \, dP(s) &= \int_{S(f)} |f + g| \, dP + \int_{S(f)} |f - g| \, dP \\ &= \int_{S(f)} (|f + g| + |f - g|) \, dP + 2 \int_{S \setminus S(f)} |g(s)| \, dP(s) \\ &= 2 + 2 \int_{S \setminus S(f)} g(s) \, dP(s) \end{aligned}$$

which implies that

$$\int_{S \setminus S(f)} |g(s)| \, dP(s) = 0$$

and this obviously implies our assertion about $S(f)$ and $S(g)$.

Since for each z, $|z| \leq 1$,

$$f = \frac{1}{2}(f + zg) + \frac{1}{2}(f - zg)$$

we can apply Lemma 3.2.4 to find a function $h_z(s)$ with the following properties:

1. h_z is positive on $S(f + zg) \cap S(f - zg)$.
2. $f + zg = hz(1 - zg)$ a.e. in $S(f + zg) \cap S(f - zg)$.

We consider now the function sign f (defined as follows:

$$\text{sign } f(s) = \begin{cases} \dfrac{f^*(s)}{f(s)} & \text{if } f(s) \neq 0 \\ 0 & \text{in the contrary case} \end{cases}$$

where z^* is the complex conjugate of z). Then, from the above relation we have that

$$z \text{ sign } f = \frac{(h_z - 1)f}{h_z + 1}$$

a.e. on $S(f + zg) \cap S(f - zg)$. Now if z is sufficiently small,

$$s \in S(f + zg)$$

whenever s is in S(f). Then for such a point, the above relation implies that

$$2f^*(s)g(s) = \frac{[h_z(s) - 1]\ f(s)^{\ 2}}{h_z(s) + 1}$$

Since the right side is real, we must have g = 0 [a.e. (P)].

Consider now the Banach algebra A of all functions defined on $\{z : |z| \leq 1\}$ which have all the Fourier coefficients $c_{-n} = 0$ for n = 1, 2, 3, ... It is known that we may consider A as the algebra of all continuous functions on the closed unit disk which are holomorphic in the open unit disk. Let H^1 be the closure of A in the set L^1 of all functions on $\{z : |z| \leq 1\}$ for which the integral

$$\frac{1}{2}\pi \int_{-\pi}^{\pi} |f(e^{is})|\ ds < \infty$$

Then we have, using the above result and Proposition 3.2.2, that the following proposition holds.

PROPOSITION 3.2.5 The Hardy space H^1 is a complex strictly convex space.

For the next result we refer to C. A. Akeman and B. Russo (1970) (Theorem 3.1).

PROPOSITION 3.2.6 If A is a von Neumann algebra with the predual A_d [see J. Dixmier (1968)] then this is a complex strictly convex Banach space.

We give now some characterizations of complex strictly convex spaces.

THEOREM 3.2.7 (I. I. Istrăţescu and V. I. Istrăţescu, 1979) The Banach space X is complex strictly convex if and only if for each pair of elements u and v of X, if for all $s \in \mathbb{R}$,

$$\|u + e^{is}v\| \le \|u\| \qquad [v,u] = 0$$

then $v = 0$.

(Here [,] means a semi-inner product on X in the sense of Lumer (1961).]

Proof: It is clear that we may suppose in the above theorem that $u = 1$. Suppose first that X is a complex strictly convex space and u, v is a pair of elements on X satisfying the conditions: for all $s \in \mathbb{R}$

$$\|u + e^{is}v\| \le \|u\| \qquad [v,u] = 0$$

In this case, for $|z| \le 1$ we have that (if $z = re^{is}$),

$$\begin{aligned}\|u\|^2 &= 1 = [u,u] \le [u + zv, u] \le \|u + zv\|\|u\| \\ &= \|re^{is}v + re^{is}(e^{-is}u) + (1 - r)e^{is}(e^{-is}u)\| \\ &\le r\|v + e^{is}u\| + 1 - r \le 1\end{aligned}$$

This implies that $u + zv = 1$ for all z with $|z| \le 1$. Then $v = 0$.

Conversely, suppose that X satisfies the condition and the assertion of the theorem that X is complex strictly convex is false. Then for some x_0 in X, $\|x_0\| = 1$, there exists y_0 in X such that for all z in $\{z : |z| \le 1\}$, $\|x_0 + zy_0\| \le 1$. We show that this implies that $[y_0,x_0] = 0$. Indeed, we have the inequalities

$$|[x_0 + zy_0, x_0]| \le |1 + z[y_0,x_0]| \le 1$$

and if $[y_0,x_0]$ is not zero, z being arbitrary in $\{z : |z| \le 1\}$, this obviously gives a contradiction. Thus $[y_0,x_0] = 0$ and then $y_0 = 0$. This completes the proof of the theorem.

The following result may be proved in a similar way.

$$\|x_n\| \le 1$$

$$\|Tx_n - Tx_m\| \ge \varepsilon \qquad n \ne m$$

then there exists an integer N and $a_1, \ldots, a_N$ in (0,1) with the following properties

$$a_1 + \cdots + a_N = 1$$

$$\|a_1x_1 + \cdots + a_Nx_N\| \le 1 - \delta(\varepsilon)$$

REMARK A.2.6 As we mentioned, there exists a nearly uniformly convex space with the property that the dual space has the Banach-Saks property and the space itself does not possess this property.

A.3 LOCALLY UNIFORMLY CONVEX OPERATORS AND RADON-RIESZ OPERATORS

As we know, an interesting class of Banach spaces, which is larger than the class of uniformly convex spaces, was defined by A. R. Lovaglia (1955) as the *locally uniformly convex spaces*. This suggests the consideration of the following class of operators, which contains the class of uniformly convex operators.

DEFINITION A.3.1 Let $T \in L(X,Y)$. Then we say that T is a locally uniformly convex operator if for each $\varepsilon > 0$ and $x \in X$, $\|x\| = 1$, there exists $\delta_T(x,\varepsilon) > 0$ such that whenever $y \in X$, $\|Tx - Ty\| \ge \varepsilon$, then

$$\|x + y\| \le 2[1 - \delta_T(x,\varepsilon)]$$

Obviously every uniformly convex operator is in this class, and since the class of locally uniformly convex spaces is strictly larger than the class of uniformly convex spaces, we see that the two classes are not equal.

We define now a class of operators suggested by the well-known result of J. Radon and Fr. Riesz about a type of convergence in L^p-spaces [p in $(1,\infty)$].

DEFINITION A.3.2 Let $T \in L(X,Y)$. Then we say that T is a Radon-Riesz operator if whenever (x_n) is a sequence in X such that

$$\|x_n\| \to \|x\| \qquad x_n \rightharpoonup x$$

then

$$\lim Tx_n = Tx$$

Now we prove the following result.

THEOREM A.3.3 If $T \in L(X,Y)$ is a locally uniformly convex operator then it is a Radon-Riesz operator.

Proof: Suppose that (x_n) is a sequence in the closed unit ball of X such that

$$x_n \rightharpoonup x \qquad \|x_n\| \to \|x\|$$

and that the assertion of the theorem is false. Then for some $\varepsilon > 0$ we have the following inequality:

$$\|Tx_n - Tx_m\| \geq \varepsilon$$

[for (x_n) or for some subsequence]. Then, since we may suppose without loss of generality that $\|x\| = 1$ we find x^* in X^* such that

$$x^*(x) = 1 \qquad \|x^*\| = 1$$

Since

$$1 = \lim x^* \left(\frac{1}{2}(x_n + x)\right) \leq \|x^*\| \limsup \left(\frac{1}{2}(\|x_n + x\|)\right) \leq 1 - \delta_T(x,\varepsilon)$$

this is clearly a contradiction and the theorem is proved.

REMARK A.3.4 From Theorem A.2.6 we see that every uniformly Kadec-Klee operator is a Radon-Riesz operator.

REMARK A.3.5 It may be of interest to know if the above classes are closed under conjugation. For example, it is not difficult to see that the class of Radon-Riesz operators is not invariant under conjugation.

A.4 GROTHENDIECK OPERATORS

Suppose that X is a Banach space. Then X is called a Grothendieck space if (and only if) the weak and weak*-sequential convergence coincide in X^*. As suggested by this important class of Banach spaces we define the following class of operators.

DEFINITION A.4.1 The operator $T \in L(X,Y)$ is called a Grothendieck operator if for every sequence (y_n^*) in Y^*, $(T^*y_n^*)$ is weak* convergent iff it is weakly convergent (in X^*).

Using a theorem of I. Leonard (1976) we can easily prove the following result.

PROPOSITION A.4.2 $T \in L(X,Y)$ iff $T_1 : \ell^p(X) \to \ell^p(Y)$ is a Grothendieck operator. Here $\ell^p(Z)$ means the Banach space of all sequences $z = (z_n)$,

z_n in Z, such that

$$z = (\Sigma\|z_n\|^p)^{1/p} < \infty \qquad T_1(z) = (Tz_n)$$

The following result gives some examples of Grothendieck operators.

THEOREM A.4.3 Suppose that X is a Grothendieck space, and for each R : $Y \to c_0$, $R \in L(Y,c_0)$, the operator

$$S = RT$$

is weakly compact, where

$$T : X \to Y$$

is a bounded linear operator. Then T is a Grothendieck operator.

Proof: Let (y_n) be a sequence in Y such that (Ty) converges weakly to zero. We define the operator $U : X \to c_0$ by the formula

$$U(x) = ((T^*y_n^*,x)) = ((y_n^*,Tx))$$

It is clear that this is a bounded linear operator from X to c_0. We identify c_0 with its image in c_0^* and we remark that

$$U^{**}X^* \subset c_0$$

Let x^{**} be an arbitrary element of X^{**}. Then we have

$$U^{**}(x^{**}) = (x^{**}(T^*y^*))$$

and thus, since $(x^{**}T^*y_n^*)$ is in c_0 for each x^{**} in X^{**}, we obtain that $T^*y_n^* \rightharpoonup 0$. Then T is a Grothendieck operator.

PROBLEM A.4.4 It may be of interest to know if this property characterizes the Grothendieck operators.

Another property of Grothendieck operators is given in the following result.

THEOREM A.4.5 Let $T \in L(X,Y)$ be a Grothendieck operator. Then for any bounded linear operator $S \in L(Y,Z)$, where Z is a WCG Banach space, the operator ST is weakly compact.

We recall that a Banach space V is said to be a WCG space if in V there exists a weakly compact set K^W whose linear span is dense in V. We note some examples of such spaces. The first example is any reflexive Banach space, which clearly satisfies the condition for a WCG space.

Second, if X is a separable Banach space then it is a WCG space. Indeed, since the space is separable the closed unit ball is also a separable metric space and thus we find a denumerable dense set, say (x_n). Then the set $K^w = \overline{(x_n/n)}$ satisfies the requirements of the definition, and thus all separable Banach spaces are WCG spaces.

Proof of Theorem A.4.5: Suppose that S is in L(Y,Z) and consider the adjoints

$$S^* : Z^* \to Y^* \qquad T^* : Y^* \to X^*$$

Let (z_n^*) be a bounded sequence in Z*. A theorem of D. Amir and J. Lindenstrauss (1972) asserts that the above sequence has a weak* convergent subsequence. Since S* and T* are weak* continuous, the above subsequence is mapped into a weakly convergent sequence of X*. Then it is itself a weakly convergent sequence. This means that ST is a weakly compact operator by Gantmacher's theorem.

We close this section with some open problems.

PROBLEM A.4.6 Is there a connection between the Grothendieck operators and the operators which admit a factorization through Banach spaces of the type C(K), with K a compact Hausdorff Stone space?

PROBLEM A.4.7 It is well known that the class of Grothendieck spaces is intimately related to measure theory for vector-valued measures. Are there connections of the Grothendieck operators with problems of measure theory for vector-valued measures?

PROBLEM A.4.8 If $T : X \to Y$ is a Grothendieck operator, is T* a Grothendieck operator?

PROBLEM A.4.9 If T and T*, $T : X \to Y$, are Grothendieck operators, is T then weakly compact?

A.5 k-UNIFORMLY CONVEX SPACES AND k-UNIFORMLY CONVEX OPERATORS

We consider now a class of Banach spaces which is larger than the class of uniformly convex spaces and contains some classes of Banach spaces studied by I. Glicksberg et al. (1958).

DEFINITION A.5.1 The Banach space X is called k-uniformly convex if for all $\varepsilon > 0$ there exists $\delta(\varepsilon) > 0$ such that whenever $(x_i),(y_i), i = 1, 2, \ldots,$

k, are elements in the closed unit ball of X and

$$\sum_{i=1}^{k} \|x_i - y_i\| \le \varepsilon$$

then

$$\|x_1 + \cdots + x_k + y_1 + \cdots + y_k\| \le 2k[1 - \delta(\varepsilon)]$$

DEFINITION A.5.2 The k-modulus of convexity of the Banach space X is the function defined by the formula

$$\delta_{k,X}(t) = \inf_t \left\{1 - \frac{1}{2} k\|x_1 + \cdots + x_k + y_1 + \cdots + y_k\| : \Sigma\|x_i - y_i\| \le t\right\}$$

It is clear just from the definition that X is k-uniformly convex iff $\delta_{k,X}(t) > 0$ for each $t > 0$.

DEFINITION A.5.3 The k-modulus of smoothness of the Banach space X is defined as follows:

$$2k\rho_{k,X}(s) = \sup\left\{\sum_{i=1}^{k} (\|x + sy_i\| + \|x - sy_i\|) - 2k : \|x\| = \|y_i\| = 1\right\}$$

The basic result for these numerical characteristics of the Banach space X is the following relation, which extends the Lindenstrauss duality to k-uniformly convex spaces.

THEOREM A.5.4 For any Banach space X the following relation holds:

$$\rho_{k,X^*}(s) = \sup\left[\frac{st}{2k} - \delta_{k,X}(t)\right]$$

Proof: Let $x_1, \ldots, x_k$ and $y_1, \ldots, y_k$ be elements in X with

$$\sum_{i=1}^{k} \|x_i - y_i\| \le t \qquad \|x_i\| = \|y_i\| = 1$$

We consider x^* and y_i^* in X^* with the following properties:

$$x^*(x_1 + \cdots + x_k + y_1 + \cdots + y_k) = \|x_1 + \cdots + x_k + y_1 + \cdots + y_k\|$$

$$y_i^*(x_i - y_i) = \|x_i - y_i\| \qquad i = 1, 2, \ldots, k$$

In this case,

$$2k\rho_{k,X^*}(s) \geq (x^* + sy_1^*)(x_1) + (x^* - sy_1^*)(y_1) + \cdots + (x^* + sy_k^*)(x_k)$$
$$+ (x^* - sy_k^*)(y_k) - 2k$$
$$= x^*(x_1 + \cdots + x_k + y_1 + \cdots + y_k)$$
$$+ s \sum_{i=1}^{k} y_i^*(x_i - y_i) = 2k$$
$$\leq \|x_1 + \cdots + x_k + y_1 + \cdots + y_k\| + st - 2k$$

which implies that

$$\delta_{k,X}(t) + \rho_{k,X^*}(s) \geq \frac{st}{2k}$$

This gives the inequality

$$\rho_{k,X^*}(s) \geq \sup \left[\frac{st}{2k} - \delta_{k,X}(t)\right]$$

Suppose now that x^* and y_i^* are given, $i = 1, 2, 3, \ldots, k$, such that

$$\|x^*\| = \|y_i^*\| = 1 \qquad i = 1, 2, 3, \ldots, k$$

Then for each arbitrary $r > 0$ we find x_i and y_i of norm 1 such that the following inequalities are satisfied:

$$\|x^* + sy_i^*\| \leq (x^* + sy_i^*)(x_i) + r$$
$$\|x^* - sy_i^*\| \leq (x^* - sy_i^*)(y_i) + r$$

for $i = 1, 2, 3, \ldots, k$. This gives further the following inequalities:

$$\|x^* + sy_1^*\| + \|x^* - sy_1^*\| + \cdots + \|x^* + sy_k^*\| + \|x^* - sy_k^*\|$$
$$\leq x^*(x_1 + \cdots + x_k + y_1 + \cdots + y_k) + s \sum_{i=1}^{n} y_i^*(x_i - y_i) + 2kr$$
$$\leq 2k + 2k \sup \left[\frac{st}{2k} - \delta_{k,X}(t)\right] + 2kr$$

Since r is arbitrary we get the inequality

$$\rho_{k,X^*}(s) \leq \sup \left[\frac{st}{2k} - \delta_{k,X}(t)\right]$$

and thus the assertion of the theorem follows.

We say that a Banach space is k-uniformly smooth if

$$\frac{\rho_{k,X}(s)}{s} \to 0$$

as $s \to 0$. From the duality result proved above we obtain that the Banach space X is k-uniformly convex iff X is k-uniformly smooth.

The notion of k-uniformly convex operator and k-smooth operator can be defined similarly and a result like Theorem A.1.5 can be proved. We leave the details to the interested reader.

A.6 FINAL REMARKS

As we noted, it is possible to define for the Banach space of operators certain classes which, in some sense, are the analogue of the identity operator for certain classes of Banach spaces. Recently the Radon-Nicodym property for Banach spaces has received much attention. Thus it may be of interest to know if such an important property can be defined for operators. Related to this is, of course, the problem of dentability. Another interesting property of Banach spaces is the so-called Schur property, and we may ask if there is an analogue of this property for operators.

Notation

References

C. Akemann and B. Russo, Geometry of the unit sphere of a C*-algebra and its dual. Pacif. J. Math. *32* (1970), 575-585. MR *41*, 5980.

N. I. Akhezer and M. G. Krein, *The Moment Problem*, Charkov, 1938 (Russian).

L. Alaoglu and G. Birkhoff, General ergodic theorems. Ann. of Math. *41* (1940), 293-309. MR *1*, 339.

Z. Altshuler, Uniform convexity in Lorentz sequence spaces. Israel J. Math. *20* (1975), 260-274. MR *52*, 6378.

D. Amir, Chebyshev centers and uniform convexity. Pacif. J. Math. *77* (1978), 1-6. MR *80h*, 46017.

——— and F. Deutsch, Approximation by certain subspaces in the Banach space of continuous vector-valued functions. J. Approx. Theory *27* (1979), 254-270. MR *81c*, 41077.

——— and J. Lindenstrauss, The structure of weakly compact sets in Banach spaces. Ann. of Math. *(2)88* (1968), 35-46. MR *37*, 4562.

E. Z. Andalafte and L. M. Blumenthal, Metric characterizations of Banach spaces and Euclidean spaces. Fund. Math. *55* (1964), 23-55. MR *29*, 2625.

K. W. Anderson, *Midpoint Local Uniform Convexity and Other Geometric Properties of Banach Spaces*. Ph.D. Thesis, Univ. of Illinois, Urbana, 1960.

R. F. Arens and J. L. Kelley, Characterizations of the space of continuous functions over a compact Hausdorff space. Trans. Amer. Math. Soc. *62* (1947), 499-508. MR *9*, 291.

N. Aronszajn and K. T. Smith, Invariant subspaces of completely continuous operators. Ann. of Math. *(2)60* (1954), 345-350. MR *16*, 468.

G. Ascoli, Sugli spazi metrici e le loro varieta lineari II. Ann. Mat. Pura Appl. *(4)10* (1932), 203-232.

E. Asplund, Averaged norms. Israel J. Math. *5* (1967), 227-233. MR *36*, 5660.

———, Fréchet differentiability of convex functions. Acta Math. *121* (1968), 31-48. MR *37*, 6754.

*MR x, xxx means Math. Reviews, Volume x, Review xxxx.

V. J. Averbuh, Duality between differentiability and rotundity of convex functions (Russian). Math. Zametki *15* (1974), 809-820. MR *50*, 8004. English Transl. Math. Notes *15* (1974), 485-491.

S. Axler, I. D. Berg, N. Jewell, and A. L. Shields, Approximation by compact operators and the space $H^{\infty} + C$. Ann. of Math. *(2)109* (1979), 601-612. MR *81h*, 30053.

J. A. Baker, Isometries in normed spaces. Amer. Math. Monthly *78* (1971), 655-658. MR *44*, 4496.

S. Banach, *Theorie des operations lineaires*. Monographie Matematiczne 1. Warszawa, 1932.

L. Baron and O. Ferguson. See B. L. Chalmers.

B. Beauzamy, *Espaces de Banach uniformement convexifiables*. Seminaire Maurey-Schwartz 1973/1974a, Exposés 13, 14. MR *53*, 11338.

———, *Operateurs convexifiants*. Seminaire Maurey-Schwartz, 1973/1974b. Exposé 17. MR *55*, 13218.

———, *Geometrie des espaces de Banach et des operateurs entre espaces de Banach*. Ph.D. Thesis, Université de Paris VII, 1976.

———, Quelques propriétés des operateurs uniformement convexifiants. Studia Math. *60* (1977), 211-222. MR *56*, 1033.

———, *Espaces d'interpolation reels: Topologie et geometrie*. Lecture Notes in Math. *666* (1978). MR *80k*, 46080.

———, *Introduction to Banach Spaces and Their Geometry*. Université de Lyon I. Notes de Matematica, Vol. 68, North-Holland, Amsterdam, 1982.

———, Propriété de Banach-Saks. Studia Math. *66* (1979/80), 227-235. MR *81i*, 46020.

———, Un cas de convergence forte des itéres d'une contraction dans un espace uniformement convexe. C. R. Acad. Sci. Paris.

——— and B. Morey, Les pointes minimums et les ensembles optimums dans les espaces de Banach. J. Funct. Anal. *24* (1977), 107-139. MR *55*, 1049.

A. Beck, A convexity condition in Banach spaces and the strong law of large numbers. Proc. A.M.S. *13* (1962), 332-334. MR *24*, A3681.

———, On the strong law of large numbers. In *Ergodic Theory*, Proc. Internat. Sympos. Tulane Univ., Tulane, New Orleans, La., 1961. Academic Press, New York, 1963, pp. 21-53. MR *28*, 3470.

P. R. Beesack, E. Hughes, and M. Ortel, Rotund complex normed spaces. Proc. A.M.S. *75* (1979), 42-44. MR *81e*, 46013.

J. L. Bell and D. H. Fremlin, A geometric form of the axiom of choice. Fund. Math. *77* (1972), 167-170. MR *48*, 5865.

L. P. Belluce and W. A. Kirk, Nonexpansive mappings and fixed point in Banach spaces. Illinois J. Math. *11* (1967), 474-479. MR *35*, 5988.

———, W. A. Kirk, and E. F. Steiner, Normal structure in Banach spaces. Pacif. J. Math. *26* (1968), 433-440. MR *38*, 1501.

I. D. Berg. See S. Axler.

E. Berkson, Some types of Banach spaces, Hermitian operators and Bade functionals. Trans. Amer. Math. Soc. *116* (1965), 376-385. MR *32*, 4554.

C. Bessaga and A. Pelczynski, On extreme points in separable conjugate spaces. Israel J. Math. *4* (1966), 262-264. MR *35*, 2126.

———, *Selected Topics in Infinite Dimensional topology.* Polish Academy of Sciences, Warszawa, 1975, Monographie Mathematiczne *58*. MR *57*, 17657.

A. Beurling and A. E. Livingston, A theorem on duality mappings. Arkiv. Math. *4* (1962), 405-411. MR *26*, 2551.

G. Birkhoff, Orthogonality in linear metric spaces. Duke Math. J. *1* (1935), 169-172.

E. Bishop and R. R. Phelps, A proof that every Banach space is subreflexive. Bull. Amer. Math. Soc. *67* (1961), 27-35. MR *23*, A503.

———, Support functionals of convex sets. In *Proc. Sympos. Pure Math. 7, Convexity*, Amer. Math. Soc., Providence, R.I., 1963, 27-35. MR *27*, 4051.

A. R. Blass and C. V. Stanojevic, Partial Mielnik spaces and characterization of uniformly convex spaces. Proc. A.M.S. *55* (1976), 75-82, MR *52*, 14926.

L. M. Blumenthal. See E. Z. Andalafte.

R. M. Blumenthal, J. Lindenstrauss, and R. R. Phelps, Extreme positive operators into C(K). Pacif. J. Math. *15* (1965), 747-756. MR *35*, 785.

R. P. Boas, Some uniformly convex spaces. Bull. A.M.S. *46* (1940), 304-311. MR *1*, 242.

S. Bochner and A. E. Taylor, Linear functionals on certain spaces of abstract-valued functions. Ann. of Math. *(2) 39* (1938), 913-944.

H. Bohnenblust and S. Karlin, Geometric properties of the unit sphere of a Banach algebra. Ann. of Math. *(2) 62* (1965), 217-229.

——— and A. Sobczyc, Extension of functionals on complex linear spaces. Bull. A.M.S. *44* (1938), 91-93.

F. F. Bonsall, J. Lindenstrauss, and R. R. Phelps, Extreme positive operators on algebras of functions. Math. Scand. *35* (1966), 161-182. MR *18*, 759.

N. Bourbaki, *Eléments de mathematique XVIII. Première partie. Les structure fondamentales de l'analyse.* Herman et Cie, Paris, 1955a. MR *17*, 1109.

———, *Espaces vectoriels topologiques*. Herman et Cie, Paris, 1955b. MR *31*, 1109.

M. S. Brodskii and D. P. Milman, On the center of a convex set. DAN SSSR *59* (1948), 837-840.

F. Browder, Multi-valued monotone nonlinear mappings and duality mappings in Banach spaces. Trans. Amer. Math. Soc. *118* (1965a), 338-351. MR *31*, 5114.

———, On a theorem of Beurling and Livinston. Canad. J. Math. *17* (1965b), 367-372. MR *31*, 595.

———, Nonexpansive, nonlinear operators in a Banach space. Proc. Nat. Acad. Sci. U.S.A. *54* (1965c), 1041-1044. MR *32*, 4574.

———, Convergence of approximants to fixed points of nonexpansive nonlinear mappings in Banach spaces. Archiv. Rational Mech. Anal. *24* (1967), 82-90. MR *34*, 6582.

A. L. Brown, A rotund reflexive Banach space having a subspace of codimension two with a discontinuous metric projection. Mich. Math. J. *21* (1974), 145-151. MR *50*, 2870.

D. R. Brown, B-convexity and reflexivity in Banach spaces. Trans. Amer. Math. Soc. *187* (1974a), 69-76. MR *48*, 12002.

———, P-convexity and B-convexity in Banach spaces. Trans. Amer. Math. Soc. *187* (1974b), 77-81. MR *48*, 12003.

M. Bruneau, Fonctions d'une variable reélle; characterisations des points extremaux de la boule unité de l'espace des fonctions a p-variation bornée. C. R. Acad. Sci. Paris, Ser. *A-B 274* (1972), A51-A54. MR *44*, 7270.

A. Brunel, H. Fong, and L. Sucheston, On ergodic superproperties of Banach spaces defined by a class of matrices. Proc. Amer. Math. Soc. *49* (1975), 373-378. MR *51*, 1433.

——— and L. Sucheston, On B-convex spaces. Math. System Theory *7* (1974), 294-299. MR *55*, 11004.

———, On J-convexity and some ergodic superproperties of Banach spaces. Trans. Amer. Math. Soc. *204* (1975), 79-90. MR *52*, 1261.

R. C. Buck, A complete characterization of extreme functionals. Bull. Amer. Math. Soc. *65* (1959), 130-133. MR *21*, 1514.

———, Approximate properties of vector-valued functions. Pacif. J. Math. *53* (1974), 85-90. MR *51*, 3892.

Bui-Min-Chi and V. I. Gurariĭ, Certain characteristic of normed spaces and their applications to the generalization of Parseval's equality in Banach spaces. Teor. Funkcii Funkcional Anal. i Priložen. Vyp. *8* (1969), 74-91. MR *42*, 3544.

J. D. Buckholtz, Sums of powers of complex numbers. Notices A.M.S. *13* (1966), 372.

———, A characterization of the exponential series. Amer. Math. Monthly *73* (1966), 121-123.

W. L. Bynum, Characterization of uniform convexity. Pacif. J. Math. *38* (1971), 577-581. MR *46*, 4164.

———, A class of spaces lacking normal structure. Compositio Math. *25* (1972), 233-236. MR *47*, 7386.

———, Normal structure of Banach spaces. Manuscripta Math. *11* (1974), 203-209. MR *49*, 3613.

———, Normal structure coefficients for Banach spaces. Pacif. J. Math. *86* (1980), 427-436. MR *81m*, 46030.

J. R. Calder, A property of ℓ_p spaces. Proc. Amer. Math. Soc. *17* (1966), 202-206. MR *32*, 8112.

———, Concerning weakly uniformly convex spaces. J. London Math. Soc. *(2)1* (1969), 116-118. MR *39*, 7394.

——— and J. B. Hill, A collection of sequence spaces. Trans. Amer. Math. Soc. *152* (1970), 107-118. MR *42*, 822.

J. Cantwell, A topological approach to extreme points in function spaces. Proc. Amer. Math. Soc. *19* (1968), 821-825. MR *37*, 4583.

S. O. Carlson, Orthogonality in normed linear spaces. Arkiv for Math. *4* (1961), 297-318. MR *25*, 5364.

P. G. Casazza and Bor-Luh-Lin, On Lorentz sequence spaces. Bull. Acad. Sinica *2* (1974), 233-240. MR *50*, 14164.

———, Some geometric properties of the Lorentz sequence spaces. Rocky Mount. J. Math. *7* (1977), 683-698. MR *56*, 12832.

B. L. Chalmers, L. Baron, and O. Ferguson, Sets of best approximation in certain classes of normed spaces. J. Approx. Theory *4* (1971), 194-203. MR *44*, 7197.

B. S. Cirel'son (= B. S. Tsirelson), Not every Banach space contains ℓ^p or c_0. Functional. Analiz. i ego Priložen. *8* (1974), 57-60. MR *50*, 2871. English Transl. Functional. Anal. Appl. *8* (1974), 138-141.

J. A. Clarkson, Uniformly convex spaces. Trans. Amer. Math. Soc. *40* (1936), 396-414.

H. B. Cohen and F. Sullivan, Projecting onto cycles in smooth, reflexive Banach spaces. Pacif. J. Math. *34* (1970), 355-364. MR *42*, 2283.

G. Converse, Extreme positive operators on C(X) which commute with a given operator. Trans. Amer. Math. Soc. *138* (1969), 149-158. MR *39*, 4693.

———, I. Namioka, and R. R. Phelps, Extreme invariant positive operators. Trans. Amer. Math. Soc. *137* (1969), 375-385. MR *39*, 4692.

H. Corson, The weak topology of a Banach space. Trans. Amer. Math. Soc. *101* (1961), 1-15. MR *24*, 2220.

——— and J. Lindenstrauss, On function spaces which are Lindelof spaces. Trans. Amer. Math. Soc. *121* (1966a), 476-491. MR *32*, 4666.

———, Continuous selections with non-metrizable range. Trans. Amer. Math. Soc. *121* (1966b), 492-504. MR *32*, 4667.

———, On weakly compact subsets of Banach spaces. Proc. Amer. Math. Soc. *17* (1966c), 407-412. MR *33*, 7812.

J. A. Crenshaw, Extreme positive linear operators. Math. Scand. *25* (1969), 195-217. MR *43*, 956.

———, Positive linear operators continuous for strict topology. Proc. Amer. Math. Soc. *46* (1974), 79-85. MR *51*, 6379.

J. Daneš, On local and global moduli of convexity. Comment. Math. Univ. Carolin. *17* (1976), 413-420. MR *54*, 583.

———, On the strict convexity of the polar operator. Comment. Math. Univ. Carolin. *18* (1977), 393-400. MR *56*, 1009.

F. K. Dashiel and J. Lindenstrauss, Some examples concerning strictly convex norms on C(K) spaces. Israel J. Math. *16* (1973), 329-342. MR *50*, 964.

W. J. Davis and W. B. Johnson, A renorming of non-reflexive Banach spaces. Proc. Amer. Math. Soc. *37* (1973), 486-488. MR *46*, 9693.

M. M. Day, The spaces L^p with $0 < p < 1$. Bull. Amer. Math. Soc. *46* (1940), 816-823. MR *2*, 102,419.

———, Reflexive Banach spaces not isomorphic to uniformly convex spaces. Bull. Amer. Math. Soc. *47* (1941a), 313-317. MR *2*, 221.

———, Some more uniformly convex spaces. Bull. Amer. Math. Soc. *47* (1941b), 504-507. MR *2*, 314.

———, Uniform convexity. III. Bull. Amer. Math. Soc. *49* (1943), 745-750. MR *5*, 146.

———, Uniform convexity in factor and conjugate spaces. Ann. of Math. *(2)45* (1944), 374-385. MR *6*, 69.

———, Strict convexity and smoothness of normed spaces. Trans. Amer. Math. Soc. *78* (1955), 516-528. MR *16*, 716.

———, Every L-space is isomorphic to a strictly convex space. Proc. Amer. Math. Soc. *8* (1957), 415-417. MR *19*, 868.

———, *Normed Linear Spaces*. Springer Verlag, New York, 1958. MR *31*, 5054.

———, A geometric proof of Asplund's differentiability theorem. Israel J. Math. *13* (1972), 277-280. MR *52*, 8882.

———, Invariant renorming. In *Fixed Point Theory and Applications*, Academic Press, New York, 1976, pp. 51-62. MR *58*, 30067.

———, R. C. James, and S. Swaminathan, Normed linear spaces that are uniformly convex in every direction. Canad. J. Math. *23* (1971), 1051-1059. MR *44*, 4492.

J. Diestel, *Geometry of Banach Spaces: Selected Topics*. Lecture Notes in Math. *485*, 1975. MR *57*, 1079.

——— and J. H. Uhl, *Vector Measures*. Amer. Math. Soc. Surveys *16* (1977). MR *56*, 12216.

C. R. Diminie and A. G. White, Strict convexity in topological spaces. Math. Japonica *22* (1977), 49-56. MR *56*, 16321.

———, Strict convexity conditions for seminorms. Math. Japonica *24* (1979/80), 489-493. MR *81g*, 46014.

J. Dixmier, Sur une theoreme de Banach. Duke Math. J. *15* (1948), 1057-1071. MR *10*, 306.

J. Dixmier, *Les algebres d'operateurs dans l'espace Hilbertien: Algebres de von Neumann*. Gauthier Villars, Paris, 1957. MR *20*, 1234. Second Ed. Paris 1969. MR *50*, 5482.

R. G. Douglas, On majorization, factorization, and range inclusion of operators on Hilbert spaces. Proc. Am. Math. Soc. *17* (1966), 413-416. MR *34*, 3315.

D. J. Downing and B. Turett, Some properties of the characteristic of convexity relating to fixed point theory. Pacif. J. Math. (to appear).

Van der Dulst, *Reflexive and Superreflexive Banach Spaces*. Math. Centre Tracts 102, Amsterdam, 1978. MR *80d*, 46019.

N. Dunford and M. Morse, Remarks on the preceding paper by James A. Clarkson. Trans. Amer. Math. Soc. *40* (1936), 415-420.

——— and J. T. Schwartz, *Linear Operators*. *I*. Interscience, New York, 1958. MR *22*, 8302.

M. Edelstein, A theorem on fixed points under isometries. Amer. Math. Monthly *70* (1963), 298-300. MR *27*, 4044.

———, On non-expansive mappings in Banach spaces. Proc. Cambridge Math. Soc. *60* (1964), 439-447. MR *29*, 1521.

———, The construction of an asymptotic center with a fixed point property. Bull. Amer. Math. Soc. *78* (1972), 206-208. MR *45*, 1005.

———, Fixed point theorems in uniformly convex spaces. Proc. Amer. Math. Soc. *44* (1974), 369-374. MR *50*, 10917.

——— and A. C. Thompson, Some results on nearest points and support properties of convex sets in c_0. Pacif. J. Math. *40* (1972), 553-560. MR *46*, 7855.

A. J. Ellis, Extreme positive operators. Quart. J. Math. Oxford Ser. *(2)15* (1964), 342-344. MR *30*, 4157.

P. Enflo, Banach spaces which can be given an equivalent uniformly convex norm. Israel J. Math. *13* (1972), 281-288. MR *49*, 1073.

———, Recent results on general Banach spaces. In *Proc. Internat. Congress of Math., Vancourver, B.C., 1974*, Vol. 2, pp. 53-55. Canad. Math. Congress, Montreal, Que., 1975. MR *54*, 11029.

M. S. Espelie, Multiplicative and extreme positive operators. Pacif. J. Math. *48* (1973), 57-66. MR *49*, 1212.

H. Fakhoury, Directions d'uniformite dans un espace norme. C. R. Acad. Sci.-Paris, Ser. *A-B 283* (1976), 473-476. MR *55*, 3750.

Ky Fan and I. Glicksberg, Fully convex normed spaces. Proc. Nat. Acad. Sci. U.S.A. *41* (1955), 947-953. MR *17*, 386.

———, Some geometrical properties of the sphere in a normed linear space. Duke Math. J. *25* (1958), 553-568. MR *20*, 5421.

T. Figiel, On moduli of smoothness and convexity. Publ. Ser. Univ. Aarhus. *24* (1974), 67-70. MR *52*, 11541.

———, On the moduli of convexity and smoothness. Studia Mathematica *56* (1976), 121-155. MR *54*, 13535.

——— and W. B. Johnson, A uniformly convex Banach space which contains no ℓ^p. Compositio Math. *29* (1974), 179-180. MR *50*, 8011.

———, and G. Pisier, Series aleatoires dans les espaces uniformement convexes ou uniformement lisses. C. R. Acad. Sci. Paris Ser. *A-B 279* (1974), A611-A614. MR *50*, 10761.

G. D. de Figueiredo, *Topics in Nonlinear Analysis*. Lecture Notes *48*, Univ. of Maryland, 1967.

S. R. Foguel, On a theorem of A. E. Taylor. Proc. Amer. Math. Soc. *9* (1958), 325. MR *20*, 219.

V. P. Fonf (see M. I. Kadets), Conditionally convergent series in a uniformly smooth Banach space. Mat. Zametki *11(2)* (1972), 201-208. MR *45*, 7446. English Transl. Math. Notes *11(2)* (1972), 129-132.

R. Fortet, Remarques sur les espaces uniformement convexes. Bull. Soc. Math. France *69* (1941), 23-46. MR *7*, 124.

D. H. Fremlin. See J. L. Bell.

A. L. Garkavi, The best possible net and the best possible cross section of a set in a normed space. Izv. Akad. Nauk SSSR *26* (1962), 87-106. MR *25*, 429. English Transl. Amer. Math. Soc. Transl. *(2)39* (1964a), 111-132.

———, On Chebyshev and almost Chebyshev subspaces. Izv. Akad. Nauk SSSR *28* (1964b), 799-818 (Russian). MR *29*, 2635.

D. J. Garling, A class of reflexive symmetric BK-spaces. Canad. J. Math. *21* (1969), 602-608. MR *53*, 14081.

———, Convexity, smoothness and martingale inequalities. Israel J. Math. *29* (1978), 189-198. MR *57*, 10819.

A. Gendler, Extreme operators in the unit ball of L(C(X),C(Y)) over the complex field. Proc. Amer. Math. Soc. *57* (1976), 85-88, MR *53*, 8967.

R. Geremia. See F. Sullivan.

D. Giesy, On a convexity condition in normed linear spaces. Trans. Amer. Math. Soc. *125* (1966), 114-146. MR *34*, 4866.

———, B-convexity and reflexivity. Israel J. Math. *15* (1973), 430-436. MR *48*, 4698.

——— and R. C. James, Uniform non-ℓ^1 and B-convex spaces. Studia Math. *48* (1973), 61-69. MR *48*, 11994.

J. R. Giles, Classes of semi-inner products. Trans. Amer. Math. Soc. *129* (1967), 436-447. MR *36*, 663.

———, A characterization of differentiability of the norm of a normed linear space. J. Austral. Math. Soc. *12* (1971), 106-114. MR *44*, 2021.

———, A non-reflexive Banach space has a non-smooth third conjugate space. Canad. Math. Bull. *17* (1974), 117-119. MR *50*, 5429.

———, Uniformly weak differentiability of the norm and a condition of Vlasov. J. Austral. Math. Soc. Ser. *A.21(4)* (1976), 393-409. MR *54*, 11030.

A. Gleit and R. McGuigan, A note on polyhedral Banach spaces. Proc. Amer. Math. Soc. *33* (1972), 398-404. MR *45*, 4123.

I. Glicksberg and Ky Fan, Some geometrical properties of the sphere in a normed linear space. Duke Math. J. *25* (1958), 553-568. MR *20*, 5421.

G. Godini, On the localization and directionalization of uniform convexity and uniform smoothness. Rev. Roum. Math. Pure Appl. *11* (1966), 1233-1239. MR *36*, 654.

———, On minimal points. Comment. Math. Univ. Carolin. *21* (1980), 407-419. MR *81j*, 46023.

K. Goebel, An elementary proof of the fixed point theorem of Browder and Kirk. Mich. Math. J. *16* (1969), 381-383. MR *40*, 4831.

———, Convexity of balls and fixed point theorems for mappings with non-expansive square. Compositio Math. *22* (1970), 269-274.

———, W. A. Kirk, and T. N. Shimi, A fixed point theorem in uniformly convex spaces. Boll. Union. Mat. Ital. *(4)7* (1973), 67-75. MR *47*, 9367.

J. P. Gossez and E. Lami Dozo, Structure normale et base de Schauder. Bull. Acad. Royal. Belgique (5e Ser.) *55* (1969), 673-681.

———, Some geometric properties related to the fixed point theory for nonexpansive mappings. Pacif. J. Math. *40* (1972), 565-573. MR *46*, 9815.

E. Granirer, *Exposed Points of Convex Sets and Weak Sequential Convergence.* Mem. Amer. Math. Soc. *123* (1972). Amer. Math. Soc., Providence, R.I. (1972). MR *51*, 1343.

F. P. Greenleaf, *Invariant Means on Topological Groups.* Van Nostrand Reinhold Co., New York.

A. Grothendieck, *Espaces vectoriels topologiques.* Inst. Mat. Pura Applicada, Universidade de Sao Paulo, Sao Paulo, 1954. MR *47*, 1110.

S. P. Gudder and D. Strawther, A characterization of strictly convex spaces. Proc. Amer. Math. Soc. *47* (1975), 268.

———, Strictly convex normed spaces. Proc. Amer. Math. Soc. *59* (1976), 263-267. MR *54*, 5806.

N. I. Gurarii and V. I. Liokumovič, The stability of orthogonal bases in uniformly convex spaces (Russian). Izv. Vysš. Učebn. Zaved. Mathematika *(10)125* (1972), 23-26. MR *47*, 9250.

V. I. Gurarii, On the moduli of rotundity and smoothness of Banach spaces. Doklady Akad. Nauk SSSR *161* (1965), 1003-1005. MR *32*, 8118. English Transl. Sov. Math. Doklady (1965), 535-539.

———, Dependence of certain geometric properties of Banach spaces on the modulus of rotundity. Teoria Functssi i Funktional. Analiz. i. Priložen. Vyp. *2* (1966), 98-107 (Russian). MR *33*, 7819.

———, Some theorems on bases in Hilbert spaces. Doklady Akad. Nauk SSSR *193* (1970), 974-977 (Russian). MR *44*, 7262. English Transl. Sov. Math. Doklady *11* (1970), 1042-1045.

——— and N. I. Gurarii, Bases in uniformly convex spaces and uniformly flattened Banach spaces (Russian). Izv. Akad. Nauk SSSR, Ser. Math. *35* (1971), 210-215. MR *44*, 780.

J. Hager and F. D. Sullivan, Smoothness and weak*-sequential compactness. Proc. Math. Soc. *78(4)* (1980), 497-503. MR *81k*, 46010.

I. Halperin, Function spaces. Canad. J. Math. *5* (1953), 273-288. MR *15*, 38.

———, Uniform convexity in function spaces. Duke Math. J. *21* (1954a), 195-204. MR *15*, 880.

———, Reflexivity of the L^{λ} function spaces. Duke Math. J. *21* (1954b), 205-208. MR *15*, 880.

——— and H. Nakano, Generalized ℓ^p spaces and the Schur property. J. Math. Soc. Japan *5* (1953), 50-58. MR *15*, 326.

O. Hanner, On the uniform convexity of L^p and ℓ^p. Arkiv for Math. *3* (1956), 244-246. MR *17*, 987.

L. A. Harris, Schwartz's lemma in normed linear spaces. Proc. Nat. Acad. Sci. U.S.A. *62* (1969), 1014-1017.

———, A continuous form of Schwarz's lemma in normed linear spaces. Pacif. J. Math. *38* (1971), 635-639. MR *46*, 4167.

———, Operator extreme points and the numerical range. Indian Math. J. *23* (1974), 937-947. MR *48*, 12085.

G. M. Henkin, Stability of unconditional bases in uniformly convex spaces. Uspechi Mat. Nauk *18* (1963), No. 6(114), 219-223. MR *29*, 451.

J. B. Hill. See J. R. Calder.

E. Hille and R. S. Phillips, *Functional Analysis and Semigroups*. Amer. Math. Soc. Colloquim Publ. Vol. 31. Amer. Math. Soc., Providence, R.I., 1957. MR *19*, 664.

W. Hintzman, Characterization of extreme functionals. Math. Japonica *19* (1974), 63-64. MR *51*, 13621.

B. Hirsberg and A. J. Lazar, Complex Lindenstrauss spaces with extreme points. Trans. Amer. Math. Soc. *186* (1973), 141-150. MR *48*, 11996.

J. Hoffman-Jørgensen, *On the Modulus of Smoothness and the G_*-Conditions in B-Spaces*. Aarhus Univeritet, Matematisk Inst. Preprint series, 1974.

R. B. Holmes, On the continuity of the best approximation operators. In *Proc. Sympos. Infinite Dimensional Topology*. Ann. of Math. Studies 69 (1972), 137-157. Princeton Univ. Press, Princeton, N.J., 1972a, MR *53*, 8757.

———, *A Course in Optimization and Best Approximation*. Lecture Notes in Math. 598. Springer-Verlag, New York, 1972b. MR *54*, 8321.

———, Smoothness of certain projections on Hilbert space. Trans. Amer. Math. Soc. *183* (1973), 87-100. MR *48*, 4596.

——— and B. Kripke, Smoothness of approximation. Mich. Math. J. *15* (1968), 225-248. MR *37*, 4483.

J. R. Holub, On the metric geometry of ideals of operators on Hilbert spaces. Math. Ann. *201* (1973), 157-163. MR *48*, 4757.

———, Rotundity, orthogonality and characterizations of inner product spaces. Bull. Amer. Math. Soc. *81* (1975), 1087-1089. MR *52*, 1263.

M. Hsieh, *Convergent and Divergent Series in Banach Spaces*. M.S. Thesis, Louisiana State Univ., 1970.

R. E. Huff, Dentability and the Radon-Nicodym property. Duke Math. J. *41* (1971), 111-114. MR *49*, 5783.

——— and P. D. Morris, Geometric characterization of the Radon-Nicodym property in Banach spaces. Studia Math. *56* (1970), 157-164. MR *54*, 897.

———, Dual spaces with the Krein-Milman property. Proc. Amer. Math. Soc. *49* (1975), 104-108. MR *50*, 14220.

E. Hughes. See P. R. Beesack.

T. Husain and B. Malviya, On semi-inner product spaces. I. Colloq. Math. *24* (1972), 235-239. MR *42*, 981.

L. Ingelstam, A vertex property for Banach algebras. Math. Scand. *11* (1962), 22-32. MR *26*, 4199.

A. Ionescu Tulcea and C. Ionescu Tulcea, On the lifting property. I. J. Math. Anal. Appl. *3* (1961), 537-546. MR *27*, 257.

———, *Topics in the Theory of Lifting*. Ergebn. der Math. Band 48, 1969. MR *43*, 2185.

———, *A Note on Extreme Points*. (Unpublished manuscript.)

V. I. Istrăţescu, Uber die Banachraume mit zahlbarer Basis. I. Rev. Roum. Math. Pure Appl. *7* (1962), 481-482. MR *31*, 2590.

———, Uber die Banachraume mit zahlbarer Basis. II. Rev. Roum. Math. Pure Appl. *9* (1964), 431-433. MR *33*, 6343.

———, *Probabilistic Metric Spaces: Theory and Applications* (Romanian). Editura Technica, Bucharest, 1975.

———, On complex strictly convex spaces. III. Rend. Ist. Mat. Univ. Trieste, Vol. 10, Fasc. I-II (1978), 48-56. MR *81k*, 46013.

———, On complex strictly convex spaces. II. J. Math. Anal. Appl. *71* (1979), 580-589. MR *80m*, 46016.

———, *Introduction to Linear Operator Theory*. Marcel Dekker, Inc., New York, 1981a.

———, *Fixed Point Theory: An Introduction*. D. Reidel Publishing Corp., Dordrecht-Boston, 1981b.

———, A simple example of a reflexive smooth Banach space which is not (F). (To appear.) 1982a

———, On some geometric properties of interpolation spaces of Loins-Peetre. (To appear.) (1983a)

———, Extreme and strongly extreme operators. (To appear.)(1983b)

———, Inner product structures: theory and applications. (To appear.)(1983c)

——— and I. I. Istrăţescu, On complex strictly convex spaces. I. J. Math. Anal. Appl. *70* (1979), 423-429. MR *80h*, 46022.

R. C. James (See M. M. Day, D. Giesy), Orthogonality in normed linear spaces. Duke Math. J. *12* (1945), 291-302. MR *6*, 273.

———, Orthogonality and linear functionals in normed linear spaces. Trans. Amer. Math. Soc. *61* (1947), 265-292. MR *9*, 42.

———, Bases and reflexivity of Banach spaces. Ann. of Math. *(2)52* (1950), 518-527. MR *12*, 616.

———, A non-reflexive Banach space isometric with its conjugate. Proc. Nat. Acad. Sci. U.S.A. *37* (1951a), 174-177. MR *13*, 356.

———, Linear functionals as differentials of a norm. Math. Mag. *24* (1951b), 237-244.

———, Reflexivity and the supremum of linear functionals. Ann. of Math. *(2)66* (1957), 159-169. MR *19*, 755.

———, Weakly compact sets. Trans. Amer. Math. Soc. *113* (1964a), 129-140. MR *29*, 2628.

———, Uniformly non-square Banach spaces. Ann of Math. *(2)80* (1964b), 542-550. MR *30*, 4139.

———, Convexity and smoothness. In *Proc. Colloq. on Convexity*, Copenhagen, 1965, 165-167. Kobenhavns Univ. Inst. Mat., 1967. MR , 657.

———, A counterexample for a sup theorem in normed spaces. Israel J. Math. *9* (1971), 511-512. MR *43*, 5287.

———, Reflexivity and the sup of linear functionals. Israel J. Math. *13* (1972a), 289-300. MR *49*, 3506.

———, Some self-dual properties of normed linear spaces. In *Proc. Sympos. Infinite Dimensional Topology*, 1972, 159-175. Ann. of Math. Studies 69, Princeton Univ. Press, Princeton, N.J., 1972b. MR *56*, 12849.

———, Superreflexive spaces with bases. Pacif. J. Math. *41* (1972c), 409-419. MR *46*, 7866.

———, Superreflexive Banach spaces. Canad. J. Math. *24* (1972d), 896-904. MR *47*, 9248.

———, A non-reflexive Banach space that is uniformly non-octahedral. Israel J. Math. *18* (1974), 145-155. MR *50*, 8012.

———, On a conjecture about ℓ^1. (To appear.)

——— and J. J. Schaefer, Super-reflexivity and the girth of spheres. Israel J. Math. *11* (1972), 398-404. MR *46*, 4175.

K. John and V. Zizler, A renorming of dual spaces. Israel J. Math. *12* (1972), 331-336. MR *49*, 9592.

———, A note on renorming of dual spaces. Bull. Acad. Polon. Sci. *21* (1973), 47-50. MR *47*, 9246.

K. John and V. Zizler, Projections in dual weakly compactly generated Banach spaces. Studia Math. *49* (1974a), 41-50. MR *49*, 1071.

———, Smoothness and its equivalence in weakly compactly generated Banach spaces. J. Funct. Anal. *15* (1974b), 1-11. MR *54*, 5807.

———, On extension of rotund norms. Bull. Acad. Polon. Sci. *24* (1976a), 705-707. MR *55*, 1041.

———, A note on renorming of Banach spaces decomposable into certain operator ranges. Acta Math. Acad. Hungarica *28* (1976b), No. 3-4, 247-251. MR *54*, 11032.

———, On rough norms on Banach spaces. Comment. Math. Univ. Carolin. *19* (1978), 335-349. MR *80d*, 46028.

———, A short proof of a version of Asplund's norm averaging theorem. Proc. Amer. Math. Soc. *73* (1979a), 277-278. MR *80b*, 46020.

———, On extension of rotund norms. Pacif. J. Math. *82* (1979b), 451-455. MR *81h*, 46009.

N. Jewell. See S. Axler.

B. E. Johnson, Norms of derivations. Pacif. J. Math. *38* (1971), 565-569. MR *46*, 6087.

W. B. Johnson (see W. J. Davis), A reflexive Banach space which is not sufficiently euclidean. Studia Math. *55* (1976), 201-205. MR *55*, 3761.

——— and J. Lindenstrauss, Some remarks on weakly generated Banach spaces. Israel J. Math. *17* (1974), 219-230. MR *54*, 5808. Correction: Ibid *32* (1979), 382-383. MR *81g*, 46015.

M. I. Kadets, On topological equivalence of uniformly convex spaces. Uspechi Mat. Nauk *10* (1955), *5*(66), 137-141. MR *17*, 511.

———, Unconditionally convergent series in a uniformly convex space. Uspechi Mat. Nauk (N.S.) *11* (1956), 185-190. MR *18*, 733.

———, On the connection between weak and strong convergence. Dopovīdī Akad. Nauk Ukrain. RSR *9* (1969), 949-952. MR *22*, 2879.(Ukrainian).

———, On weak and norm convergence. Doklady Akad. Nauk SSSR *122* (1958), 13-16. MR *20*, 5422.

———, On spaces isomorphic to locally uniformly rotund spaces. Izv. Vysš. Učebn. Zaved. Matematika *6* (1959), 51-57. MR *23*, A3987.

———, Condition for the differentiability of the norm of a Banach space. Uspechi Mat. Nauk *20* (1965a), 183-188. MR *32*, 2883.

———, The topological equivalence of certain cones in a Banach space. Doklady Akad. Nauk SSSR *162* (1965b), 1241-1244. MR *31*, 3824. English Transl. Sov. Math. Doklady *6* (1965), 847-850.

———, The geometry of Banach spaces. Math. Analysis (Russian) Akad. Nauk SSSR, Inst. Naučn. i. Techn. Inform. Moscow, 1965c. MR *58*, 30064. English Transl. J. Soviet Math. *7* (1977), 953-973.

———, A proof of the equivalence of all separable infinite-dimensional Banach spaces. Functional. Analiz. i Priložen. *1* (1967), 61-70. MR *35*, 700.

———, The method of equivalent norms in the theory of abstract almost periodic functions. Studia Math. *31* (1968), 89-94. MR *38*, 3692.

——— and A. Pelczynski, Basic sequences, biorthogonal systems and norming sets in Banach and Fréchet spaces. Studia Math. (1965), 297-323. MR *31*, 6112.

——— and V. P. Fonf, Some properties of extreme points of the unit ball of a Banach space (Russian). Mat. Zametki *20* (1976), No. 3, 315-320. MR *55*, 3759. English Transl. Math. Notes *20* (1976), No. 3-4, 737-739.

S. Kakutani, Weak convergence in uniformly convex spaces. Tohoku Math. J. *45* (1938), 188-193.

S. Kaniel, Construction of a fixed point for contractions in a Banach space. Israel J. Math. *9* (1971), 535-540. MR *44*, 2110.

R. Kannan, Some results on fixed points, II. Amer. Math. Monthly *76* (1969), 405-408. MR *41*, 2487.

———, Fixed point theorems in reflexive Banach spaces. Proc. A.M.S. *38* (1973), 111-118. MR *47*, 2448.

J. L. Kelley. See R. Arens.

C. W. Kim, Extreme contraction operators on ℓ_∞. Math. Z. *151* (1976), 101-110. MR *54*, 413.

W. A. Kirk. See L. P. Belluce, K. Goebel.

S. V. Kisliakov, p-absolutely summing operators. In *Geometry of Linear Spaces and the Theory of Operators* (Russian). B. S. Mitiagin, Ed., Jaroslavl, 1977, pp. 114-175.

V. Klee, Some characterizations of reflexivity. Revista Ci. Lima *52* (1950), 15-23. MR *13*, 250.

———, Convex bodies and periodic homeomorphisms in Hilbert spaces. Trans. Amer. Math. Soc. *74* (1953), 10-43. MR *14*, 989.

———, A note on topological properties of normed linear spaces. Proc. Amer. Math. Soc. *7* (1956), 673-674. MR *17*, 1227.

———, Some new results on smoothness and rotundity in normed linear spaces. Math Ann. *139* (1959), 51-63. MR *22*, 5879.

———, Mappings into normed linear spaces. Fund. Math. *49* (1960a), 25-34. MR *23*, A3985.

———, Relative extreme points. In *Proc. Sympos. Linear Spaces.* Jerusalem, 1960b, 283-289. Jerusalem Academic Press, Pergamon Press, Oxford, 1961. MR *24*, A2825.

———, Extreme points of convex sets without completeness of the scalar field. Mathematika *11* (1964), 59-63. MR *29*, 2712.

———, Remarks on nearest points in normed linear spaces. In *Proc. Sympos. on Convexity*, Copenhagen, København, 1965, 168-176.

———, Two renorming constructions related to a question of Anselone. Studia Math. *33* (1969), 231-242. MR *40*, 1756.

G. Köthe, *Topological Linear Spaces*. Springer-Verlag, Heidelberg-Berlin-New York. I, 1969; II, 1980.

C. A. Kottman, Packing and reflexivity in Banach spaces. Trans. Amer. Math. Soc. *150* (1970), 565-570. MR *42*, 827.

S. N. Kračkovskii and A. A. Vinogradov, On a criterion of uniform convexity of space of type (B). Uspechi Mat. Nauk (N.S.) *7*, No. 3(49), 131-134. MR *14*, 55.

B. Kripke. See R. B. Holmes.

S. S. Kuteladze, Extremal operators (Russian). DAN SSSR *234* (1977), 291-293. MR *56*, 3673.

———, Extremal operators (Russian). DAN SSSR *234* (1977), 291-293. MR *56*, 3673.

S. Kwapien, Some remarks on (p,q)-absolutely summing operators in ℓ_p-spaces. Studia Math. *29* (1968), 327-337.

J. Kyle, Norms of derivations. J. London Math. Soc. *(2)16* (1977), 297-312. MR *58*, 7113.

K. S. Lau, Extreme operators on Choquet simplexes. Pacif. J. Math. *52* (1974), 129-142. MR *50*, 14176.

A. J. Lazar. See B. Hirsberg.

E. B. Leach and J. H. M. Whitfield, Differentiable functions and rough norms on Banach spaces. Proc. Amer. Math. Soc. *33* (1972), 120-126. MR *45*, 2471.

I. E. Leonard, Banach sequences spaces. J. Math. Anal. Appl. *54* (1976), 245-265. MR *54*, 8230.

——— and K. Sundaresan, A note on smooth Banach spaces. J. Math. Anal. Appl. *43* (1973a), 450-454. MR *48*, 4695.

———, Geometry of Lebesgue-Bochner function spaces, smoothness. Bull. Amer. Math. Soc. *79* (1973b), 546-549. MR *48*, 4720.

———, Geometry of Lebesgue-Bochner function spaces, smoothness. Trans. Amer. Math. Soc. *198* (1974a), 229-251. MR *51*, 3894.

———, Smoothness and duality in $L^p(E)$. J. Math. Anal. Sppl. *46* (1974b), 513-522. MR *49*, 9608.

E. A. Lifschitz, Fixed point theorems for operators in strongly convex spaces (Russian). Voronez. Gosud. Univ. Trudy Mat. Fak. *16* (1975), 23-28.

T. C. Lim, A fixed point theorem for multi-valued nonexpansive mappings in uniformly convex spaces. Bull. Amer. Math. Soc. *80* (1974), 1123-1126. MR *52*, 15136.

———, Fixed point theorems for mappings of nonexpansive type. Proc. Amer. Math. Soc. *66* (1977), 69-74. MR *57*, 13593.

Bor-Luh-Lin. See P. G. Casazza.

J. Lindenstrauss (see D. Amir, R. M. Blumenthal, F. F. Bonsall, H. Corson, K. Dashiel, W. B. Johnson), On the modulus of smoothness and divergent series in Banach spaces. Mich. Math. J. *10* (1963a), 241-252. MR *29*, 6316.

———, On operators which attain their norm. Israel J. Math. *1* (1963b), 139-148. MR *28*, 3308.

———, *Extension of Operators*. Mem. Amer. Math. Soc. *48* (1964). MR *31*, 3828.

———, On reflexive Banach spaces having the metric approximation property. Israel J. Math. *3* (1965), 199-204. MR *34*, 4874.

———, Notes on Klee's paper "Polyhedral sections of convex bodies." Israel J. Math. *4* (1966a), 235-242. MR *35*, 704.

———, On nonseparable reflexive Banach spaces. Bull. Amer. Math. Soc. *72* (1966b), 967-970. MR *34*, 4875.

———, On extreme points in ℓ^1. Israel J. Math. *41* (1966c), 59-61. MR *34*, 589.

———, Some aspects of the theory of Banach spaces. Advances in Math. *5* (1970), 159-180. MR *43*, 5288.

———, Factoring compact operators. Israel J. Math. *9* (1971), 337-345.

———, *Weakly-Compact Sets: Their Topological Properties and the Banach Space They Generate*. Ann. of Math. Studies *69* (1972), 235-273. MR *54*, 5809.

——— and W. B. Johnson, Some remarks on weakly compactly generated spaces. Israel J. Math. *17* (1974), 219-230. MR *54*, 5808.

——— and A. Pelczynski, Absolutely summing operators in L^p-spaces and their applications. Studia Math. *29* (1968), 275-326. MR *37*, 6743.

———, R. R. Phelps, and J. V. Ryff, Extreme nonmultiplicative operators. Unpublished manuscript.

——— and L. Tzafriri, *Classical Banach Spaces*. I. Ergenb. Math. Bd. *92* (1977). MR *58*, 17766.

V. I. Liokumovič (see N. I. Gurarii), The existence of B-spaces with non-convex modulus of convexity. Izv. Vyss. Ucebn. Zaved. Matematika (1973), 43-49 (Russian). MR *50*, 949.

———, Certain theorems from the geometry of Minkowski spaces (Russian). Teor. Funkcii Funkcional Anal. i Priložen. Vyp. *14* (1981), 117-126. MR *47*, 5557.

J. L. Lions and J. Peetre, Sur une classe d'espaces d'interpolation. Inst. Hautes Etudes Sci. Publ. Math. *19* (1964), 5-68. MR *29*, 2627.

L. A. Liusternik and V. I. Sobolev, *The Elements of Functional Analysis* (Russian). Gosud. Izdat. Techn. Teor. Lit. Moscow-Leningrad, 1951. MR *14*, 54.

A. R. Lovaglia, Locally uniformly convex spaces. Trans. Amer. Math. Soc. *78* (1955), 225-238. MR *16*, 596.

G. Lumer, Semi-inner product spaces. Trans. Amer. Math. Soc. *100* (1961), 29-43. MR *24*, A2860.

R. P. Mallev and S. L. Troyanski (S. L. Trojanski), On moduli of convexity and smoothness in Orlicz spaces. Studia Math. *54* (1975), 131-141. MR *XX*, 528.

R. de Marr, Common fixed points for commuting contraction mappings. Pacif. J. Math. *13* (1963), 1139-1141. MR *28*, 2446.

S. Mazur, Uber konvexe Mengen in linearen normierten Rauman. Studia Math. *4* (1933), 70-84.

C. A. McCarthy, c_p. Israel J. Math. (1967), 249-271. MR *37*, 735.

R. McGuigan (see A. Gleit), Strongly extreme points in Banach spaces. Manuscripta Math. *5* (1971), 113-122. MR *46*, 7858.

E. J. McShane, Linear functionals on certain Banach spaces. Proc. Amer. Math. Soc. *1* (1950), 402-408. MR *12*, 110.

M. Menarer, *Semi-inner Product Spaces*. Ph.D. Thesis. Univ. of Maryland, 1969.

K. Menger, Statistical metrics. Proc. Nat. Acad. Sci. U.S.A. *28* (1942), 535-537, MR *4*, 163.

E. Michael, Continuous selections. I. Ann. of Math. *(2)63* (1956a), 361-382. MR *17*, 990.

———, Continuous selections. II. Ann. of Math. *(2)64* (1956b), 562-580. MR *18*, 3254.

———, A linear mapping between function spaces. Proc. Amer. Math. Soc. *15* (1964a), 407-409. MR *28*, 5327.

———, Three mapping theorems. Proc. Amer. Math. Soc. *15* (1964b), 410-415. MR *28*, 5328.

———, A selection theorem. Proc. Amer. Math. Soc. *17* (1966), 1404-1406. MR *34*, 3551.

P. Miles, B*-algebras unit ball extremal points. Pacif. J. Math. *14* (1964), 627-637. MR *30*, 3385.

D. P. Milman, Certain tests for the regularity of spaces of type (B). Doklady Akad. Nauk SSSR *20* (1938), 243-246.

V. D. Milman, The infinite dimensional geometry of the unit sphere of a Banach space (Russian). DAN SSSR *177* (1967), 514-517. MR *39*, 3289. English Transl. Sov. Math. Doklady *8* (1967), 1440-1444.

———, On a transformation of convex functions and the duality of the and characteristics of a B-space. Doklady Akad. Nauk SSSR *187* (1969), 33-35. MR *41*, 797. English Transl. Sov. Math. Doklady *10* (1969), 789-792.

———, Facial characterization of convex sets: extremal elements. Trudy Moscow. Math. Obscestva *22* (1970a), 63-126. English Transl. Transact. Moscow Math. Soc. *22* (1970), 69-139.

———, Geometric theory of Banach spaces. I. Uspechi Math. Nauk *25(3)* (1970b), 113-173. MR *43*, 6704. English Trans. Russian Math. Surveys *25:3* (1970), 111-170.

———, Geometric theory of Banach spaces. II. Uspechi Math. Nauk *26* (1971), 73-149. MR *54*, 8240. English Trans. Russian Math. Surveys *26* (1971), 79-163.

R. L. Moore and T. T. Trent, A strict maximum modulus theorem for certain Banach spaces. Mh. Math. *92* (1981), 197-201.

J. J. Moreau, Un cas de convergence des iterées d'une contraction d'un espace Hilbertien. C. R. Acad. Sci.-Paris Ser. *A-B 286* (1978), 143-144. MR *57*, 7272.

P. D. Morris (see R. E. Huff) and R. R. Phelps, Theorems of Krein-Milman type for certain convex sets of operators. Trans. Amer. Math. Soc. *150* (1970), 183-200. MR *41*, 7409.

S. E. Mosiman and R. F. Wheeler, The strict topology in a completely regular setting; relations to topological measure theory. Canad. J. Math. *24* (1972), 873-890. MR *48*, 6909.

D. H. Muštari, The notion of a normed space of cotype p2 (Russian). Functional. Analiz. i Priložen. *13* (1979), 75-76. MR *81i*, 46012.

I. Namioka (see G. Converse), Neighborhoods of extreme points. Israel J. Math. *5* (1967), 145-152. MR *36*, 4323.

H. Nakano (see I. Halpern), *Topology and Linear Topological Spaces*. Maruzen Co., Maruzen, Tokyo, 1951. MR *13*, 753.

V. N. Nikoskii, The best approximation and bases in Fréchet spaces. DAN SSSR *59* (1948), 639-642 (Russian). MR *10*, 128.

G. Nordlander, The modulus of convexity in normed linear spaces. Arkiv. för Math. *4* (1960), 15-17. MR *25*, 4329.

———, On sign-independent and almost sign-independent convergence in normed linear spaces. Arkiv. för Math. *4* (1962), 287-296. MR *25*, 4330.

V. P. Odinets (= V. P. Odinec), On normal bases in Banach spaces (Russian). Izv. Vysš. Učebn. Zaved. Matematika *18* (1974), 67-69. MR *51*, 8797.

W. Orlicz, Über unbedingte Konvergenz in Funktionen-Räumen. I. Studia Math. *4* (1933a), 33-37.

———, Über unbedingte Konvergenz in Funktionen-Räumen. II. Studia Math. *4* (1933b), 41-47.

M. Ortel. See P. R. Beesack.

E. Oshman, On continuity of metric projection onto some classes of subspaces in a Banach space. DAN SSSR *195* (1970), 555-557 (Russian). MR *42*, 8251. English Transl. Sov. Math. Dokl. *11* (1970), 1521-1523.

C. L. Outlaw and C. W. Groetsch, Averaging iteration in Banach spaces. Not. Amer. Math. Soc. *15* (1968), 180.

———, Averaging iteration in Banach spaces. Bull. Amer. Math. Soc. *75* (1969), 430-432. MR *39*, 835.

J. R. Partington, Equivalent norms on spaces of bounded functions. Israel J. Math. *35* (1980), 205-209. MR *81h*, 46013.

N. T. Peck, Representation of functions in C(X) by means of extreme points. Proc. Amer. Math. Soc. *18* (1967), 133-135. MR *34*, 8167.

J. Peetre. See J. L. Lions.

A. Pelczynski. See C. Bessaga, M. I. Kadets.

W. V. Petryshyn, A characterization of strict convexity of Banach spaces and other use of duality mappings. J. Funct. Anal. *6* (1970), 282-291. MR *44*, 3396.

——— and W. T. Williamson, Strong and weak convergence of the sequence of successive approximations for quasi-nonexpansive mappings. J. Math. Anal. Appl. *43* (1973), 459-497. MR *48*, 4854.

B. J. Pettis, A proof that uniformly convex space is reflexive. Duke Math. J. *5* (1939), 249-253.

R. R. Phelps (see R. M. Blumenthal, G. Converse, P. D. Morris), A representation theorem for bounded convex sets. Proc. Amer. Math. Soc. *11* (1960a), 976-983. MR *23*, A501.

———, Uniqueness of Hahn-Banach extensions and unique best approximation. Trans. Amer. Math. Soc. *95* (1960b), 238-255. MR *22*, 3964.

———, Extreme positive operators and homeomorphisms. Trans. Amer. Math. Soc. *108* (1963), 265-274. MR *27*, 6153.

———, Extreme points in function algebras. Duke Math. J. *32* (1965), 267-277. MR *31*, 3890.

———, *Lectures on Choquet's Theorem*. Van Nostrand, Princeton, N.J., 1966. MR *33*, 1690.

———, Dentability and extreme points in Banach spaces. J. Funct. Anal. *16* (1974), 78-90. MR *50*, 5427.

R. S. Phillips. See E. Hille.

G. Pisier, *Martingales a valeurs dans les espaces uniformement convexes*. Mimeographed Notes, Paris, 1974a.

———, Martingales a valeurs dans les espaces uniformement convexes. C. R. Acad. Sci. Paris, Ser. *A-B 279* (1974b), 647-649. MR *50*, 10764.

———, Martingales with values in uniformly convex spaces. Israel J. Math. *20* (1975), 326-350. MR *52*, 14940.

W. Pranger, Extreme points of convex sets. Math. Ann. *205* (1973), 299-302. MR *54*, 8447.

J. Rainwater, Local uniform convexity of Day's norm on $c_0(\Gamma)$. Proc. Amer. Math. Soc. *22* (1969), 335-339.

S. A. Rakov, The Banach-Saks property for a Banach space (Russian). Mat. Zametki *26* (1979), 823-834. MR *81i*, 46018. English Transl. Math. Notes (1979), 909-916.

M. M. Rao, Smoothness in Orlicz spaces. I., II. Proc. Amsterdam Acad. Sci. *28* (1965), 671-680; 681-690. MR *32*, 8116.

———, Characterizing Hilbert spaces by smoothness. Proc. Wetenschp. Ser. *A*, *LXX* (1967), 132-135. MR *35*, 2131.

———, Notes on characterizing Hilbert spaces by smoothness and smoothness in Orlicz spaces. J. Math. Anal. Appl. *37* (1972a), 228-234. MR *52*, 8042.

———, Prediction sequences in smooth Banach spaces. Ann. Inst. H. Poincaré, Sect. B (N.S.) *8* (1972b), 319-332. MR *50*, 8042.

S. Reich, Extreme invariant operators. Atti Accad. Nazionale dei Lincei, Rend. *(8)55* (1973), 31-36. MR *51*, 3967.

J. Reif and V. Zizler, On strongly extreme points. Comment. Math. Univ. Carolin. *18* (1974, 1975), 63-70. MR *50*, 10759.

G. Restrepo, Differentiable norms in Banach spaces. Bull. Amer. Math. Soc. *70* (1964), 413-414. MR *28*, 4338.

J. R. Retherford, On Chebyshev subspaces and unconditional bases in Banach spaces. Bull. Amer. Math. Soc. *76* (1967), 238-241. MR *34*, 8152.

———, Schauder bases and best approximation. Colloq. Math. *22* (1970), 91-110. MR *42*, 8253.

B. Reznik, Banach spaces which satisfy linear identities. Pacif. J. Math. *74* (1979a), 221-233. MR *58*, 7045.

———, Banach spaces with polynomial norms. Pacif. J. Math. *82* (1979b), 223-235.

C. E. Rickart, *General Theory of Banach Algebras*. Van Nostrand, Princeton, N.J., 1960. MR *22*, 5903.

J. R. Ringrose, A note on uniformly convex spaces. J. London Math. Soc. *34* (1959), 92. MR *20*, 6648.

W. H. Ruckle, The extend of the sequence space associated with a basis. Canad. J. Math. *24* (1972), 636-641. MR *45*, 9105.

A. F. Ruston, A note on the convexity in Banach spaces. Proc. Cambridge Philos. Soc. *45* (1949a), 157-159. MR *10*, 197.

———, A short proof of a theorem on reflexive spaces. Proc. Cambridge Philos. Soc. *45* (1949b), 674. MR *11*, 186.

B. N. Sadovskii, On a fixed point principle. Funktional. Analiz i Prilozen. *1* (1967), 74-76. MR *35*, 2184. English Transl. Functional Anal. and Appl. *1* (1967), 151-153.

———, Limit compact and condensing operators (Russian). Uspekhi Mat. Nauk *27* (1972), No. 1(163), 81-146. English Trans. Russian Math. Surveys *27* (1972), No. 1, 85-156.

A. G. Sangna, B_p-convex spaces (Russian; Georgian and English summaries). Soobsc. Akad. Nauk Gruz. SSR *94* (1979), 553-556. MR *81g*, 46020.

Yu. Saskin, Convex sets, extreme points and simplexes. Akad. Nauk SSSR, Vsesoiuz. Inst. Naučn. Techn. Inform. Moscow. Math. Analysis, Vol. 11, 1-50. English Transl. J. Soviet Math. *4* (1975), 625-655. MR *53*, 8861.

J. J. Schaefer. See R. C. James.

R. Schatten, *Norm Ideals of Completely Continuous Operators*. Ergebnisse der Math. 27, Springer Verlag, New York, 1966. MR *22*, 9878.

A. N. Šerstnev, On the notion of random normed space. Doklady Akad. Nauk SSSR *149* (1963a), 280-283. MR *27*, 4268.

———, Best approximation problems in random normed spaces (Russian). DAN SSSR *149* (1963b), 539-542. MR *24*, 4269.

———, Some problems of the best approximation in random normed spaces. Rev. Roum. Math. Pure Appl. *9* (1964), 771-789. MR *34*, 3622.

M. Sharir, Characterization and properties of extreme operators in C(Y). Israel J. Math. *12* (1972), 174-183. MR *47*, 5574.

———, Extremal structure in operator spaces. Trans. Amer. Math. Soc. *186* (1973), 91-111. MR *48*, 12151.

———, A note on the extremal elements in $A_0(K,E)$. Proc. Amer. Math. Soc. *46* (1974), 244-246. MR *51*, 11075.

———, A counterexample on extreme operators. Israel J. Math. *24* (1976), 320-328. MR *55*, 11096.

———, A non-nice extreme operator. Israel J. Math. *26* (1977), 306-312. MR *56*, 12970.

A. L. Shields. See S. Axler.

T. N. Shimi. See K. Goebel.

I. Singer, On the set of best approximation of an element in a normed linear space. Rev. Roum. Math. Pure Appl. *5* (1960), 383-402. MR *24*, A1629.

———, *Best Approximation in Normed Linear Spaces by Elements of Linear Subspaces* (Romanian). Bucharest, 1967. MR *38*, 3677. English Transl. Bucharest, 1970. MR *42*, 4937.

———, *The Theory of Best Approximation and Functional Analysis*. CBMS Conf. 13, SIAM, Philadelphia, 1974. MR *51*, 10967.

M. A. Smith, A smooth non-reflexive second conjugate space. Bull. Austral. Math. Soc. *15* (1976a), No. 1, 129-131. MR *54*, 11039.

———, Rotundity and smoothness in conjugate spaces. Proc. Amer. Math. Soc. *61* (1976b), 232-234. MR *55*, 8763.

———, Banach spaces that are uniformly rotund in weakly compact sets of directions. Canad. J. Math. *29* (1977), 963-970. MR *56*, 9232.

———, Some examples concerning rotundity in Banach spaces. Math. Ann. *233* (1978a), 155-161. MR *58*, 2168.

———, Products of Banach spaces that are uniformly rotund in every direction. Pacif. J. Math. *70* (1977b), 215-219, MR *57*, 3830.

———, A reflexive Banach space which is LUR and not 2R. Canad. J. Math. *21* (1978), 251-252. MR *80d*, 46032.

———, and F. Sullivan, *Extremely Smooth Banach Spaces*. Lecture Notes in Math. 604 (1980), pp. 125-137. MR *56*, 6328.

———, and B. Turett, Rotundity in Lebesgue-Bochner function spaces. Trans. Amer. Math. Soc. *257* (1980), 105-118. MR *80m*, 46031.

V. L. Smulian, On some geometrical properties of the sphere in spaces of type (B). Mat. Sbornik *6* (1939a), 77-94. MR *1*, 242.

———, On the principle of inclusion in the space of type (B). Math. Sbornik (Rec. Math.) *5* (1939b), 317-328. MR *1*, 335.

———, Some geometrical properties of the sphere in spaces of type (B). DAN SSSR *24* (1939c), 648-652. MR *1*, 242.

———, On the differentiability of the norm of a Banach space. DAN SSSR *27* (1940), 643-648. MR *2*, 102.

———, Sur quelques proprietes geometriques de la sphere dans les espaces lineaires semi-ordonées de Banach. C. R. (Doklady) Acad. Sci. USSR (N.S.) *30* (1941), 394-398. MR *2*, 314.

S. L. Sobolev, *Applications of Functional Analysis in Mathematical Physics*. Translated from the Russian. Transl. Vol. 7, Amer. Math. Soc., Providence, R.I., 1963. MR *29*, 2624.

V. I. Sobolev. See L. A. Liusternik.

P. S. Soltan, *Extremal Problems on Convex Sets* (Russian). Izdat. Stiinţa, Kishinev, 1976. MR *58*, 18164.

C. V. Stanojevič (see A. R. Blass), Mielnik's probability spaces and characterization of inner product spaces. Trans. Amer. Math. Soc. *183* (1973), 441-448. MR *48*, 6904.

E. F. Steiner. See L. P. Belluce.

D. Strawther (see S. P. Gudder) and S. P. Gudder, A characterization of strictly convex spaces. Proc. Amer. Math. Soc. *47* (1975), 268. MR *50*, 8016.

L. Sucheston. See A. Brunel.

F. Sullivan (see H. B. Cohen, M. A. Smith), A generalization of best approximation operators. Annali Mat. Pura Appl. *4(57)* (1975), 245-261. MR *53*, 3564.

———, An approximation-theoretic characterization of uniformly rotund spaces. J. Approx. Theory *16* (1976), 281-286. MR *53*, 3655.

———, Geometrical properties determined by the higher duals of a Banach space. Illinois J. Math. *21* (1977), 315-331. MR *56*, 16327.

———, A generalization of uniformly rotund spaces. Canad. J. Math. *31* (1979), 628-636. MR *80h*, 46023.

——— and B. Altestam, Descent methods in smooth, rotund spaces with application to approximation in L^p. J. Math. Anal. Appl. *48* (1974), 155-164. MR *50*, 5318.

——— and R. Geremia, Multi-dimensional volumes and moduli of convexity in Banach spaces. Annali di Mat. Pura Appl. *127* (1981), 197-201.

K. Sundaresan (see I. Leonard), On the strict and uniform convexity of certain Banach spaces. Pacif. J. Math. *15* (1965), 1083-1086. MR *32*, 4522.

———, Smooth Banach spaces. Bull. Amer. Math. Soc. *72* (1966a), 520-521. MR *32*, 8117

———, Orlicz spaces isomorphic to strictly convex spaces. Proc. Amer. Math. Soc. *17* (1966b), 1353-1356. MR *34*, 601.

———, Smooth Banach spaces. Math. Ann. *173* (1967), 191-199. MR *36*, 1960.

———, Uniform convexity of Banach spaces (p_i). Studia Math. *39* (1971), 227-231. MR *46*, 7859.

S. Swaminathan. See M. M. Day.

G. Szegö, Über eine Eigenschaft der Exponentialreihe. Sitzungsberichte der Berlin Gesselschaft Math. *23* (1924), 50-64.

D. G. Tacon, The conjugate of a smooth Banach space. Bull. Austral. Math. Soc. *2* (1970), 415-425. MR *41*, 5942.

J. Tate, On the relation between extremal points of convex sets and homomorphisms of algebras. Comm. Pure Appl. Math. *4* (1951), 31-32. MR *13*, 361.

A. E. Taylor, The extension of linear functionals. Duke Math. J. *5* (1939), 538-547. MR *1*, 58.

E. Throp and R. Witley, The strong maximum modulus theorem for analytic functions into a Banach space. Proc. Amer. Math. Soc. *18* (1967), 640-646. MR *35*, 5643.

N. Tomczak-Jaegermann, The moduli of smoothness and convexity and the Rademacher averages of the trace class S_p. Studia Math. *50* (1974), 163-182. MR *50*, 8141.

———, Finite dimensional subspaces of uniformly convex and uniformly smooth Banach lattices and the trace classes S_p. Studia Math. *66* (1979/80), 261-281. MR *81j*, 46026.

F. A. Toranzos, Metric betweeness in normed linear spaces. Colloq. Math. *23* (1971), 99-102. MR *46*, 4171.

E. Torrance, Strictly convex spaces via semi-inner product spaces orthogonality. Proc. Amer. Math. Soc. *26* (1970), 108-111. MR *41*, 5943.

S. L. Troyanski (= S. L. Trojanski) (see R. P. Maleev), The connection between weak uniform convexity and reflexivity (Russian). Teoria Functsii Funckional. Analiz. i Priložen. Vyp. *8* (1969), 92-96, MR *42*, 2287.

———, An example of a smooth space whose dual is not strictly convex (Russian). Studia Math. *35* (1970), 305-309. MR *42*, 6589.

———, On locally uniformly convex spaces and differentiable norms in certain non-separable Banach spaces. Studia Math. *37* (1971), 173-180. MR *46*, 5995.

———, On equivalent norms and minimal systems in nonseparable Banach spaces. Studia Math. *43* (1972), 125-138. MR *48*, 2734.

———, with unconditional basis (Russian). C. R. A. Bulgare Sci. *30* (1977), 1243-1246.

———, Locally uniformly convex norms. C. R. Acad. Bulgare Sci. *32* (1979), 1167-1169. MR *81b*, 46015.

B. Turett (see D. J. Downing, M. A. Smith), Rotundity of Orlicz spaces. Nederl. Akad. Wetensch. Proc. Ser. *A79* (= Indag. Math. *38*) (1976), No. 5, 462-469. MR *55*, 1058.

———, *Frenchel-Orlicz Spaces*. Dissertationes Math. 181, Warzawa, 1981.

———, A dual view of a theorem of Baillon. In *Proc. Conf. Nonlinear Analysis and Applications*, St. John's, Newfoundland, 1981. (To appear.)

A. J. Vasiliev, Cebysev sets and strict convexity of linear metric spaces (Russian). Mat. Zametki *25* (1979), 653-664. MR *81j*, 46007.

E. Veits (= E. Veic), Orthogonal systems of elements in smooth Banach spaces. DAN SSSR *192* (1970), 9-12. MR *42*, 8257. English Transl. Sov. Math. Doklady *11* (1970), 553-557.

E. Vesentini, Maximum h theorems for vector valued holomorphic functions. Rend. del Sem. Mat. e Fis. di Milano *40* (1970a), 24-55. MR *44*, 4506.

———, Maximum theorems for spectra. In *Essays in Topology and Related Topics*. Memoires dediés a G. de Rham, 111-114. Springer Verlag, New York, 1970b. MR *42*, 6612.

A. A. Vinogradov. See S. N. Kračkovskiĭ).

L. P. Vlasov, Approximate convex sets in Banach spaces. DAN SSSR *163* (1965), 18-21. MR *34*, 6497.

———, On Chebyshev subsets. DAN SSSR *173* (1967a), 491-494. MR *35*, 5903.

———, Chebyshev sets and approximatively convex sets (Russian). Mat. Zametki *2* (1967b), 191-200. MR *35*, 5904.

———, Approximate convex sets in uniformly smooth spaces (Russian). Mat. Zametki *1* (1967c), 443-450. MR *35*, 4706.

———, Almost convex and Chebyshev sets (Russian). Mat. Zametki *8* (1970), 545-550. MR *43*, 2476.

———, Some theorems on Chebyshev sets. Mat. Zametki *11*(1972), 135-144. MR *45*, 9046. Mat. Notes *11* (1972), 87-92.

———, Approximate properties of sets in normed linear spaces. Uspechi Mat. Nauk *28* (1973), 3-66. MR *53*, 8761. Russian Math. Surveys *28* (1973), 1-66.

R. Vyborny, On the weak convergence in locally uniformly convex spaces (Czech). Casopis Pest. Math. CSSR *81* (1956), 352-353.

J. Wada, Strict convexity and smoothness of normed spaces. Osaka Math. J. *10* (1958), 221-230. MR *20*, 7203.

S. Watanabe, Note on positive linear maps of Banach algebras with an involution. Sci. Rep. Nagata Univ. Ser. A, *7* (1969), 17-21. MR *42*, 855.

———, Positive linear maps of Banach algebras with an involution. Sci. Rep. Nagata Univ. Ser. A, *8* (1971), 1-6. MR *44*, 2041.

G. Weiss, A note on Orlicz spaces. Portugaliae Math. *29* (1967), 35-47.

R. F. Wheeler. See S. E. Mosiman.

A. G. White. See C. R. Diminie.

W. T. Williamson. See W. V. Petryshyn.

A. Wilansky, *Functional Analysis*.

R. Witley. See E. Throp.

W. A. Woyczynsky, *Geometry and Martingales in Banach Spaces*. Lecture Notes in Math. 472, 299-275. MR *52*, 14936.

A. C. Yorke, Weak rotundity in Banach spaces. J. Austral. Math. Soc. *A24* (1977), 224-233. MR *58*, 2171.

———, Differentiability and local rotundity. J. Austral. Math. Soc. *A28* (1979a), 205-213. MR *81m*, 46033.

———, Locally uniformly rotund norms and Markkushevic bases. J. Austral. Math. Soc. *A27* (1979b), 371-377. MR *80j*, 46034.

P. P. Zabreiko, A theorem on subadditive functionals (Russian). Funckional. Analiz. i Priložen. *3* (1969), 86-88. MR *39*, 3283.

V. Zizler (see K. J. John, J. Reif), One theorem on rotundity and smoothness of separable Banach spaces. Comment. Math. Univ. Carolin. *9* (1968a), 637-640. MR *41*, 798.

———, Banach spaces with differentiable norms. Comment. Math. Univ. Carolin. *8* (1968b), 415-440. MR *39*, 7427.

———, Some notes on various rotundity and smoothness properties of separable Banach spaces. Comment. Math. Univ. Carolin. *10* (1969), 195-205. MR *39*, 7401.

———, On some rotundity and smoothness properties of Banach spaces. Dissert. Math. (Rozsprawy) *87* (1971), 5-33. MR *45*, 9108.

Additional Bibliography

T. J. Abatzoglu, Norm derivatives on spaces of operators. Mathematical Annalen *239* (1979), 129-135.

C. C. Ahuja, T. D. Narang, and S. Trehan, Best approximation on convex sets in a metric space. J. Approx. Theory *12* (1974), 94-97. MR *50*, 10642.

L. Alaoglu and G. Birkhoff, General ergodic theorems. Proc. Nat. Acad. Sci. U.S.A. *25* (1939), 628-630.

———, General ergodic theorems, Annals of Math. *(2) 41* (1940), 293-309.

Z. Altshuler, The modulus of convexity of Lorentz and Orlicz spaces. In *Notes on Banach Spaces*, Univ. of Texas, Austin, Tex., 1980, pp. 359-370. MR *82e*, 46020.

D. Amir, Characterization of relative Chebyshev centers. In *Approximation Theory. III*. Proc. Conf. Texas Univ. Texas, Austin, 1980, pp. 157-162. Academic Press, 1980. MR *82c*, 41018.

V. J. Andreev, The existence of discontinuous metric projections (Russian). Mat. Zametki *16* (1974), 857-864. MR *51*, 6364. English Transl. Math. Notes *16* (1974), 1107-1111 (1975).

J. Arazy, More on convergence in unitary matrix spaces. Proc. Amer. Math. Soc. *81* (1981), 44-48. MR *82f*, 46009.

H. Auerbach, S. Mazur, and S. Ulam, Sur une propriete de l'ellipsoide. Monath. Math. Phys. *42* (1935), 45-48.

D. Avis, On the extreme rays of the metric cone. Canad. J. Math. *32* (1980), 126-144. MR *82b*, 52012.

R. Bantegnie, Sur certains espaces métriques. Proc. Koninkl. Akad. Nederl. Wetensch. A. *70* (1967), 74-75.

E. Blanc, Les espaces métriques quasiconvexes. Ann. Sci. Ecole Norm. Sup. 3 *55* (1938), 1-82.

A. Bosznay, A remark on uniquely remotal sets in C(K,X). Period. Hung. *12* (1981), 11-14. MR *82f*, 46016.

J. Bourgain, An averaging result for c_0-sequences. Bull. Soc. Math. Belgique *30* (1978), 83-87. MR *81d*, 46013.

———, A new class of L^1-spaces. Israel J. Math. *39* (1981), 113-126. MR *82f*, 46020.

———, ℓ^{∞}/c_0 has no equivalent strictly convex norm. Proc. Amer. Math. Soc. *78* (1980), 225-226. MR *81b*, 46029.

———, A Hausdorff-Young inequality for B-convex Banach spaces. Preprint (to appear).

——— and F. Delbaen, A class of L_{∞} spaces. Acta Math. *145* (1980), 155-176. MR *82h*, 46023.

J. Buir, A propos des points extremaux de la some de deux ensembles convexes (English summary). Bull. Soc. Math. Roy. Liege *48* (1979), 262-264. MR *81m*, 52004.

I. I. Ceitlin, Extremal points of the unit ball of certain operator spaces. Mat. Zametki *20* (1976), 521-527 (Russian). MR *55*, 13220. English Transl. Math. Notes *20* (1976), 848-852 (1977).

———, Strongly extremal points in the space of I-integral mappings (Russian). In *Functional Anal. No. 8: Operator Theory* (Russian), pp. 149-161. Ul'janovsk Gos. Ped. Inst. Ulyanovsk, 1977. MR *58*, 30491. Entire Coll. MR *57*, 7206.

M. M. Day, A geometric proof of Asplund's differentiability theorem. Israel J. Math. *13* (1972), 277-280.

C. R. Diminie and A. G. White, Strict convexity conditions for seminorms. Math. Japon. *24* (1979/80), 489-493. MR *81g*, 46014.

D. van Dulst, Flat spots on the unit sphere. J. Austral. Math. Soc. *A27* (1979), 289-304.

———, Equivalent norms and the fixed point property for nonexpansive mappings. J. London Math. Soc. *(2)25* (1982a), 139-144.

———, *Some More Banach Spaces with Normal Structure*. Preprint, 1982b.

——— and B. Sims, *Fixed Points of Nonexpansive Mappings and Chebyshev Centers in Banach Spaces with Norms of Type (KK)*. Preprint, 1982.

——— and A. J. Pach, A renorming of Banach spaces. Rev. Roum. Math. Pure Appl. *26* (1981), 843-847. MR *82j*, 46074.

T. Figiel, Uniformly convex norms on Banach lattices. Studia Math. *68* (1980), 215-247. MR *82a*, 46024.

V. P. Fonf, Some properties of polyhedral Banach spaces. Funkcional Anal. i Prilozen. *14* (1980), No. 4, 89-90. MR *82a*, 46017.

C. Franchetti, On the radial projections in Banach spaces. In *Approxima-Theory III*. Proc. Conf. Texas Univ. Texas, Austin, 1980, pp. 425-428. Academic Press, 1980. MR *82c*, 46017.

R. Freese and G. Murphy, The CMP and Banach spaces with unique metric lines. J. Geom. *14* (1980), 257-278. MR *81g*, 58007.

M. Frunza, Analytic characterizations of quasi-conformality in normed spaces. An. Stiint. Univ. Al. I. Cuza, Sect. I a Mat. (N.S.) *25* (1979), 273-278. MR *81c*, 30030.

M. and S. Frunza, La quasi-conformite et la geometrie des espaces normes. C. R. Acad. Sci. Paris, Ser.

B. V. Gordun, The relation of modulus of convexity to the characteristic of normed subspaces (Russian). Sibirsk. Mat. Z. *22* (1981), 229-231. MR *82e*, 46023.

R. Grzaslewicz, Extreme operators on 2-dimensional ℓ_p spaces. Colloq. Math. *44* (1980), 309-315.

———, Extreme contractions on real Hilbert spaces. Math. Ann. *261* (1982), 463-466.

K. Hoffman, *Banach Spaces of Analytic Functions*. Prentice Hall, Englewood Cliffs, N.J., 1962. MR *24a*, 12844.

A. Hopenwasser, R. L. Moore, and V. Paulsen, C*-extreme points. Trans. Amer. Math. Soc. *266* (1981), 291-307. MR *82f*, 46065.

R. Huff, Banach spaces which are nearly uniformly convex. Rocky Mountain J. Math. *10* (1980), 743-749. MR *82b*, 46016.

N. A. Il'jasov, Differential and smoothness properties of functions in various metrics (Russian). Azerbaidzan Gosud. Univ. Ucen. Zap., 1979, No. 6, 86-98. MR *82c*, 46040.

V. I. Istrăţescu, On classes of metric spaces related to uniformly convex metric spaces and some applications to fixed point theory. (To appear, 1984).

K. P. Kirčev and S. Troyansky, On a characteristic property of spaces with an inner product. C. R. Acad. Bulgare Sci. *28* (1975), 445-447 (Russian).

W. A. Kirk, Fixed point theorems for nonexpansive mappings, in *Nonlinear Functional Analysis*, Proc. Sympos. Pure Math. 13, Part 1, Amer. Math. Soc., Providence, R.I., 1970, pp. 162-168. MR *42*, 6677.

B. L. Lin, On generalized James quasi-reflexive Banach space. Bull. Inst. Math. Acad. Šinica *8* (1980), No. 2-3, Part 2, 389-399. MR *82a*, 46020.

R. Loewy, Extreme points of a convex set of the cone of positive semidefinite matrices. Math. Ann. *253* (1980), 227-238. MR *82b*, 52011.

K. Menger, Untersuchungen uber allgemeine Metrik, I, II, III. Math. Ann. *100* (1928), 75-163.

D. P. Milman, Facial characterization of convex sets. Extremal elements (Russian). Trudy Moskov. Mat. Obsc. *22* (1970), 63-126. English Transl. Trans. Moscow Math. Soc. *22* (1970), 69-139. MR *51*, 6362.

V. D. Milman, Duality of certain geometric characteristics of Banach spaces (Russian). Teor. Funkcii Funkcional. Analiz i Priložen. Vyp. *18* (1973), 120-137. MR *48*, 12004.

B. Mitiagin, Ed., *Theory of Linear Operators and the Geometry of Banach Spaces* (Russian). Jaroslavl, 1977.

S. V. R. Naidu and K. P. R. Sastry, Convexity conditions in normed linear spaces. J. Reine Angew. Math. *297* (1978), 35-53.

———, Different types of isometries in metric linear spaces. Jnanabha *7* (1977), 17-20. MR *81c*, 46003.

T. D. Narang, On Chebyshev sets in metric linear spaces. Math. Sem. Notes Kobe Univ. *7* (1979), 593-595. MR *81c*, 41062.

T. D. Narang, Convexity of Chebyshev sets. Nieuw Arch. Wisk. *(3)25* (1977), 377-402. MR *81f*, 41037.

V. P. Odinec, On a lemma of Banach on peak functions (Russian). Comment. Math. Prace Mat. *21* (1980), 195-201. MR *81m*, 46044.

———, On a property of reflexive Banach spaces. Bull. Acad. Polon. Sci. Ser. Sci. Math. *30* (1982), No. 7-8, 353-357.

P. L. Papini and I Singer, Best coapproximation in normed linear spaces. Monatsh. Math. *88* (1979), 27-44. MR *81c*, 41080.

J. R. Partington, On nearly uniformly convex spaces. Math. Proc. Cambridge Philos. Soc. *93* (1983), 127-129.

———, Almost sure summability of subsequences in Banach spaces. Studia Math. *71* (1982), 27-36.

———, Extremal symmetric norms on a two-dimensional space. Preprint, 1983 (to appear).

———, On the Banach-Saks property. Math. Proc. Cambridge Philos. Soc. *82* (1977), 369-374.

———, Personal communications.

A. Pelczynski, Certain problems of Banach (Polish). Wiadom. Mat. *(2)15* (1972), 3-11. MR *58*, 12298.

J. C. B. Perrot, Transfinite duals of Banach spaces and ergodic superproperties equivalent to super-reflexivity. Quart. J. Math. Oxford *(2)30* (1979), 99-111. MR *82b*, 46015.

A. T. Plant, A note on the modulus of convexity. Glasgow Math. J. *22* (1981), 157-158. MR *82h*, 46030.

S. Ramaswami, A simple proof of Clarkson's inequality. Proc. Amer. Math. Soc. *68* (1978), 249-250.

S. Reich, A remark on a problem of Asplund. Atti Accad. Nazionale dei Lincei, Rend. *(8)67* (1979), 204-205.

M. Smith, A Banach space which is MLUR and not HR. Math. Ann. *256* (1981), 277-279. MR *82h*, 46032.

F. Sullivan, Dentability, smoothability and stronger properties of Banach spaces. Indiana Univ. Math. Journal *26* (1977), 545-553. MR *55*, 11007.

———, Nearly smooth norms on Banach spaces. Rev. Roum. Math. Pure Appl. *26* (1981), 1053-1057. MR *82i*, 46031.

V. Tokarev, On the Banach-Saks property in lattices (Russian). Sibirskii Matematicheskii Journal *24* (1983), 187-188.

S. Troyansky, Equivalent norms that are uniformly convex and uniformly differentiable in every direction (Russian). C. R. Acad. Bulgare Sci. *32* (1979), 1461-1464. MR *81d*, 46016.

N. N. Vakhania, *Probability Distributions on Linear Spaces*. North-Holland Publ. Co., 1981. Transl. from the Russian. MR *82i*, 60015.

J. E. Valentine, Banach-Euclidean four-point properties. Fund. Math. *106* (1980), 227-230. MR *81i*, 52002.

J. S. W. Wong, Remarks on metric spaces. Proc. Koninkl. Nederl. Akad. Wetensch. *A69* (1966), 70-73.

A. C. Yorke, Differentiability and local rotundity. J. Austral. Math. Soc. *A28* (1979), 205-213. MR *81m*, 46033.

Author Index

Subject Index